T0360443

LIMIT THEOREMS FOR NONLINEAR COINTEGRATING REGRESSION

NONLINEAR TIME SERIES AND CHAOS (NTSC)

Series Editors: Howell A M Tong *(London School of Economics, UK)*
Jiti Gao *(Monash University, Australia)*

Nonlinear Time Series and Chaos – Vol. 5

LIMIT THEOREMS FOR NONLINEAR COINTEGRATING REGRESSION

Qiying Wang

The University of Sydney, Australia

 **World Scientific**

NEW JERSEY · LONDON · SINGAPORE · BEIJING · SHANGHAI · HONG KONG · TAIPEI · CHENNAI

Published by

World Scientific Publishing Co. Pte. Ltd.
5 Toh Tuck Link, Singapore 596224
USA office: 27 Warren Street, Suite 401-402, Hackensack, NJ 07601
UK office: 57 Shelton Street, Covent Garden, London WC2H 9HE

Library of Congress Cataloging-in-Publication Data
Wang, Qiying, author.
    Limit theorems for nonlinear cointegrating regression / Qiying Wang, The University of Sydney,
Australia.
        pages cm -- (Nonlinear time series and chaos ; volume 5)
        Includes bibliographical references and index.
        ISBN 978-9814675628
        1. Limit theorems (Probability theory). 2. Nonlinear systems. I. Title. II. Series: Nonlinear time
series and chaos ; vol. 5.
        QA273.67.W36 2015
        519.5'36--dc23

                                                                                    2015017266

British Library Cataloguing-in-Publication Data
A catalogue record for this book is available from the British Library.

Printed in Singapore

Preface

The past decade has witnessed great progress in the development of non-linear cointegrating regression. Unlike linear cointegration and nonlinear regression with stationarity where the traditional and classical methods are widely used in practice, estimation and inference theory in nonlinear cointegrating regression produce new mechanisms involving local time, a mixture of normal distributions and stochastic integrals. This book aims to introduce the machinery of the theoretical developments in nonlinear cointegrating regression, providing up-to-date results on convergence to local time, extended martingale limit theorems and weak convergence to stochastic integrals for econometric applications.

The core context of this book comes from the current research by the author and his collaborators in recent years. The majority of the original works, however, has been significantly extended in line with current development in the area. In particular, a new limit theorem for martingale is established under the convergence in distribution for the conditional variance. This extension removes a main barrier in applications of the classical martingale limit theorem to estimation and inference theory in nonlinear cointegrating regression. This book is intended for use as a reference book. The materials are mainly self-contained, but may not have sufficient details for a text book.

I acknowledge grant support for my research project related to this book from the Australian Research Council (DP130102408). I sincerely thank my collaborator Peter Phillips for his continuous support and many encouraging discussions over the years. Without his help, I may never have touched the project related to this book. I thank my other collaborators and colleagues Nigel Chan, Xiaohong Chen, Jiti Gao, Degui Li, Qi Li, Hanying Liang, Oliver Linton, Weidong Liu, Joon Park, John Robinson,

Dongsheng Wu and Weibiao Wu for working with me on related projects and for their helpful comments. In particular, Dongsheng Wu has read most of the manuscript and provided invaluable feedback. Jiti Gao suggested that I write a book on this topic. I also thank Ms Zhang Ji in World Scientific for her help in the writing of this book.

Qiying Wang, April, 2015

Contents

Chapter 1

Introduction

Consider a nonlinear regression model:

$$y_t = f(x_t) + u_t, \quad t = 1, 2, ..., n, \tag{1.1}$$

where $f : \mathrm{R} \to \mathrm{R}$ is an unknown function, x_t and u_t are regressors and regression errors, respectively. Let $K(x)$ be a kernel function and $h \equiv h_n \to 0$ be a bandwidth. The conventional kernel estimator of $f(x)$ is given by

$$\hat{f}(x) = \frac{\sum_{t=1}^{n} y_t K[(x_t - x)/h]}{\sum_{t=1}^{n} K[(x_t - x)/h]}.$$

Let x_0 be fixed. We may split $\hat{f}(x_0) - f(x_0)$ as

$$\frac{\sum_{t=1}^{n} u_t \, K[(x_t - x_0)/h]}{\sum_{t=1}^{n} K[(x_t - x_0)/h]} + \frac{\sum_{t=1}^{n} \left[f(x_t) - f(x_0) \right] K[(x_t - x_0)/h]}{\sum_{t=1}^{n} K[(x_t - x_0)/h]}. \tag{1.2}$$

The second term of (1.2) is easy to handle with and is negligible under mild conditions. The limit behavior of $\hat{f}(x_0)$ is determined by the first term of (1.2), which follows from the asymptotics (jointly) of S_{n1} and S_{n2} defined by

$$S_{n1} = \sum_{t=1}^{n} K[(x_t - x_0)/h], \quad S_{n2} = \sum_{t=1}^{n} u_t \, K[(x_t - x_0)/h].$$

If (x_t, y_t) is a stationary process, we may prove that $\frac{1}{nh} S_{n1}$ converges to a constant C under mild conditions on x_t and h. Moreover, a standard central limit theorem can be employed to show

$$\frac{1}{\sqrt{nh}} S_{n2} \to_D N(0, \sigma^2), \tag{1.3}$$

where $\sigma^2 > 0$ is a constant. As a consequence, $\sqrt{nh} \left[\hat{f}(x_0) - f(x_0) \right]$ converges to a normal distribution, an important result on the estimation and

1

inference theory in nonlinear regression. There are extensive researches in the area. See, e.g., Härdle (1990), Fan and Gijbels (1996) and the references therein.

The situation becomes very different if (x_t, y_t) includes a non-stationary component. To illustrate, assume that $x_t = \sum_{j=1}^n \epsilon_j$, where ϵ_j is a sequence of i.i.d. $N(0, 1)$ random variables. We further assume $u_t = \epsilon_{t+1}$ and $K(x)$ is a uniform kernel, i.e., $K(x) = \frac{1}{2}I(|x| \le 1/2)$. In this case, it is not true to say $\frac{1}{nh}S_{n1} \to_P C$. Instead, we may prove

$$\frac{1}{\sqrt{nh}}S_{n1} \to_D L_B(1, 0), \tag{1.4}$$

where $B = \{B_t\}_{t \ge 0}$ is a Brownian motion and $L_B(1, 0)$ is a local time of B. Furthermore, S_{n2} forms a martingale, but limit behavior of S_{n2} cannot be investigated by the classical martingale limit theorem that is stated in Theorem 3.2 of Hall and Heyde (1980). Indeed, as a martingale, the conditional variance σ_n^2 of S_{n2} has the property:

$$\sigma_n^2 = \sum_{t=1}^n K^2[(x_t - x_0)/h] = \frac{1}{2}S_{n1} \to_D \frac{1}{2}L_B(1, 0), \tag{1.5}$$

by (1.4). Since the convergence in distribution in (1.5) for the conditional variance σ_n^2 cannot be extended to the convergence in probability, the classical martingale limit theorem is no longer useful to provide the asymptotics of S_{n2}.

When the regressor x_t is a nonstationary time series, in view of the obvious link to cointegration when $f(\cdot)$ is linear, the model (1.1) is conventionally called a nonlinear cointegrating regression model in econometrics even though nonlinear functions may well affect the memory and persistence properties of the regressor x_t. This example indicates that estimation theory in nonlinear cointegrating regression produces new mechanisms involving local time and its development relies on new limit results for the sample covariances between nonstationary and stationary random elements.

This book intends to provide the limit theorems that can be used in the development of non-linear cointegrating regression. Toward the aims, Chapter 2 investigates the convergence to local time for nonlinear functionals of nonstationary time series, providing two frameworks: one was developed in Wang and Phillips (2009a) by the conditional density arguments and another was given in Wang (2015) by using characteristic functions. This chapter introduces the definition of local time for various processes, presents a number of examples on convergence to local time and also considers the uniform convergence to local time.

Chapter 3 deals with the convergence to a mixture of normal distributions for the sample covariances between nonstationary and stationary random elements. This chapter extends the classical martingale limit theorem. Unlike previous books, for a certain class of martingales, weak convergence to a mixture of normal distributions is established under the convergence in distribution for the conditional variance. This extension is essential in econometrics, removing a main barrier in applications of the classical martingale limit theorem to estimation and inference theory in nonlinear cointegrating regression. Except the extended martingale limit theorem, this chapter establishes the convergence to a mixture of normal distributions without assuming martingale array structure. Many applications in econometrics involve a cointegration framework where endogeneity and nonlinearity play a major role, the results in this chapter provide a foundation in developing nonlinear cointegrating regression. This chapter also considers uniform convergence for a class of martingales.

Chapter 4 contributes to the convergence to stochastic integrals for the sample covariances between nonstationary and stationary random elements. This chapter collects some of the most useful results in the area and explores the limit theory without assuming semi-martingale structure. In particular, this chapter establishes a novel decomposition theorem that can be used as a platform for future research.

Finally, Chapter 5 investigates estimation and inference theory in nonlinear cointegrating regression, which illustrates the use of the techniques developed in Chapters 2-4.

Notation

The following notation is used throughout the book.

a.s.	almost surely
a.e.	almost everywhere
$=_D$	the same in distribution
$\to_D$	convergence in distribution
$\to_P$	convergence in probability
$\to_{a.s.}$	convergence almost surely
$\Rightarrow$	weak convergence of probability measures
$[a]$	integer part of the real number a
$a_n \sim b_n$	$\lim_{n\to\infty} a_n/b_n = 1$
$a_n = o(b_n)$	$\lim_{n\to\infty} a_n/b_n = 0$
$a_n = O(b_n)$	$\limsup_{n\to\infty} a_n/b_n < \infty$
$a_n \asymp b_n$	$a_n = O(b_n)$ and $b_n = O(a_n)$
$X_n = o_P(b_n)$	$X_n/b_n \to_P 0$, as $n \to \infty$
$X_n = O_P(b_n)$	$X_n/b_n < \infty$, in probability
$X_n = o(b_n)$, a.s.	$X_n/b_n \to_{a.s.} 0$, as $n \to \infty$
$X_n = O(b_n)$, a.s.	$X_n/b_n < \infty$, almost surely
$\|a\| = \sum_{j=1}^d |a_j|$	if $a = (a_1, \cdots, a_d)$
$\|x\| = \sum_{i=1}^l \sum_{j=1}^m |x_{ij}|$	if x is an $l \times m$ matrix with components x_{ij}
i.i.d.	independent and identically distributed
$N(\mu, \sigma^2)$	normally distributed random variables with mean μ and variance σ^2

The transpose of a vector or matrix a is denoted by a'. The complement of an event A is denoted by A^c and the indicator function of A by $I_A(x)$ or $I(A) = I(A)(x)$, i.e.,

$$I_A(x) = I(A) = \begin{cases} 1, \text{ if } x \in A, \\ 0, \text{ otherwise.} \end{cases}$$

$R = (-\infty, \infty)$, $R^+ = [0, \infty)$ and $R^d = R \times R \times \cdots \times R$. $\mathbb{Z} = \{0, \pm 1, \pm 2, \cdots\}$.

For $x = (x_1, ..., x_d)$, we define

$$\int_{-\infty}^{\infty} f(x)dx = \int_{-\infty}^{\infty} \cdots \int_{-\infty}^{\infty} f(x_1, ..., x_d)dx_1 \cdots dx_d.$$

$g(x)$ is called locally integrable (locally bounded) if, for all compact set $K \subset R^d$, $\int_K |g(x)|dx < \infty$ ($\sup_{x \in K} |g(x)|$ is bounded by a constant). The integral is understood in the sense of Lebesgue, except explicitly mentioned as Riemann. We let $\Delta X_t = X_t - X_{t-}$.

Chapter 2

Convergence to local time

Consider an array $\{X_{nk}\}_{k\geq 1, n\geq 1}$ constructed from some underlying nonstationary time series and assume that there is a continuous limiting process $X = \{X_t\}_{t\geq 0}$ to which $X_{n,[nt]}$ converges weakly. Typically, $X_{nk} = d_n^{-1} X_k$ where X_k is a nonstationary time series such as an $I(1)$ process and d_n is an appropriate standardization factor. In many applications, we encounter a functional of the form

$$S_n(x) = \frac{c_n}{n} \sum_{k=1}^{n} g\big[c_n\,(X_{nk} + x)\big],$$

where g is a real integrable function on R and c_n is a certain sequence of positive constants satisfying that $c_n \to \infty$ and $n/c_n \to \infty$. For instance, such functionals arise in nonlinear regression with integrated time series and nonparametric estimation in relation to nonlinear cointegration models. See Chapter 4 for more details. In such cases, g may be a kernel function K or a squared kernel function K^2, and the sequence c_n depends on the bandwidth used in the nonparametric regression.

This chapter is concerned with asymptotic behaviors of $S_n(x)$ and other related statistics. Under certain conditions on g, X and X_{nk}, this chapter will prove

$$S_n(x) \to_D \tau\, L_X(1, x),$$

where τ is a constant and $L_X = \{L_X(t, s)\}_{t\geq 0, s\in\mathrm{R}}$ is a local time of X. Convergence of $S_n(x)$ to stochastic integrals and uniform convergence for $S_n(x)$ are also investigated.

2.1 Local time: definition and existence

Let $X = \{X_s\}_{s\geq 0}$ be a real-valued stochastic process. A measurable process $L_X = \{L_X(t,x)\}_{t\geq 0, x\in R}$ is said to be a *local time* of X if it satisfies

$$\int_0^t h(X_s)ds = \int_{-\infty}^{\infty} h(x)L_X(t,x)dx, \tag{2.1}$$

for all positive (or bounded) Borel functions $h : R \to R$. The concept of local time has its origins from the occupation measure of X, defined by

$$\mu_t(A) = \int_0^t I_A(X_s)ds = \lambda(s \in [0,t] : X_s \in A), \quad A \in \mathcal{R},$$

where $\lambda(\cdot)$ is a Lebesgue measure and $\mathcal{R}$ is a Borel σ-field. $\mu_t(A)$ represents "the amount of time" spent by X in the set A during $[0,t]$. If μ_t is absolutely continuous with respect to λ, the Radon-Nikodym derivative $\frac{d\mu_t(x)}{d\lambda}$ (written as $L_X(t,x)$) satisfies the equation

$$\int_0^t I_A(X_s)ds = \int_A L_X(t,x)dx, \tag{2.2}$$

for any $A \in \mathcal{R}$. Using (2.2), for each $t \geq 0$, there exists a version of $L_X(t,x)$ that is $\mathcal{R}$-measurable and satisfies (2.1).

We call (2.1) *occupation time formula* that has a generalization:

$$\int_0^t g(s, X_s)ds = \int_{-\infty}^{\infty} \int_0^t g(s,x)L_X(ds,x)dx, \tag{2.3}$$

for all positive Borel functions $g : R^+ \times R \to R$. Due to (2.1) and (2.3), we also call $L_X(t,x)$ the *occupation density* of X over $[0,t]$.

For the purpose of this book, we consider existence of local time (occupation density) using Fourier analysis and the differentiation of measures, which was first treated somewhat systematically by Berman (1969). See also Geman and Horwicz (1976, 1980). The following theorem is of fundamental importance.

Theorem 2.1. *(i) X has a local time L_X if, for all $0 \leq v \leq T < \infty$*

$$\liminf_{\epsilon\downarrow 0} \epsilon^{-1} \int_0^T P(|X_u - X_v| \leq \epsilon)du < \infty. \tag{2.4}$$

(ii) X has a local time L_X satisfying $\int_{-\infty}^{\infty} L_X^2(t,x)dx < \infty$, a.s., if and only if, for all $T > 0$,

$$\liminf_{\epsilon\downarrow 0} \epsilon^{-1} \int_0^T \int_0^T P(|X_u - X_v| \leq \epsilon)dudv < \infty, \tag{2.5}$$

if and only if, for all $T > 0$,

$$\int_{-\infty}^{\infty} \mathrm{E}\Big| \int_0^T e^{i\theta X_u} du\Big|^2 d\theta < \infty. \tag{2.6}$$

(iii) Suppose that, for all $T > 0$,

$$\int_{-\infty}^{\infty}\int_{-\infty}^{\infty}\int_0^T\int_0^T \big|\mathrm{E}e^{i\theta X_u + i\eta X_v)}\big| du\, dv\, d\theta\, d\eta < \infty. \tag{2.7}$$

Then X has a local time L_X which satisfies $\int_{-\infty}^{\infty} L_X^2(t,x)dx < \infty$, a.s., and which can be represented as

$$L_X(t,x) = \frac{1}{2\pi}\int_{-\infty}^{\infty} e^{-i\,ux}\int_0^t e^{iuX_s} ds\, du, \tag{2.8}$$

where the right hand is understood in the framework of L^2-theory, i.e., as a limit of $L^N(t,x)$ in the sense that

$$\lim_{N\to\infty}\sup_x \mathrm{E}\,|L^N(t,x) - L_X(t,x)|^2 = 0,$$

where

$$L^N(t,x) = \frac{1}{2\pi}\int_{-N}^{N} e^{-i\,ux}\int_0^t e^{iuX_s} ds\, du.$$

Remark 2.1. It follows from the occupation time formula (2.1) that

$$L_X(t,x) = \lim_{\epsilon\to 0}\frac{1}{2\epsilon}\int_0^t \mathrm{I}(|X_s - x| \le \epsilon)ds, \quad a.s. \tag{2.9}$$

For any Borel function X_s, the limit (2.9) actually exists (finite) for a.e x, and $L_X(t,x)$ is a non-decreasing Borel function in t and the points of increase of the function $t \to L_X(t,x)$ are contained in the level set $\{t : X_t = x\}$. This suggests the nature of local times (occupation densities).

A measurable process $L_X = \{L_X(t,x)\}_{t\ge 0, x\in\mathrm{R}}$ satisfying (2.9) is usually set as an alternative definition of local time.

Remark 2.2. The limit $V(t,v) := \lim_{\epsilon\to 0}\frac{1}{2\epsilon}\int_0^t \mathrm{I}(|X_u - X_v| \le \epsilon)du$ exists $(\le \infty)$ for $v \in \mathrm{R}$ except a Lebesgue measure zero set. Under (2.4), we have $V(t,v) < \infty$ and

$$V(t,v) = L_X(t, X_v), \quad a.s.$$

If (2.5) or (2.6) is satisfied, then

$$
\begin{aligned}
V(t) &:= \lim_{\epsilon\to 0}\frac{1}{2\epsilon}\int_0^t\int_0^t \mathrm{I}(|X_u - X_v| \le \epsilon)du\, dv\\
&= \int_0^t L_X(t, X_u)du = \int_{-\infty}^{\infty}\int_0^t L_X(t,x)\, L_X(ds,x)\, dx\\
&= \int_{-\infty}^{\infty} L_X^2(t,x)\, dx < \infty, \quad a.s.
\end{aligned}
$$

Furthermore, whence $L_X(t, x)$ has expression (2.8), we may write

$$V(t) = \frac{1}{2\pi} \lim_{N \to \infty} \int_{-N}^{N} \int_0^t \int_0^t e^{iu(X_s - X_v)} ds dv\, du,$$

where, as in (2.8), the limit is understood in the framework of L^2-theory. The process $V = \{V(t)\}_{t \geq 0}$ is called the *self-intersection local time* of X.

2.1.1   *Local time of Gaussian process*

Let $X = \{X_s\}_{s \geq 0}$ be a Gaussian process with mean zero. The condition of Theorem 2.1 on the existence of local time L_X can be written in terms of the covariance function $R(u, v) = \mathrm{E}X_u X_v$. Note that $X_u - X_v \sim N(0, \Delta(u, v))$, where

$$\Delta(u, v) = \mathrm{E}(X_u - X_v)^2 = R(u, u) + R(v, v) - 2R(u, v).$$

Simple calculations imply the following results.

(i) X has a local time L_X if, for all $0 \leq v \leq T < \infty$,

$$\int_0^T [\Delta(u, v)]^{-1/2} du < \infty. \tag{2.10}$$

(ii) X has a local time L_X satisfying $\int_{-\infty}^{\infty} L_X^2(t, x) dx < \infty, a.s.$, if and only if, for all $T > 0$,

$$\int_0^T \int_0^T [\Delta(u, v)]^{-1/2} du dv < \infty. \tag{2.11}$$

(iii) Suppose that, for all $T > 0$,

$$\int_0^T \int_0^T [R(u, u)R(v, v) - \mathrm{R}^2(u, v)]^{-1/2} du dv < \infty. \tag{2.12}$$

Then X has a local time L_X satisfying (2.8) and $\int_{-\infty}^{\infty} L_X^2(t, x) dx < \infty, a.s.$

As indicated in Geman (1976), (2.10) is implied by (2.11) or (2.12) for a continuous covariance. If X has stationary increments and $R(u, u + s) = r(s), s \geq 0$, the conditions (2.10), (2.11) and (2.12) all are equivalent to

$$\int_0^T [1 - r(s)/r(0)]^{-1/2} ds < \infty, \tag{2.13}$$

which is satisfied if $r(s) \sim 1 + c|s|^\alpha, 0 < \alpha < 2$ as $s \to 0$, where c is a constant.

$L_X(t, x)$ is continuous in t for a.e. x if (2.10) or (2.12) holds. Joint continuity of $L_X(t, x)$ is more involved, requiring the concept of locally nondeterministic that was first introduced by Berman (1973).

A Gaussian process $X = \{X_s\}_{s \geq 0}$ is called *locally nondeterministic* (LND) on $[0, T]$ if, for each $k \geq 2$, there exist positive constants $c > 0$ and $\delta > 0$ (may depend on k) such that

$$\text{Var}\Big(\sum_{j=1}^{k} u_j(X_{t_j} - X_{t_{j-1}})\Big) \geq c \sum_{j=1}^{k} u_j^2 \Delta(t_j, t_{j-1}) \qquad (2.14)$$

for all $0 \leq t_0 < t_1 < ... < t_k$ in $[0, T]$ with $t_j - t_{j-1} \leq \delta$ and all $u_j \in \mathbb{R}$.

The property (2.14) says that X has locally approximately independent increments. It was shown in Nolan (1989) that (2.14) is equivalent to the original definition of LND by Berman (1973). See also Geman and Horwicz (1980).

Theorem 2.2. *Let X be LND on $[0, T]$ with $X_0 = 0$ and $\text{E}X_t = 0$. Suppose that, for some $\gamma > 0$,*

$$\sup_{v \in [0, T]} \int_0^T [\Delta(u, v)]^{-1/2 - \gamma} du < \infty. \qquad (2.15)$$

Then X has a local time L_X which is jointly continuous on $[0, T] \times \mathbb{R}$ and which satisfies

$$|L_X(t, x) - L_X(t, y)| \leq C |x - y|^\beta, \quad a.s., \qquad (2.16)$$

for any $\beta < \gamma$, $t \in [0, T]$ and $x, y \in \mathbb{R}$. Furthermore, $L_X(t, x)$ can be represented as that of (2.8).

Theorem 2.2 is a corollary of Theorem 26.1 in Geman and Horwicz (1980). In applications, the condition (2.15) is easy to verify. Suppose $X_0 = 0$ and X has stationary increments with incremental variance function $\sigma^2(t) = \text{E}(X_{t+s} - X_s)^2, t \geq 0$. If

(i) $\sigma(t) \to 0$ as $t \to 0$ and $\sigma^2(t)$ is concave on $(0, \delta)$ for some $\delta > 0$; or
(ii) the spectral measure F of X has a density $f(x)$ which satisfies

$$0 < \liminf_{x \to \infty} \frac{xf(x)}{\int_x^\infty f(y)dy} < \limsup_{x \to \infty} \frac{xf(x)}{\int_x^\infty f(y)dy} < 2, \qquad (2.17)$$

then X is LND on $\mathbb{R}^+$. See Berman (1978, 1987).

Consider a Gaussian process $X = \{X_t\}_{t \geq 0}$ defined by

$$X_t = \int_{-\infty}^t F(t, u)dB_u, \qquad (2.18)$$

where $F(t, u)$ is a measurable function of (t, u) and $B = \{B_u\}_{u \in \mathrm{R}}$ is a Brownian motion. If, for $0 \le t \le T$, $\int_{-\infty}^{t} F^2(t, u) du < \infty$ and

$$\lim_{\delta \to 0} \inf_{\substack{0 < t - s < \delta \\ 0 \le s, t \le T}} \frac{\int_s^t F^2(t, u) du}{\int_{-\infty}^s [F(t, u) - F(s, u)]^2 du} > 0, \qquad (2.19)$$

then X is LND on $[0, T]$. See Theorem 3.2 of Berman (1973).

More sufficient conditions for X to be LND can be found in Berman (1973, 1988, 1991), Cuzick (1978), Pitt (1978) and Xiao (2006).

Example 2.1. One of the most important examples for Gaussian processes is a fractional Brownian motion $B_H = \{B_H(t)\}_{t \ge 0}$, having a representation:

$$B_H(t) = \kappa_H \int_{-\infty}^{t} \left[((t - u)_+)^{H-1/2} - ((-u)_+)^{H-1/2} \right] dB_u,$$

where $a_+ = \max\{a, 0\}$, $0 < H < 1$, $B = \{B_u\}_{u \in \mathrm{R}}$ is a Brownian motion and $\kappa_H > 0$ is a constant such that $EB_H^2(1) = 1$.

Let $F(t, u) = \kappa_H \left[((t - u)_+)^{H-1/2} - ((-u)_+)^{H-1/2} \right]$. We have, for all $0 \le s < t \le T$,

$$\int_s^t F^2(t, u) du = \kappa_H^2 \int_s^t (t - u)^{2H-1} du = \frac{\kappa_H^2}{2H} (t - s)^{2H},$$

$$\int_{-\infty}^s [F(t, u) - F(s, u)]^2 du = \kappa_H^2 \int_0^\infty [(t - s + u)^{H-1/2} - u^{H-1/2}]^2 du$$

$$= c_0 (t - s)^{2H},$$

where $0 < c_0 = \kappa_H^2 \int_0^\infty [(1 + u)^{H-1/2} - u^{H-1/2}]^2 du < \infty$. Hence (2.19) holds for any $T > 0$ and B_H is LND on R^+.

On the other hand, by noting $\Delta(u, v) = E(X_u - X_v)^2 = |u - v|^{2H}$, we have (2.15) for any $0 < \gamma < 2^{-1}(1/H - 1)$. A simple application of Theorem 2.2 now shows that B_H has a local time $L_{B_H} = \{L_{B_H}(t, x)\}_{t \ge 0, x \in \mathrm{R}}$ which is continuous on $\mathrm{R}^+ \times \mathrm{R}$ and which satisfies

$$\left| L_{B_H}(t, x) - L_{B_H}(t, y) \right| \le C |x - y|^\beta, \quad a.s.,$$

for any $0 < \beta < 2^{-1}(1/H - 1)$, $t \ge 0$ and $x, y \in \mathrm{R}$. Furthermore, $L_{B_H}(t, x)$ can be represented as

$$L_{B_H}(t, x) = \frac{1}{2\pi} \int_{-\infty}^\infty e^{-iux} \int_0^t e^{iuB_H(s)} ds\, du.$$

Example 2.2. Consider a fractional Ornstein-Uhlenbeck process U defined by

$$U_t = \int_0^t e^{-\beta(t-s)} dB_H(s) = B_H(t) + \beta \int_0^t e^{-\beta(t-s)} B_H(s) ds,$$

where $0 < H < 1$, $\beta > 0$ and B_H is a fractional Brownian motion. As in Example 2.1, U has a local time $L_U = \{L_U(t, x)\}_{t \geq 0, x \in \mathbb{R}}$ which is continuous on $\mathbb{R}^+ \times \mathbb{R}$ and which satisfies

$$\left| L_U(t, x) - L_U(t, y) \right| \leq C \left| x - y \right|^\beta, \quad a.s.,$$

for any $0 < \beta < 2^{-1}(1/H - 1)$, $t \geq 0$ and $x, y \in \mathbb{R}$. Furthermore, $L_U(t, x)$ can be represented as

$$L_U(t, x) = \frac{1}{2\pi} \int_{-\infty}^{\infty} e^{-i u x} \int_0^t e^{i u U_s} \, ds \, du.$$

2.1.2 Local time of Lévy process

Let $X = \{X_s\}_{s \geq 0}$ be a Lévy process with characteristics (a, σ^2, ν), i.e., $\mathrm{E} e^{i \theta X_t} = e^{-t \psi(\theta)}$, where

$$\psi(\theta) = -ia\theta + \frac{1}{2}\sigma^2 \theta^2 - \int_{-\infty}^{\infty} \left(e^{i\theta y} - 1 - i\theta y \mathbf{I}_{(|y|<1)} \right) \nu(dy).$$

Here ν is a measure on $\mathbb{R}$ satisfying $\int_{-\infty}^{\infty} (1 \wedge x^2) \nu(dx) < \infty$ and $\nu(\{0\}) = 0$.

Theorem 2.3. *X has a local time L_X satisfying $\int_{-\infty}^{\infty} L_X^2(t, x) dx < \infty, a.s.,$ if and only if*

$$\int_{-\infty}^{\infty} \mathrm{Re} \frac{1}{1 + \psi(\theta)} d\theta < \infty, \tag{2.20}$$

where $\mathrm{Re}(x)$ denotes the real part of x.

Theorem 2.3 is a well-known theorem of Hawkes (1986). Also see Khoshnevisan, et al. (2003a, b). "If" part is easy to prove. Since, for any $\theta \in \mathbb{R}$,

$$\left| \int_0^T e^{i\theta X_u} du \right|^2 \leq C \left| \int_0^{\infty} e^{-u + i\theta X_u} du \right|^2,$$

where $C > 0$ is a constant, it follows from the definition of Lévy process that, for any $T > 0$,

$$\int_{-\infty}^{\infty} \mathrm{E} \left| \int_0^T e^{i\theta X_u} du \right|^2 d\theta \leq C \int_{-\infty}^{\infty} \mathrm{E} \left| \int_0^{\infty} e^{-u + i\theta X_u} du \right|^2 d\theta$$

$$= C \int_{-\infty}^{\infty} \int_0^{\infty} \int_0^{\infty} e^{-u - v - |u - v| \psi[\theta \, \mathrm{sgn}(u - v)]} du \, dv \, d\theta$$

$$= C \int_{-\infty}^{\infty} \mathrm{Re} \frac{1}{1 + \psi(\theta)} d\theta < \infty.$$

Theorem 2.1 now implies the claim.

Let $\gamma = \sup\{\beta \geq 0 : \lim_{|\theta| \to \infty} |\theta|^{-\beta} \operatorname{Re} \psi(\theta) = \infty\}$. γ is called a *lower index* of Lévy process X. We always have $0 < \gamma \leq 2$, and $\gamma = \alpha$ if X is a stable Lévy process with an index α.

Theorem 2.4. *If $\gamma > 1$, then X has a local time L_X which is jointly continuous on $\mathrm{R}^+ \times \mathrm{R}$ and which satisfies*

$$|L_X(t, x) - L_X(t, y)| \leq C |x - y|^\beta, \quad a.s., \tag{2.21}$$

for any $\beta < (\gamma - 1)/2$, $t \geq 0$ and $x, y \in \mathrm{R}$.

Proof.   Under $\gamma > 1$, we have

$$\begin{aligned}
\phi^2(x) &\equiv \int_{-\infty}^{\infty} \left(1 - \cos(\theta x)\right) \operatorname{Re} \frac{1}{1 + \psi(\theta)} d\theta \\
&\leq C \int_{-\infty}^{\infty} |1 - \cos(\theta x)|(1 + |\theta|^{-\gamma}) d\theta \\
&\sim C |x|^{\gamma - 1},
\end{aligned}$$

as $x \to 0$. The result follows from Theorem 2 of Barlow (1988).   □

Example 2.3. Let $\xi_\alpha = \{\xi_\alpha(t)\}_{t \geq 0}$ be a stable Lévy process. If $1 < \alpha \leq 2$, then ξ_α has a jointly continuous local time $L_{\xi_\alpha} = \{L_{\xi_\alpha}(t, x)\}_{t \geq 0, x \in \mathrm{R}}$ satisfying:

$$|L_{\xi_\alpha}(t, x) - L_{\xi_\alpha}(t, y)| \leq C |x - y|^\beta, \quad a.s.,$$

for any $\beta < (\alpha - 1)/2$, and

$$L_{\xi_\alpha}(t, x) = \frac{1}{2\pi} \int_{-\infty}^{\infty} e^{-i u x} \int_0^t e^{i u \xi_\alpha(s)} ds \, du,$$

where the representation is understood as in (2.8).

It is well-known that, except $\alpha = 2$, the process ξ_α has infinitely many jumps in finite time and has no continuous part. Moreover, ξ_α has a local time L_{ξ_α} if and only if $\alpha > 1$. See, e.g., Kesten (1969).

2.1.3   *Local time of semimartingale*

Different notion of local time is used in this section. For a semimartinagle X, we define its local time by using the "occupation time formula" with respect to the continuous part of $[X, X]$, rather than (2.1). There are some advantages for this definition of local time in the theory of semimartingale.

In particular, an explicit formula of local time can be established for a certain class of semimartingales.

Let $\{\Omega, \mathcal{F}, \mathbf{F}, \mathrm{P}\}$ be a stochastic basis and $X = \{X_s\}_{s \geq 0}$ be a semimartingale. The quadratic variation $[X, X]$ of X is a cádlág, increasing, adapted process satisfying

$$[X, X]_t = [X, X]_t^c + X_0^2 + \sum_{0 < s \leq t} (\Delta X_s)^2,$$

where $\Delta X_s = X_s - X_{s-}$ and $[X, X]^c$ denotes the path-by-path continuous part of $[X, X]$. A measurable process $L_X = \{L_X(t, a)\}_{t \geq 0, a \in \mathbb{R}}$ is said to be a *local time* of X if it satisfies

$$\int_0^t h(X_s)d[X, X]_s^c = \int_{-\infty}^{\infty} h(a)L_X(t, a)da, \tag{2.22}$$

for all positive (bounded) Borel functions $h : \mathbb{R} \to \mathbb{R}$.

Define the sgn function by

$$\mathrm{sgn}(x) = \begin{cases} 1, & \text{if } x > 0, \\ -1, & \text{if } x \leq 0. \end{cases}$$

Note that $\mathrm{sgn}(x)$ is the left derivative of $|x|$ and $|x - a|$ is a convex function for any $a \in \mathbb{R}$. For a semimartingale X, the Itô's formula II (see Chapter 3) states that

$$|X_t - a| = |X_0 - a| + \int_0^t \mathrm{sgn}(X_{s-} - a)dX_s + A_t^a, \tag{2.23}$$

where $A^a = \{A_t^a\}_{t \geq 0}$ is an adapted, right continuous, increasing process and

$$\Delta A_t^a = |X_t - a| - |X_{t-} - a| - \mathrm{sgn}(X_{s-} - a)\Delta X_t.$$

Set $l_t^a = A_t^a - \sum_{0 < s \leq t} \Delta A_s^a$. For any $a \in \mathbb{R}$, the process $l^a = \{l_t^a\}_{t \geq 0}$ is the continuous part of A^a. In particular, $l_t^a = A_t^a$, if X is a continuous semimartingale.

Theorem 2.5. *Let $X = \{X_s\}_{s \geq 0}$ be a semimartingale. Then,*

(i) $\{l_t^a\}_{t \geq 0, a \in \mathbb{R}}$ is a local time of X;

(ii) if in addition $\sum_{0 < s \leq t} |\Delta X_s| < \infty$, a.s. for each $t > 0$, then there exists a version of $\{l_t^a\}_{t \geq 0, a \in \mathbb{R}}$ which is continuous in t and right continuous with left limit in a:

$$\lim_{\lim_{s \to t, b \downarrow a}} l_s^b = l_t^a, \qquad \lim_{\lim_{s \to t, b \uparrow a}} l_s^b = l_t^{a-} \quad \text{exists.}$$

Moreover,

$$l_t^a - l_t^{a-} = 2 \int_0^t I_{(X_s=a)} dA_s = 2 \int_0^t I_{(X_s=a)} dX_s,$$

where $X_t = X_0 + M_s + A_s + J_t$ with $J_t = \sum_{0<s\leq t} \Delta X_s$ and we let

$$X - J = M + A \qquad (2.24)$$

be the (unique) canonical decomposition (see Appendix A), i.e., M is a continuous local martingale and A is a continuous finite variation (FV) process with $M_0 = A_0 = 0$.

Proof.   (i) Let $g(x)$ be a positive continuous real function. Let

$$f(x) = \frac{1}{2} \int_{-\infty}^{\infty} |x - y| \, g(y) dy.$$

Then $f'(x) = \frac{1}{2} \int_{-\infty}^{\infty} \text{sgn}(x-y) g(y) dy$ and

$$f''(x) = \int_{-\infty}^{\infty} \delta(x-y) g(y) dy = g(x),$$

where $\delta(x)$ is the Dirac δ-function at 0. By noting

$$\int_0^t f'(X_{s-}) dX_s = \frac{1}{2} \int_{-\infty}^{\infty} \int_0^t \text{sgn}(X_{s-} - a) dX_s \, g(a) da,$$

due to the Fubini's theorem, it follows from Itô's formula I (see Chapter 3) that

$$\frac{1}{2} \int_{-\infty}^{\infty} g(a) \, l_t^a \, da = \frac{1}{2} \int_{-\infty}^{\infty} \Big[A_t^a - \sum_{0<s\leq t} \Delta A_s^a \Big] g(a) da$$

$$= f(X_t) - f(X_0) - \int_0^t f'(X_{s-}) dX_s$$

$$- \sum_{0<s\leq t} \big\{ f(X_s) - f(X_{s-}) - f'(X_{s-}) \Delta X_s \big\}$$

$$= \frac{1}{2} \int_0^t f''(X_{s-}) d[X, X]_s^c$$

$$= \frac{1}{2} \int_0^t g(X_{s-}) d[X, X]_s^c. \qquad (2.25)$$

Since (2.25) holds for any positive, continuous function g, by using a monotone class argument, we have (2.22) for all positive (bounded) Borel functions $h : \text{R} \to \text{R}$. This proves the claim.

   (ii) See Theorem IV.76 and its Corollary 1 of Protter (2005).   □

It follows from Theorem 2.5 that $\int_0^t I_{(X_s=a)} dM_s = 0$ for every $t \geq 0, a \in$ R, and $\{l_t^a\}_{t \geq 0, a \in R}$ is jointly continuous in (t, a) if and only if

$$\int_0^\infty I_{(X_s=a)} |dA_s| = 0.$$

For a continuous local martingale X, as $J = A = 0$ in (2.24), there exists a version of $\{l_t^a\}_{t \geq 0, a \in R}$ which is jointly continuous in $t \geq 0$ and $x \in R$. Furthermore, the map $a \to l_t^a$ is Hölder continuous of order α for every $\alpha < 1/2$ and uniformly in t on every compact interval, as stated in Corollary 1.8 of Revuz and Yor (2003).

Theorem 2.6. *Let* $X = \{X_s\}_{s \geq 0}$ *be a continuous local martingale. For every* $\alpha < 1/2$, *we have*

$$\lambda\{a \geq 0 : \limsup_{\epsilon \downarrow 0} \epsilon^{-\alpha} \sup_{t \in I} |l_t^{a+\epsilon} - l_t^a| > 0\} = 0, \quad a.s.,$$

where λ *is the Lebesque measure on* R^+ *and* I *is an arbitrary compact interval of* R.

It is straightforward by Theorems 2.5–2.6 and the definition of local time that

$$l_t^a = \lim_{\epsilon \to 0} \frac{1}{\epsilon} \int_0^t I(a \leq X_s \leq a + \epsilon) d[X, X]_s^c, \quad a.s. \quad (2.26)$$

$$\frac{l_t^a + l_t^{a-}}{2} = \lim_{\epsilon \to 0} \frac{1}{2\epsilon} \int_0^t I(|X_s - a| \leq \epsilon) d[X, X]_s^c, \quad a.s. \quad (2.27)$$

for every $t \geq 0, a \in R$, and if X is a continuous local martingale

$$l_t^a = \lim_{\epsilon \to 0} \frac{1}{2\epsilon} \int_0^t I(|X_s - a| \leq \epsilon) d[X, X]_s. \quad (2.28)$$

The equalities (2.26)-(2.28) suggest the nature of local time $\{l_t^a\}_{t \geq 0, a \in R}$. The lack of symmetry in (2.26) is due to the definition of sgn function, which is not symmetric in this chapter.

A measurable process $\{l_t^a\}_{t \geq 0, a \in R}$ satisfying (2.26) is usually set as an alternative definition of local time for a semimartingale X. We may use (2.28) as a definition of local time if X is a continuous local martingale.

Example 2.4. Let $B = \{B_t\}_{t \geq 0}$ be a Brownian motion. The process $L_B = \{L_B(t, a)\}_{t \geq 0, a \in R}$, where

$$L_B(t, a) = |B_t - a| - |a| - \int_0^t \text{sgn}(B_s - a) dB_s, \quad (2.29)$$

is a local time of B.

Example 2.4 follows straightforward from Theorem 2.5. Due to $[B, B]_t = t$, $L_B(t, x)$ satisfies both (2.1) and (2.22). Consequently, there is no difference for the local time of B under two different notions.

Example 2.5. Let $B = \{B_t\}_{t \geq 0}$ be a Brownian motion. $|B| = \{|B_t|\}_{t \geq 0}$ is a continuous semimartingale with $[|B|, |B|]_t = t$. A local time L_X of $X = |B|$ is given by

$$L_X(t, a) = \begin{cases} L_B(t, a) + L_B(t, -a), & \text{if } a \geq 0, \\ 0, & \text{if } a < 0, \end{cases} \quad (2.30)$$

where $L_B(t, a)$ is defined as in (2.29). Indeed it follows from (2.26) that

$$L_X(t, a) = \lim_{\epsilon \to 0} \frac{1}{\epsilon} \int_0^t \mathrm{I}(a \leq |B_s| \leq a + \epsilon) ds.$$

Clearly $L_X(t, a) = 0$ if $a < 0$. When $a \geq 0$, we have

$$\begin{aligned} L_X(t, a) &= \lim_{\epsilon \to 0} \frac{1}{\epsilon} \int_0^t \big[\mathrm{I}(a \leq B_s \leq a + \epsilon) + \mathrm{I}(-a - \epsilon \leq B_s \leq -a) \big] ds \\ &= L_B(t, a) + L_B(t, -a), \end{aligned}$$

which yields (2.30). Once again, $L_X(t, x)$ satisfies both (2.1) and (2.22) and hence there is no difference for the local time of $X = |B|$ under two different notions.

It is interesting to notice that $L_X(t, a)$ is not continuous at $a = 0$. Example 2.5 provides an example of continuous semimartingale having a discontinuous local time.

Example 2.6. Let $\{\Omega, \mathcal{F}, \mathbf{F}, \mathrm{P}\}$ be a stochastic basis. Assume that
(i) μ_s is adapted to $\mathbf{F}_s$ and $\int_0^T |\mu_s| ds < \infty$ a.s., and
(ii) σ_s is predictable with respect to $\mathbf{F}$ and $\int_0^T \sigma_s^2 ds < \infty$, a.s.
Let $B = \{B_t\}_{t \geq 0}$ be a Brownian motion and $X = \{X_t\}_{0 \leq t \leq T}, T > 0$, be an Ito's process defined by

$$X_t = X_0 + \int_0^t \mu_s ds + \int_0^t \sigma_s dB_s. \quad (2.31)$$

X is a semimartingale with $[X, X]_t^c = \int_0^t \sigma_s^2 ds$. Using Theorem 2.5, there exists a local time $L_X = \{L_X(t, x)\}_{0 \leq t \leq T, x \in \mathrm{R}}$ satisfying (2.22), i.e.,

$$\int_0^t h(X_s) d[X, X]_s^c = \int_0^t h(X_s) \sigma_s^2 ds = \int_{-\infty}^{\infty} h(a) L_X(t, a) da,$$

for all positive Borel functions $h : \mathrm{R} \to \mathrm{R}$.

If $\sigma_s^2 = \sigma^2(X_s)$, then $l_X = \{l_X(t,x)\}_{0 \le t \le T, x \in \mathbf{R}}$, where $l_X(t,x) = L_X(t,x)/\sigma^2(x)$, satisfying (2.1), i.e.,

$$\int_0^t g(X_s)ds = \int_{-\infty}^\infty g(a)l(t,a)da,$$

for all positive Borel functions $g : \mathbf{R} \to \mathbf{R}$. Hence l_X is a local time of X in the notion based on (2.1). In this situation, there is a relationship $l_X = \sigma^2(x)L_X$ between two different notions of local time.

Example 2.7. Let $X = \{X_s\}_{s \ge 0}$ be a Lévy process with characteristics (a, σ^2, ν). Then X is semimartingale with $[X, X]_t^c = \sigma^2 t$.

If $\sigma^2 \ne 0$, there exist local times in both different notions of local time. The difference is only up to a constant σ^2.

If X is a stable Lévy process with index $1 < \alpha < 2$ (hence $\sigma^2 = 0$ and $[X, X]_t^c \equiv 0$), there is no process $L_X = \{L_X(t,x)\}_{t \ge 0, x \in \mathbf{R}}$ satisfying (2.22) except meaningless $L_X(t,x) \equiv 0$. On the other hand, Example 2.3 shows that there exists a non-zero local time of X in the notion based on (2.1).

Remark 2.3. The notion of local time for a semimartingale based on (2.22) has advantages in Stochastic Calculus, Stochastic Differential Equations and other related areas. Although the class of semimartingales is rather large, the requirement $[X, X]^c \ne 0$ in (2.22) is a restriction. On the other hand, there are processes which are not semimartingales. Examples include fractional Brownian motion and $\{|B_t|^\alpha\}_{t \ge 0}$, where $0 < \alpha < 1$.

The notion of local time based on (2.1) arises naturally in density-like kernel estimation, nonlinear cointegrating regression and other nonparametric estimation with non-stationary time series.

2.2 Convergence to integral functionals of stochastic processes

Let $\{X_{nk}\}_{k \ge 1, n \ge 1}$ be a vector random array and assume that there is a limiting process $X_t = (X_1(t), \cdots, X_d(t))$ such that

$$X_{n,[nt]} \Rightarrow X_t, \quad \text{on } D_{\mathbf{R}^d}[0, \infty). \tag{2.32}$$

This section provides sufficient conditions in such a way that

$$\left(X_{n,[nt]}, \frac{1}{n} \sum_{k=1}^n g(X_{nk}) \right) \Rightarrow \left(X_t, \int_0^1 g(X_s)ds \right), \tag{2.33}$$

on $D_{\mathbf{R}^{d+1}}[0, \infty)$, where $g(x) = g(x_1, ..., x_d)$ is a Borel function on $\mathbf{R}^d$.

2.2.1   Existence of $\int_0^1 g(X_s)ds$

Without loss of generality, assume $g \geq 0$. Otherwise, we consider $g^+ = g\mathrm{I}(g \geq 0)$ and $g^- = g\mathrm{I}(g \leq 0)$, separately. Let $X = \{X_t\}_{t \geq 0}$ be a d-dimensional cádlág process. We say $\int_0^1 g(X_s)ds$ exists if

$$\mathrm{P}\{\omega : \int_0^1 g(X_s)ds < \infty\} = 1.$$

Let $g_N(x) = g(x)\mathrm{I}(\|x\| \leq N)$ and $S^N = \int_0^1 g_N(X_s)ds$. Since

$$\mathrm{P}\big(S^N \neq \int_0^1 g(X_s)ds\big) \leq \mathrm{P}\big(\sup_{0 \leq t \leq 1} \|X_t\| > N\big) \to 0,$$

the existence of $\int_0^1 g(X_s)ds$ is determined by that of S^N for each $N \geq 1$.

Theorem 2.7. $\int_0^1 g(X_s)ds$ *exists if $g(x)$ is a locally bounded function.*

Proof.   For each $N \geq 1$, we have

$$S^N \leq \sup_{\|x\| \leq N} g(x) \int_0^1 \mathrm{I}\big(|X_s| \leq N\big)ds < \infty, \quad a.s.$$

implying the result.   □

Example 2.8. Let $B = \{B_t\}_{t \geq 0}$ be a Brownian motion. Then $\int_0^1 B_s^{-1}ds = \infty$ and $\int_0^1 |B_s|^{-\alpha}ds < \infty$, a.s. if $\alpha < 1$, indicating that, if there are more smooth conditions on X_t, the locally bounded condition on $g(x)$ for the existence of $\int_0^1 g(X_s)ds$ can be reduced.

Theorem 2.8. *Suppose that, for all $t \in [0,1]$, X_t has a density $p_t(x)$ with respect to a d-dimensional Lebesgue measure. Then $\int_0^1 g(X_s)ds$ exists if one of the following condition sets holds.*

(i) $g(x)$ is locally integrable and $\sup_x \int_0^1 p_s(x)ds < \infty$.
(ii) There exists $p > 1$ such that $g^p(x)$ is locally integrable and

$$\int_0^1 \Big(\int_{-\infty}^{\infty} p_s^q(x)dx\Big)^{1/q}ds < \infty,$$

where q satisfies $1/p + 1/q = 1$.

Proof.   Fix $N \geq 1$. Under the condition set (i), it is readily seen that

$$\mathrm{E}\,S^N \leq \int_0^1 \mathrm{E}\,g(X_s)\mathrm{I}(\|X_s\| \leq N)ds$$

$$\leq \int_0^1 \Big(\int_{\|x\| \leq N} g(x)p_s(x)dx\Big)ds$$

$$\leq \sup_{\|x\| \leq N} \int_0^1 p_s(x)ds \int_{\|x\| \leq N} g(x)dx < \infty.$$

Similarly, it follows from Hölder's inequality and the condition set (ii) that

$$\mathrm{E}\, S^N \leq \left(\int_{\|x\| \leq N} g^p(x) dx \right)^{1/p} \int_0^1 \left(\int_{-\infty}^{\infty} p_s^q(x) dx \right)^{1/q} ds < \infty.$$

Hence S^N is finite a.s., indicating that $\int_0^1 g(X_s) ds$ exists. □

Remark 2.4. The condition that X_t has a density $p_t(x)$ is quite natural in many applications, including Gaussian process, stable Lévy process, certain diffusion process and many others.

Example 2.9. Let X_t be a stationary d-dimensional Gaussian process with mean zero and covariance matrix $\sum_t$. Denote by σ_t^2 the minimal eigenvalue of $\sum_t$. Then $\int_0^1 g(X_s) ds$ exists if one of the following condition sets holds.

   (i) $d = 1$, $\int_0^1 \sigma_t^{-1} dt < \infty$ and $g(x)$ is locally integrable.
   (ii) $d \geq 2$, $\int_0^1 \sigma_t^{-d/p} dt < \infty$ and $g^p(x)$ is locally integrable for some
      $p > 1$.

Indeed, due to $X_t \sim N(0, \sum_t)$, its density $p_s(x)$ satisfies:
   if $d = 1$ and $\int_0^1 \sigma_t^{-1} dt < \infty$, then $\sup_x \int_0^1 p_s(x) ds < \infty$;
   if $d \geq 2$ and $\int_0^1 \sigma_t^{-d/p} dt < \infty$, then

$$\int_0^1 \left(\int_{-\infty}^{\infty} p_s^q(x) dx \right)^{1/q} ds \leq C \int_0^1 \sigma_s^{-d/p} ds < \infty,$$

for any $p > 1$ and $1/p + 1/q = 1$. See, e.g., Christopeit (2009). Theorem 2.8 now implies the results.

If $X = \{X_t\}_{t \geq 0}$ is one-dimensional cádlág process, the definition of local time can be used to find alternative conditions imposed on X so that $\int_0^1 g(X_s) ds$ exists.

Definition 2.1. A semimartingale $X = \{X_t, \mathcal{F}_t\}_{t \geq 0}$ is called *regular* if $\sum_{0 < s \leq 1} |\Delta_s| < \infty$, a.s., where $\Delta_s = X_s - X_{s-}$, and $[X, X]_t^c = \int_0^t \sigma^2(s, Z_s) ds$ with $\Delta = \inf_{0 \leq t \leq 1} \sigma^2(s, Z_s) > 0$ a.s., where Z_s is adapted to $\mathcal{F}_s$ and $[X, X]^c$ denotes the path-by-path continuous part of $[X, X]$.

Theorem 2.9. *Suppose* $X = \{X_t\}_{t \geq 0}$ *is one-dimensional cádlág process. Then* $\int_0^1 g(X_s) ds$ *exists if one of the following condition sets holds:*

   (i) $g(x)$ *is locally integrable and* X *has a local time* L_X *so that* $L_X(1, x)$
      is continuous on $\mathbb{R}$.
   (ii) $g^2(x)$ *is locally integrable and* $\int_{-\infty}^{\infty} \mathrm{E} \left| \int_0^1 e^{i\theta X_u} du \right|^2 d\theta < \infty$.

(iii) $g(x)$ is locally integrable and $X = \{X_t, \mathcal{F}_t\}_{t \geq 0}$ is a regular semi-martingale

Proof. It suffices to show $S^N < \infty$, a.s. for each $N \geq 1$. This is simple under the condition sets (i) and (ii). In fact, by the occupation formula (2.1), we have

$$S^N = \int_{-N}^{N} g(x) L_X(1, x) dx.$$

It is readily seen that, under the condition set (i),

$$S^N \leq \sup_{|x| \leq N} L_X(1, x) \int_{-N}^{N} g(x) dx < \infty, \quad a.s.$$

for any $N \geq 1$. Similarly, if the condition set (ii) holds, we have $\int_{-\infty}^{\infty} L_X^2(1, x) dx < \infty$, a.s. by Theorem 2.1, This, together with Hölder's inequality and the local integrability of $g^2(x)$, implies that

$$S^N \leq \left(\int_{-N}^{N} g(x)^2 dx \right)^{1/2} \left(\int_{-N}^{N} L_X^2(1, x) dx \right)^{1/2} < \infty, \quad a.s.,$$

for any $N \geq 1$, as required.

Under the condition set (iii), it follows from the notion of local time in Section 2.1.3 [recalling (2.22)] that

$$S^N = \int_0^1 \frac{g(X_s) \mathrm{I}(|X_s| \leq N)}{\sigma^2(s, Z_s)} d[X, X]_s^c$$

$$\leq \Delta^{-1} \int_{-N}^{N} g(x) L_X(1, x) dx$$

$$\leq \Delta^{-1} \sup_{|x| \leq N} L_X(1, x) \int_{-N}^{N} g(x) dx. \qquad (2.34)$$

As $\sum_{0 < s \leq 1} |\Delta_s| < \infty$, a.s., Theorem 2.5 shows that $L_X(1, x)$ can be chosen continuous in t and cádlág in x. As a consequence, $\sup_{|x| \leq N} L_X(1, x)$ is finite a.s. This, together with the local integrability of $g(x)$ and $\Delta > 0$, a.s., yields $S^N < \infty$, a.s. □

Remark 2.5. Examples for X_t satisfying $\int_{-\infty}^{\infty} \mathrm{E} \left| \int_0^1 e^{i\theta X_u} du \right|^2 d\theta < \infty$ can be found in Section 2.1. The Itô's process X_t defined by

$$X_t = \int_0^t \mu_s ds + \int_0^t \sigma_s dB_s,$$

where $\inf_{0 \leq s \leq 1} \sigma_s^2 > 0$, is a regular semimartingale.

2.2.2 Convergence to $\int_0^1 g(X_s)ds$

As in Section 2.2.1, let $g_N(x) = g(x)\mathrm{I}(\|x\| \leq N)$, $S_{nN} = \frac{1}{n}\sum_{k=1}^n g_N(X_{nk})$ and $S^N = \int_0^1 g_N(X_s)ds$. For $\forall \epsilon > 0$, if (2.32) holds, i.e.,

$$X_{n,[nt]} \Rightarrow X_t, \quad \text{on } D_{\mathrm{R}^d}[0,\infty),$$

(indicating $X = \{X_t\}_{t\geq 0}$ is a d-dimensional cádlág process), then there exist n_0 and N_0 such that

$$\mathrm{P}\big(S^N \neq \int_0^1 g(X_s)ds\big) \leq \mathrm{P}\big(\sup_{0\leq t\leq 1} \|X_t\| \geq N\big) \leq \epsilon,$$

$$\mathrm{P}\Big(S_{nN} \neq \frac{1}{n}\sum_{k=1}^n g(X_{nk})\Big) \leq \mathrm{P}\big(\max_{1\leq k\leq n} \|X_{nk}\| \geq N\big) \leq 2\epsilon,$$

for all $n \geq n_0$ and $N \geq N_0$. Hence (2.33) will follow if we prove

$$\big(X_{[nt]},\ S_{nN}\big) \Rightarrow \big(X_t,\ S^N\big), \quad \text{on } D_{\mathrm{R}^{d+1}}[0,\infty), \tag{2.35}$$

for each $N \geq 1$, as $n \to \infty$.

Theorem 2.10. *Let $g(x)$ be a locally Riemann integrable function. Then (2.32) implies (2.35), and hence (2.33).*

Proof. Without loss of generality, assume $g(x) \geq 0$. Fix $N \geq 1$. Recalling the Riemann integrability of $g_N(x) = g(x)\mathrm{I}(\|x\| \leq N)$, for $\forall \delta > 0$, there exist two positive continuous functions $g_{1N}(x)$ and $g_{2N}(x)$ such that

$$g_{1N}(x) \leq g_N(x) \leq g_{2N}(x),$$

$g_{1N}(x) = g_{2N}(x) = 0$ for $\|x\| \geq (d+1)N$ and

$$g_{2N}(x) - g_{1N}(x) \leq \delta, \quad \text{for all } x \in \mathrm{R}^d. \tag{2.36}$$

Set, for $i = 1$ and 2, $S_i^N = \int_0^1 g_{iN}(X_s)ds$ and

$$S_{nN}^{(i)} = \frac{1}{n}\sum_{k=1}^n g_{iN}\big(X_{nk}\big)$$

$$= \int_0^1 g_{iN}(X_{n,[ns]})ds - \frac{1}{n}g_{iN}(0) + \frac{1}{n}g_{iN}(X_{nn}).$$

Since the mapping $H : D_{\mathrm{R}^d}[0,1] \to \mathrm{R}$ defined by $H(x) = \int_0^1 g_{iN}[x(s)]ds$ is continuous in Skorokhod topology, the continuous mapping theorem and (2.32) imply that, for $i = 1$ and 2,

$$\big(X_{[nt]},\ S_{nN}^{(i)}\big) \Rightarrow \big(X_t,\ S_i^N\big), \quad \text{on } D_{\mathrm{R}^{d+1}}[0,\infty),. \tag{2.37}$$

This implies (2.35) since, for $i = 1$ and 2,

$$|S_i(N) - S^N| \leq S_2^N - S_1^N \leq \delta, \quad |S_{nN}^{(i)} - S_{nN}| \leq S_{nN}^{(2)} - S_{nN}^{(1)} \leq \delta,$$

and δ can be chosen arbitrarily small. $\qquad\square$

Example 2.10. Let $X_{nk} = \sum_{j=1}^{k} \epsilon_j / \sqrt{n}$, where ϵ_j are i.i.d. random variables taking integer values with $E\epsilon_1 = 0$ and $0 < \sigma^2 = E\epsilon_1^2 < \infty$. Let $g(x) = 1$ if x is irrational and $g(x) = 0$ if x is rational. It is well-known that $g(x)$ is locally (Lebesgue) integrable (not locally Riemann integrable),

$$X_{n,[nt]} \Rightarrow \sigma B_t, \quad \text{on } D[0, \infty),$$

where $B = \{B_t\}_{t \geq 0}$ is a Brownian motion. Furthermore, if $n = k^2$ where k is an integer, $\frac{1}{n} \sum_{k=1}^{n} g(X_{nk}) = 0$. However, $\int_0^1 g(\sigma B_s)ds = 1, a.s.$ Hence (2.33) cannot be true.

Example 2.10 indicates that some additional smooth conditions on X_{nk} are required to extend Theorem 2.10.

Definition 2.2. $\{X_{nk}\}_{k \geq 1, n \geq 1}$ is called a *smooth vector array (smooth array for d = 1)* if there exists a sequence of positive constants d_{nk} such that $\sup_{n \geq 1} \frac{1}{n} \sum_{k=1}^{n} \frac{1}{d_{nk}} < \infty$ and X_{nk}/d_{nk} has a density $p_{nk}(x)$ that is uniformly bounded by a constant K for all $x, k \geq 1, n \geq 1$.

Let $f(s, x)$ be a measurable function and $A_0 = \sup_{n \geq 1} \frac{1}{n} \sum_{k=1}^{n} \frac{1}{d_{nk}}$. For a smooth (vector) array X_{nk}, routine calculation shows that, for any $c_n > 0$,

$$\frac{c_n}{n} \sum_{k=1}^{n} E \left| f\left(\frac{k}{n}, c_n X_{nk}\right) \right| \leq A_0 K \int_{-\infty}^{\infty} \sup_{0 \leq s \leq 1} |f(s, x)| dx. \qquad (2.38)$$

Concept of smooth (vector) array and the result (2.38) is vital for the extension of Theorem 2.10 to locally integrable $g(x)$ and other functionals.

Theorem 2.11. *Let $g(x)$ be a locally integrable function and $\{X_{nk}\}_{k \geq 1, n \geq 1}$ be a smooth (vector) array. Then (2.32) implies (2.35), and hence (2.33), if one of the following condition sets holds:*

   (i) *for all $t \in [0, 1]$, X_t has a density $p_t(x)$ with respect to d-dimensional Lebesgue measure satisfying $\sup_x \int_0^1 p_s(x)ds < \infty$;*
   (ii) *(2.32) holds with $d = 1$ and X has a local time L_X so that $L_X(1, x)$ is continuous on $\mathbb{R}$;*
   (iii) *(2.32) holds with $d = 1$ and $X = \{X_t, \mathcal{F}_t\}_{t \geq 0}$ is a regular semimartingale.*

Proof. Fix $N \geq 1$. For any $\delta > 0$, by noting

$$\int_{-\infty}^{\infty} |g_N(x)| dx = \int_{\|x\| \leq N} g(x)dx < \infty,$$

Lusin's theorem implies that there is a continuous function $\widetilde{g}_N(x)$ such that $\int_{-\infty}^{\infty} |\widetilde{g}_N(x)| dx < \infty$ and

$$\int_{-\infty}^{\infty} |\widetilde{g}_N(x) - g_N(x)| dx \leq \delta. \tag{2.39}$$

Set $\widetilde{S}^N = \int_0^1 \widetilde{g}_N(X_s) ds$ and $\widetilde{S}_{nN} = \frac{1}{n} \sum_{k=1}^n \widetilde{g}_N(X_{nk})$. As in (2.37), it follows from the continuous mapping theorem that

$$\left(X_{[nt]}, \widetilde{S}_{nN}\right) \Rightarrow \left(X_t, \widetilde{S}^N\right), \quad \text{on } D_{\mathbb{R}^{d+1}}[0, \infty).$$

On the other hand, by (2.38) and (2.39),

$$\mathrm{E}\left|\widetilde{S}_{nN} - S_{nN}\right| \leq \frac{1}{n} \sum_{k=1}^n \mathrm{E}\left|\widetilde{g}_N\left(X_{nk}\right) - g_N\left(X_{nk}\right)\right| \leq C\,\delta.$$

Hence, to establish (2.35), it suffices to show

$$\left|\widetilde{S}^N - S^N\right| = O_P(\delta),$$

under one of the condition sets (i)–(iii). This follows from (2.39) and the similar arguments as in the proofs of Theorems 2.8 and 2.9. We omit the details. $\square$

2.2.3 Supplement and generalization

Using the similar arguments as that in previous sections, this section investigates the asymptotic behaviors of functionals that are somewhat different but of a sufficient similar form. We only consider one-dimensional cádlág process $X = \{X_s\}_{s \geq 0}$ on $D[0, \infty)$. Extension to multidimensional settings is straightforward.

Theorem 2.12. *(i) For any locally Riemann integrable function* $g(x_1, \cdots, x_d)$, *we have*

$$S_d = \int_0^1 \cdots \int_0^1 g(X_{t_1}, \cdots, X_{t_d}) dt_1 \cdots dt_d,$$

exists, and (2.32) implies

$$n^{-d} \sum_{k_1=1}^n \cdots \sum_{k_d=1}^n g(X_{n,k_1}, ..., X_{n,k_d}) \to_D S_d. \tag{2.40}$$

(ii) Suppose that, for all $t_1, ..., t_d \in [0, 1]$, $(X_{t_1}, \cdots, X_{t_d})$ *has a density* $p_{t_1,...,t_d}(x)$ *satisfying*

$$\sup_x \int_0^1 \cdots \int_0^1 p_{t_1,...,t_d}(x) dt_1 \cdots dt_d < \infty.$$

For any locally integrable function $g(x_1, \cdots, x_d)$, *we have* S_d *exists, and if in addition* $\{X_{nk}\}_{k \geq 1, n \geq 1}$ *is a smooth array, then (2.32) implies (2.40).*

Proof. The proof is along the same line as that in Theorems 2.10 and 2.11. ☐

Theorem 2.13. *Let $g(x, y)$ be a locally Riemann integrable function on $[0, 1] \times \mathrm{R}$. Then $\int_0^1 g(s, X_s) ds$ exists, and (2.32) implies*

$$\left(X_{n, [nt]}, \frac{1}{n} \sum_{k=1}^{n} g\left(\frac{k}{n}, X_{nk}\right) \right) \Rightarrow \left(X_t, \int_0^1 g(s, X_s) ds \right), \qquad (2.41)$$

on $D_{\mathrm{R}^2}[0, \infty)$.

Proof. Let $g_N(x, y) = g(x, y) \mathrm{I}(|y| \leq N)$. As in Section 2.2.2, it suffices to prove (2.41) with $g(x, y)$ being replaced by $g_N(x, y)$. The remaining of the proof is similar to that of Theorem 2.10. We omit the details. ☐

Definition 2.3. A real function $g(s, x) : [0, 1] \times \mathrm{R} \to \mathrm{R}$ is called P-*locally integrable* if, for each $N \geq 1$ and $\forall \delta > 0$, there exists a continuous function $\widetilde{g}_N(s, y)$ such that $\int_{-\infty}^{\infty} \sup_{0 \leq s \leq 1} |\widetilde{g}_N(s, y)| dy < \infty$ and

$$\int_{-\infty}^{\infty} \sup_{0 \leq s \leq 1} |g(s, y) \mathrm{I}(|y| \leq N) - \widetilde{g}_N(s, y)| dy \leq \delta. \qquad (2.42)$$

It is easy to verify that if $a(s)$ is uniformly bounded on $[0, 1]$ and $v(x)$ is locally integrable, then $g(s, x) = a(s) v(x)$ is P-locally integrable.

Theorem 2.14. *Suppose that, for all $t \in [0, 1]$, X_t has a density $p_t(x)$ satisfying $\sup_x \int_0^1 p_s(x) ds < \infty$.*

(i) *For any* P-*locally integrable function $g(s, x)$, we have $\int_0^1 g(s, X_s) ds$ exists, and if $\{X_{nk}\}_{k \geq 1, n \geq 1}$ is a smooth array, then (2.32) implies (2.41).*

(2) *If in addition $g_n(s, y)$ is a real function such that*

$$\int_{-N}^{N} \sup_{0 \leq s \leq 1} |g_n(s, y) - g(s, y)| dy = o(1), \qquad (2.43)$$

for all $N \geq 1$, then

$$\left(X_{n, [nt]}, \frac{1}{n} \sum_{k=1}^{n} g_n\left(\frac{k}{n}, X_{nk}\right) \right) \Rightarrow \left(X_t, \int_0^1 g(s, X_s) ds \right), \quad (2.44)$$

on $D_{\mathrm{R}^2}[0, \infty)$.

Proof. The proof for $\int_0^1 g(s, X_s)ds < \infty$, a.s. is similar to that of Theorem 2.9. Let $g_N(x, y) = g(x, y)\mathrm{I}(|y| \leq N)$. For the $\widetilde{g}_N(s, x)$ in (2.42), it follows from (2.38) and Theorem 2.13 that

$$\frac{1}{n}\sum_{k=1}^n \mathrm{E}\left|\widetilde{g}_N\left(\frac{k}{n}, X_{nk}\right) - g_N\left(\frac{k}{n}, X_{nk}\right)\right| \leq C\,\delta, \tag{2.45}$$

$$\frac{1}{n}\sum_{k=1}^n \widetilde{g}_N\left(\frac{k}{n}, X_{nk}\right) \to_D \int_0^1 \widetilde{g}_N(s, X_s)ds. \tag{2.46}$$

On the other hand, by the smooth condition on X_t, we have

$$\int_0^1 \mathrm{E}\left|\widetilde{g}_N(s, X_s) - g_N(s, X_s)\right|ds$$

$$\leq \int_0^1 \int_{-\infty}^\infty \left|\widetilde{g}_N(s, y) - g_N(s, y)\right|p_s(y)dy\,ds$$

$$\leq \sup_x \int_0^1 p_s(x)ds \int_{-\infty}^\infty \sup_{0 \leq s \leq 1}|g(s, y)\mathrm{I}(|y| \leq N) - \widetilde{g}_N(s, y)|dy$$

$$\leq C\,\delta. \tag{2.47}$$

Due to the arbitrarity of δ, (2.45)-(2.47) imply (2.41) whenever $g(x, y)$ is replaced by $g_N(x, y)$. Now, by using the same arguments as in Section 2.2.2, (2.41) holds for any P-locally integrable function $g(s, x)$.

In order to prove (2.44), it suffices to show

$$\frac{1}{n}\sum_{k=1}^n \sup_{0 \leq s \leq 1}|g_n(s, X_{nk}) - g(s, X_{nk})| = o_\mathrm{P}(1),$$

which follows from that, for all $N \geq 1$,

$$\frac{1}{n}\sum_{k=1}^n \mathrm{E}\sup_{0 \leq s \leq 1}|g_{nN}(s, X_{nk}) - g_N(s, X_{nk})| = o(1),$$

where $g_{nN}(x, y) = g_n(x, y)\mathrm{I}(|y| \leq N)$, as $n \to \infty$. This is a consequence of (2.38) and (2.43). The details are omitted. □

Remark 2.6. It is readily seen that $g(s, t) = s^{-\alpha}t^{-1/2}$, where $1/2 < \alpha < 1$, is locally integrable, but $\int_0^1 s^{-\alpha}|B_s|^{-1/2}ds$ does not exist, where $B = \{B_t\}_{t \geq 0}$ is a Brownian motion. This example indicates, in Theorem 2.14, P-local integrability of $g(x, y)$ cannot be reduced to the condition that $g(x, y)$ is locally integrable.

Definition 2.4. A real function $g(s,x) : [0,1] \times \mathrm{R} \to \mathrm{R}$ is called P-*asymptotically homogeneous* if

$$g(s,\lambda x) = \nu(\lambda)H(s,x) + R(s,x,\lambda)$$

where $H(s,x)$ is locally Riemann integrable and $\mathrm{R}(s,x,\lambda)$ is such that

(i) $\sup_{0 \leq s \leq 1} |R(s,x,\lambda)| \leq a(\lambda)\,T_1(x)$, where $\lim_{\lambda \to \infty} a(\lambda)/\nu(\lambda) = 0$ and $\overline{T_1}(x)$ is locally bounded; or

(ii) $\sup_{0 \leq s \leq 1} |R(s,x,\lambda)| \leq b(\lambda)\,T_2(\lambda x)$, where $\lim_{\lambda \to \infty} b(\lambda)/\nu(\lambda) < \infty$ and $\overline{T_2}(x)$ is bounded satisfying $T_2(x) \to 0$ as $x \to \infty$.

Typical examples for P-asymptotically homogeneous functions $g(s,x)$ include: $s(x^k + a_1 x^{k-1} + ... + a_k)$, where $a_1, ..., a_k$ are real constants;

$$s \log |x|; \quad (s+x)^2; \quad x[1 + (1+s)x]^{-1}\mathrm{I}(x \geq 0).$$

Theorem 2.15. *Suppose (2.32) holds and $\int_0^1 \mathrm{I}(|X_s| \leq x)ds = o_P(1)$, as $x \to 0$. Then, for any $0 < c_n \to \infty$ and any* P-*asymptotically homogeneous function $g(s,x)$, we have*

$$\left(X_{n,[nt]}, \ \frac{1}{n\nu(c_n)} \sum_{k=1}^n g\left(\frac{k}{n}, c_n X_{nk}\right)\right) \Rightarrow \left(X_t, \ \int_0^1 H(s,X_s)ds\right), \quad (2.48)$$

on $D_{\mathrm{R}^2}[0,\infty)$.

Proof. Due to Theorem 2.13, it suffices to show that

$$\frac{1}{n}\sum_{k=1}^n T_1(X_{nk}) = O_P(1), \quad (2.49)$$

where $T_1(x)$ is locally bounded, and for any $0 < c_n \to \infty$,

$$\frac{1}{n}\sum_{k=1}^n T_2(c_n X_{nk}) = o_P(1), \quad (2.50)$$

where $T_2(x)$ is bounded satisfying $T_2(x) \to 0$ as $|x| \to \infty$.

We only prove (2.50) as (2.49) is similar. For any $\delta > 0$, there exists $K > 0$ such that $|T_2(x)| \leq \delta$ whenever $|x| \geq K$. This, together with the boundedness of $T_2(x)$, implies that, for any $\delta > 0$,

$$\frac{1}{n}\sum_{k=1}^n |T_2(c_n X_{nk})| \leq \delta + \frac{1}{n}\sum_{k=1}^n |T_2(c_n X_{nk})|I(c_n|X_{nk}| \leq K)$$

$$\leq \delta + \frac{C}{n}\sum_{k=1}^n I(|X_{nk}| \leq K/c_n). \quad (2.51)$$

Since $g(s) = I(|s| \leq x)$ is locally Riemann integrable, it follows from Theorem 2.13 that, for any $x > 0$,

$$\frac{1}{n} \sum_{k=1}^{n} I(|X_{nk}| \leq x) \rightarrow_D \int_0^1 I(|X_s| \leq x)ds. \qquad (2.52)$$

Recalling $\int_0^1 I(|X_s| \leq x)ds \rightarrow_P 0$, as $x \rightarrow 0$, (2.50) follows from (2.51)-(2.52) and $c_n \rightarrow \infty$. $\qquad \square$

Theorem 2.15 can be strengthened if we impose more smooth conditions on X_{nk} and X_t.

Theorem 2.16. *Suppose (2.32) holds, $\{X_{nk}\}_{k \geq 1, n \geq 1}$ is a smooth array and, for all $t \in [0,1]$, X_t has a density $p_t(x)$ satisfying $\sup_x \int_0^1 p_s(x)ds < \infty$. Suppose there exists a P-locally integrable function $H(s,x)$ such that, for all $N \geq 1$,*

$$\lim_{\lambda \to \infty} \int_{-N}^{N} \sup_{0 \leq s \leq 1} \left| \frac{g(s, \lambda x)}{\nu(\lambda)} - H(s, x) \right| dx = 0, \qquad (2.53)$$

where $\nu(\lambda)$ is a positive function away from zero. Then, for any $0 < c_n \to \infty$, (2.48) holds.

Proof. Due to Theorem 2.14, it suffices to show that

$$\frac{1}{n} \sum_{k=1}^{n} \sup_{0 \leq s \leq 1} \left| \frac{g(s, c_n X_{nk})}{\nu(c_n)} - H(s, X_{nk}) \right| = o_P(1).$$

This is true by (2.38) and (2.53). $\qquad \square$

2.3 Convergence to local time

Let $g(s,x) : [0,1] \times \mathbb{R} \to \mathbb{R}$ be a real function satisfying $\int_{-\infty}^{\infty} \tilde{g}(x)dx < \infty$, where $\tilde{g}(x) = \sup_{0 \leq s \leq 1} |g(s,x)|$. Let $X = \{X_t\}_{t \geq 0}$ be a process having a continuous local time L_X. By the occupation time formula (2.3) and the dominate convergence theorem, we have

$$c_n \int_0^1 g(t, c_n X_t)dt = \int_0^1 \int_{-\infty}^{\infty} g(t, x) L_X(dt, x/c_n)dx$$

$$\rightarrow \int_0^1 G(t) L_X(dt, 0), \quad a.s., \qquad (2.54)$$

for any $0 < c_n \to \infty$, where $G(t) = \int_{-\infty}^{\infty} g(t,x)dx$. In particular, if $g(s,x) = f(x)$ is integrable, then

$$c_n \int_0^1 f(c_n X_t)dt \rightarrow \tau L_X(1,0), \quad a.s.,$$

where $\tau = \int_{-\infty}^{\infty} f(x)dx$. This section is devoted to the extension of (2.54) to random arrays $\{X_{nk}\}_{k\geq 1, n\geq 1}$ in such a way that

$$\frac{c_n}{n} \sum_{k=1}^{n} g\Big(\frac{k}{n}, c_n X_{nk}\Big) \to_D \int_0^1 G(t) L_X(dt, 0),$$

for any $c_n \to \infty$ and $c_n/n \to 0$.

We introduce the definition of strong smooth array in Section 2.3.1, Sections 2.3.2 and 2.3.5 provide frameworks for convergence to local time. Some examples that satisfy certain smooth conditions are given in Sections 2.3.3 and 2.3.6.

2.3.1   *Strong smooth array: definition and examples*

Smooth conditions on X_{nk}, required for establishing the convergence to local time, were originally introduced in Wang and Phillips (2009a).

Definition 2.5. A random array $\{X_{nk}\}_{k\geq 1, n\geq 1}$ is called *strong smooth* if there exists a sequence of constants $d_{nkj} > 0$ such that

(i) for any $0 < \delta \leq 1$, some $0 < \alpha < 1$ and $\beta > 0$,

$$\limsup_{n\to\infty} \frac{\delta^{-\beta}}{n} \max_{0\leq j<k\leq \delta n} \sum_{k=j+1}^{\delta n} (d_{nkj})^{-1} < \infty, \qquad (2.55)$$

$$\liminf_{n\to\infty} \delta^{-\alpha/2} \inf_{j+\delta n\leq k\leq n} (d_{nkj})^{-1} > 0; \qquad (2.56)$$

(ii) for $k \geq 1$ and $n \geq 1$, X_{nk}/d_{nk0} has a density $h_{nk0}(x)$ which is uniformly bounded by a constant K;

(iii) for $j \geq 1, n \geq 1$ and $k \geq j + n_0$ where n_0 is some positive integer, conditional on $\mathcal{F}_{nj} = \sigma(X_{n1}, ..., X_{nj})$, $(X_{nk} - X_{nj})/d_{nkj}$ has a density $h_{nkj}(x)$ which is uniformly bounded by a constant K and

$$\sup_{u\in\mathrm{R}} \big|h_{nkj}(u+t) - h_{nkj}(u)\big| \leq C \min\{|t|, 1\}, \qquad (2.57)$$

for $t \in \mathrm{R}$, where C is a constant.

In many applications, the d_{nkj} forms a numerical sequence such that, conditional on $\mathcal{F}_{nj}$, $(X_{nk} - X_{nj})/d_{nkj}$ has a limit distribution as $k - j \to \infty$. Typically we may have $d_{nkj} \sim [(k-j)/n]^d$, where $0 < d < 1$, which enables the verifications of (2.55) and (2.56) trivial. The integer n_0 in (iii) may be taken to be sufficiently large, but not depending on n. Due to the conditional arguments, this may be necessary for certain problems.

If $\{X_{nk}\}_{k\geq 1, n\geq 1}$ is a strong smooth array, it is trivial to have (2.38). We further have the following bounds for conditional moments of X_{nk}, which are vital in establishing the convergence to local time.

Theorem 2.17. *Suppose* $\{X_{nk}\}_{k\geq 1, n\geq 1}$ *is a strong smooth array. Then, for any real function* $p(x)$ *on* R *with* $\int_{-\infty}^{\infty} |p(x)| dx < \infty$, *we have*

(i) for any $k \geq 1$ *and* $n \geq 1$,

$$\mathrm{E}\, |p(X_{nk})| \leq K d_{nk0}^{-1} \int_{-\infty}^{\infty} |p(x)| dx; \tag{2.58}$$

(ii) for $j \geq 1, n \geq 1$ *and* $k \geq j + n_0$,

$$\mathrm{E}\left\{ |p(X_{nk})| \mid \mathcal{F}_{nj} \right\} \leq K\, d_{nkj}^{-1} \int_{-\infty}^{\infty} |p(x)| dx. \tag{2.59}$$

If in addition $\int_{-\infty}^{\infty} p(x) dx = 0$, *then*

(iii) for $j \geq 1, n \geq 1$ *and* $k \geq j + n_0$,

$$\left| \mathrm{E}\left[p(X_{nk}) \mid \mathcal{F}_{nj} \right] \right|$$
$$\leq C\, d_{nkj}^{-1} \min\left\{ d_{nkj}^{-1} \inf_t \int_{-\infty}^{\infty} |p(y-t)|\, |y|\, dy,\ 1 \right\}, \tag{2.60}$$

for some constant $C > 0$.

Proof. We only prove (2.60). Others are simple. In fact, uniformly for $t \in$ R, it follows from $\int_{-\infty}^{\infty} p(x) dx = 0$ and (2.57) that

$$\left| \mathrm{E}\left[p(X_{nk}) \mid \mathcal{F}_{nj} \right] \right| = \left| \int_{-\infty}^{\infty} p(X_{nj} + d_{nkj}\, y)\, h_{nkj}(y)\, dy \right|$$

$$\leq d_{nkj}^{-1} \int_{-\infty}^{\infty} p(y-t)\, \Big| h_{nkj}\big[(y - t - X_{nj})/d_{nkj} \big]$$
$$\qquad\qquad - h_{nkj}\big[-(t + X_{nj})/d_{nkj} \big] \Big| dy$$

$$\leq C\, d_{nkj}^{-1} \int_{-\infty}^{\infty} |p(y-t)|\, \min\{ d_{nkj}^{-1} |y|, 1 \}\, dy$$

$$\leq C\, d_{nkj}^{-1} \min\left\{ d_{nkj}^{-1} \int_{-\infty}^{\infty} |p(y-t)|\, |y|\, dy,\ 1 \right\},$$

as required. $\qquad\qquad\qquad\qquad\qquad\qquad\qquad\qquad\qquad\qquad\qquad\square$

Example 2.11. Let $\{\epsilon_j, j \geq 1\}$ be a stationary sequence of Gaussian random variables with $\mathrm{E}\epsilon_1 = 0$ and covariances $\gamma(j - i) = \mathrm{E}\epsilon_i \epsilon_j$ satisfying the following condition, for some $0 < \alpha < 2$,

$$d_n^2 := \sum_{1 \leq i, j \leq n} \gamma(j - i)\ \sim\ n^\alpha h(n),$$

as $n \to \infty$, where $h(n)$ is a slowly varying function at ∞.

Set $d_{nkj} = d_{k-j}/d_n$ and $X_{nk} = S_k/d_n$, where $S_k = \sum_{j=1}^{k} \epsilon_j$. It is readily seen that (2.55) and (2.56) are satisfied and $X_{nk}/d_{nk0} = S_k/d_k$ has a bounded density function $h_{nk0}(x)$. On the other hand, by noting

$$(S_k, S_l - S_k) \sim N(0, \textstyle\sum), \quad \text{where } \textstyle\sum = \begin{pmatrix} d_k^2 & \tilde{\gamma}_{lk} \\ \tilde{\gamma}_{lk} & d_{l-k}^2 \end{pmatrix}$$

and $\tilde{\gamma}_{lk} = cov(S_k, S_l) - d_k^2$, the conditional distribution of $S_l - S_k$ given S_k is $N(\tilde{\gamma}_{lk} S_k/d_k^2, \ d_{lk}^{*2})$, where $d_{lk}^{*2} = d_{l-k}^2 - \tilde{\gamma}_{lk}/d_k^2$. This implies that, conditional on $\epsilon_1, ..., \epsilon_k$,

$$(X_{nl} - X_{nk})/d_{nlk} = (S_l - S_k)/d_{l-k} \sim N\big(*, \ 1 - \tilde{\gamma}_{lk}/(d_k^2 d_{l-k}^2)\big).$$

Therefore there exists an n_0 such that, conditional on $\epsilon_1, ..., \epsilon_k$ and $l - k \geq n_0$, $(X_{nl} - X_{nk})/d_{nlk}$ has a bounded density $h_{nlk}(x)$, which satisfies

$$\sup_x \big| h_{nlk}(x) - h_{nlk}(x + u) \big|$$

$$\leq C \sup_x |e^{-(u+x)^2/2} - e^{-x^2/2}| \leq A \min\{|u|, 1\}.$$

Combining these facts, $\{X_{nk}\}_{k \geq 1, n \geq 1}$ is a strong smooth array.

Example 2.12. Suppose $\epsilon_j, j \in \mathbb{Z}$ is a sequence of i.i.d. random variables with $E\epsilon_0 = 0$ and $E\epsilon_0^2 = 1$. Consider a linear process $\xi_j, j \geq 1$, defined by

$$\xi_j = \sum_{k=0}^{\infty} \phi_k \, \epsilon_{j-k},$$

where the coefficients $\phi_k, k \geq 0$, satisfy one of the following conditions:

LM. $\phi_k \sim k^{-\mu} \rho(k)$, where $1/2 < \mu < 1$ and $\rho(k)$ is a function slowly varying at ∞.

SM. $\sum_{k=0}^{\infty} |\phi_k| < \infty$ and $\phi \equiv \sum_{k=0}^{\infty} \phi_k \neq 0$.

Let $\tau \geq 0$ be a constant, $\gamma = 1 - \tau/n$, $y_{n0} = 0$,

$$y_{nk} = \gamma \, y_{n,k-1} + \xi_k \quad \text{and} \quad Z_{nk} = y_{nk}/d_n, \tag{2.61}$$

where $d_n^2 = var(\sum_{j=1}^{n} \xi_j)$. It is well-known that

$$d_n^2 = \begin{cases} c_\mu \, n^{3-2\mu} \, \rho^2(n), & \text{under } \textbf{LM}, \\ \phi^2 \, n, & \text{under } \textbf{SM}, \end{cases} \tag{2.62}$$

where $c_\mu = \frac{1}{(1-\mu)(3-2\mu)} \int_0^\infty x^{-\mu}(x+1)^{-\mu} dx$. See, Wang, et al. (2003), for instance.

Theorem 2.18. *Suppose* $\lim_{|t| \to \infty} |t|^\eta |E \, e^{it\epsilon_0}| < \infty$ *for some* $\eta > 0$. *Then* $\{Z_{nk}\}_{k \geq k_0, n \geq 1}$ *is a strong smooth array for some* $k_0 \geq 1$.

Proof. Letting $a_k = \sum_{j=0}^{k} \gamma^{k-j} \phi_j$ and $\widetilde{y}_n = \sum_{k=0}^{n} a_k \, \epsilon_k$, we first claim the fact:

F. there exists an $n_0 \geq 1$ such that, for $n \geq n_0$, $\widetilde{y}_n/d_n$ has a density $h_n(x)$ that is uniformly bounded by a constant.

Indeed, by recalling $\gamma = 1 - \tau/n$, there exist $0 < c_1 < c_2 < \infty$ and $n_1 \geq 1$ such that

$$c_1/\sqrt{n} \leq |a_k|/d_n \leq c_2/\sqrt{n}, \quad c_1 d_n^2 \leq \sum_{i=1}^{n} a_i^2 \leq c_2 d_n^2, \qquad (2.63)$$

for all $n \geq n_1$ and for $n/2 \leq k \leq n$. On the other hand, due to $\lim_{|t|\to\infty} |t|^{\eta} |\mathrm{E}\, e^{it\epsilon_0}| < \infty$ for some $\eta > 0$, there exists a $\delta_0 \geq \max\{2, 1/c_2\}$ such that

$$|\mathrm{E}\, e^{it\epsilon_0}| \leq t^{-\eta/2}, \quad \text{for } |t| \geq \delta_0.$$

This, in turn, implies that there exist $\gamma > 0$ and $\delta_1 \geq \max\{\delta_0, 2, 1/c_1\}$ such that

$$|\mathrm{E}\, e^{it\epsilon_0}| \leq e^{-\gamma t^2}, \quad \text{for } |t| \leq \delta_1.$$

See, e.g., Feller (1971, Chapter 8). Consequently, we have

$$\int_{-\infty}^{\infty} (1 + |t|) |\mathrm{E}\, e^{it\widetilde{y}_n/d_n}| dt$$

$$\leq \left(\int_{|t| \leq \delta_1 \sqrt{n}} + \int_{|t| > \delta_0 \sqrt{n}} \right) \prod_{j=[n/2]+1}^{n} |\mathrm{E}\, e^{ita_j \epsilon_j/d_n}| dt$$

$$\leq \int_{-\infty}^{\infty} (1 + |t|) e^{-\gamma t^2/2} dt + \int_{|t| \geq 2} t^{-\eta n/2} dt < \infty, \qquad (2.64)$$

uniformly for $n \geq \max\{n_1, 1 + 2/\eta\}$. This yields the fact F by chosen $n_0 = \max\{n_1, 1 + 2/\eta\}$ (see, e.g., Lukács, 1970, Theorem 3.2.2).

We now prove Theorem 2.18 by verifying (i)-(iii) of Definition 2.5 with $d_{nkj} = d_{k-j}/d_n$. (i) is trivial. Note that

$$
\begin{aligned}
y_{nk} - y_{nj} &= \sum_{i=j+1}^{k} \gamma^{k-i} \xi_i + \sum_{i=1}^{j} (\gamma^{k-i} - \gamma^{j-i}) \xi_i \\
&= \sum_{i=j+1}^{k} \gamma^{k-i} \left(\sum_{u=j+1}^{i} + \sum_{u=-\infty}^{j} \right) \epsilon_u \phi_{i-u} + \sum_{i=1}^{j} (\gamma^{k-i} - \gamma^{j-i}) \xi_i \\
&:= y_{nkj}^{(1)} + y_{nkj}^{(2)}, \qquad (2.65)
\end{aligned}
$$

where $y_{nkj}^{(1)} = \sum_{i=j+1}^{k} \epsilon_i a_{k,i}$ with

$$a_{k,i} = \sum_{u=i}^{k} \gamma^{k-u} \phi_{u-i} = a_{k-i},$$

and $y_{nkj}^{(2)}$ depends only on $\epsilon_j, \epsilon_{j-1}, \ldots$ To see (iii) of Definition 2.5, take $k - j$ sufficiently large, say $k - j \geq n_0$, where n_0 is as in the fact $\mathbb{F}$. Since $y_{nkj}^{(1)} =_D \tilde{y}_{k-j}$, it follows from the fact $\mathbb{F}$ and the independence of ϵ_j that, conditional on $\mathcal{F}_{nj} = \sigma(\epsilon_i, -\infty < i \leq j)$,

$$(Z_{nk} - Z_{nj})/d_{nkj} =_D \tilde{y}_{k-j}/d_{k-j} + y_{nkj}^{(2)} \qquad (2.66)$$

has a density $h_{k-j}(x - y_{nkj}^{(2)}/d_{k-j})$, which is uniformly bounded by a constant K. Furthermore it follows from (2.64) and the inversion formula of $\mathrm{E}\, e^{it\tilde{y}_n/d_n}$ that, for any $u \in \mathrm{R}$ and $k - j \geq n_0$,

$$\sup_x \left| h_{k-j}(x - y_{nkj}^{(2)}/d_{k-j} + u) - h_{k-j}(x - y_{nkj}^{(2)}/d_{k-j}) \right|$$

$$\leq \sup_x |h_{k-j}(x + u) - h_{k-j}(x)|$$

$$\leq C \sup_x \left| \int_{-\infty}^{\infty} \left(e^{-it(x+u)} - e^{-itx} \right) \mathrm{E}\, e^{it\tilde{y}_{k-j}/d_{k-j}} dt \right|$$

$$\leq C \min\{|u|, 1\} \int_{-\infty}^{\infty} (1 + |t|) \left| \mathrm{E}\, e^{it\tilde{y}_{k-j}/d_{k-j}} \right| dt$$

$$\leq C_1 \min\{|u|, 1\},$$

as required. As for (ii) of Definition 2.5, it only needs to take $y_{nj} = 0$ in (2.65) and then follows the verification of (iii). Depending on Definition 2.5, $\{Z_{nk}\}_{k \geq k_0, n \geq 1}$ is a strong smooth array. This completes the proof. $\square$

2.3.2   Convergence to local time: framework I

Throughout this section, suppose that $\{X_{nk}\}_{k \geq 1, n \geq 1}$ is a strong smooth array and there exists a stochastic process $X = \{X_t\}_{t \geq 0}$ having a continuous local time L_X such that $X_{n,[nt]} \Rightarrow X_t$ on $D[0, \infty)$.

Theorem 2.19. *Let $\tilde{g}(x) = \sup_{0 \leq s \leq 1} |g(s, x)|$. Then, for any real function $g(s, x)$ such that $G(t) = \int_{-\infty}^{\infty} g(t, x) dx$ is Riemann integrable on $[0, 1]$ and $\int_{-\infty}^{\infty} [\tilde{g}(x) + \tilde{g}^2(x)] dx < \infty$, we have*

$$\left(X_{n,[nt]}, \frac{c_n}{n} \sum_{k=1}^{n} g\left(\frac{k}{n}, c_n X_{nk} \right) \right) \Rightarrow \left(X_t, \int_0^1 G(t) L_X(dt, 0) \right), \qquad (2.67)$$

on $D_{\mathrm{R}^2}[0, \infty)$, for any $c_n \to \infty$ and $c_n/n \to 0$.

Proof. Write $L_n = \frac{c_n}{n} \sum_{k=1}^n g\left(\frac{k}{n}, c_n X_{nk}\right)$ and

$$L_{n,\epsilon} = \frac{c_n}{n} \sum_{k=1}^n \int_{-\infty}^{\infty} g\left[\frac{k}{n}, c_n \left(X_{nk} + z\epsilon\right)\right] \phi(z) dz,$$

where $\phi(x) = \phi_1(x)$ with $\phi_\epsilon(x) = \frac{1}{\epsilon\sqrt{2\pi}} \exp\left\{-\frac{x^2}{2\epsilon^2}\right\}$. It suffices to show that, for any $\epsilon > 0$,

$$L_{n,\epsilon} - \frac{1}{n} \sum_{k=1}^n G(k/n)\, \phi_\epsilon(X_{nk}) = o_P(1), \tag{2.68}$$

and

$$\lim_{\epsilon \to 0} \lim_{n \to \infty} E\,|L_n - L_{n,\epsilon}| = 0. \tag{2.69}$$

Indeed, by noting that $G(s)\phi_\epsilon(x)$ is Riemann integrable on $[0,1] \times R$, it follows from Theorem 2.13, the occupation time formula (2.3) and the continuity of $L_X(t,x)$ that

$$\left(X_{n,[nt]}, \frac{1}{n} \sum_{k=1}^n G(k/n)\phi_\epsilon(X_{nk})\right)$$

$$\Rightarrow \left(X_t, \int_0^1 G(t)\,\phi_\epsilon(X_t)dt\right), \quad (\text{as } n \to \infty)$$

$$= \left(X_t, \int_0^1 \int_{-\infty}^{\infty} G(t)\,\phi(x)L_X(dt, \epsilon x)dx\right)$$

$$\to_{a.s.} \left(X_t, \int_0^1 G(t)L_X(dt, 0)\right), \quad \text{as } \epsilon \to 0. \tag{2.70}$$

This, together with (2.68)-(2.69), yields (2.67).

Note that, by Theorem 2.17 (i),

$$E\,\left|\phi_\epsilon(y/c_n - X_{nk}) - \phi_\epsilon(X_{nk})\right|$$

$$\leq K\,(d_{nk0})^{-1} \int_{-\infty}^{\infty} |\phi_\epsilon(y/c_n - x) - \phi_\epsilon(x)|\,dx$$

$$\leq C\,(d_{nk0})^{-1} \left[I(|y| \geq \sqrt{c_n}) + (\epsilon c_n)^{-1/2}\right].$$

Result (2.68) follows from, for any $\epsilon > 0$,

$$E\,\left|L_{n,\epsilon} - \frac{1}{n} \sum_{k=1}^n G(k/n)\,\phi_\epsilon(X_{nk})\right|$$

$$\leq \frac{1}{n} \sum_{k=1}^n \int_{-\infty}^{\infty} g\left(\frac{k}{n}, y\right) E\,|\phi_\epsilon(y/c_n - X_{nk}) - \phi_\epsilon(X_{nk})|\,dy$$

$$\leq \frac{C}{n} \sum_{k=1}^n (d_{nk0})^{-1} \left[\int_{|y| \geq \sqrt{c_n}} \tilde{g}(y)dy + (\epsilon c_n)^{-1/2} \int_{-\infty}^{\infty} \tilde{g}(y)dy\right]$$

$$\to 0, \quad \text{as } n \to \infty,$$

due to $c_n \to \infty$, $\int_{-\infty}^{\infty} \widetilde{g}(y)dy < \infty$ and (2.55) with $\delta = 1$.

We next prove (2.69), starting with some preliminaries. Write

$$Y_{nk}(z) = g[\frac{k}{n}, c_n X_{nk}] - g[\frac{k}{n}, c_n(X_{nk} + z\epsilon)].$$

As in the proof of Theorem 2.17 (iii), for $j \geq 1$ and $j + n_0 \leq k \leq n$, we have

$$\left| \mathrm{E}\left(Y_{nk}(z) \mid \mathcal{F}_{nj}\right) \right| = \Big| \int_{-\infty}^{\infty} \Big(g\big[\frac{k}{n}, c_n X_{nj} + c_n d_{nkj}\, y\big]$$

$$- g\big[\frac{k}{n}, c_n(X_{nj} + z\epsilon) + c_n d_{nkj}\, y\big]\Big) h_{nkj}(y)dy \Big|$$

$$\leq (c_n\, d_{nkj})^{-1} \int_{-\infty}^{\infty} \widetilde{g}(y) \left| V(y, c_n X_{nk})\right| dy$$

$$\leq C\, (c_n\, d_{nkj})^{-1} \min\{1, |z|\epsilon/d_{nkj}\}, \tag{2.71}$$

where $V(y, t) = h_{nkj}\left(\frac{y-t}{c_n d_{nkj}}\right) - h_{nkj}\left(\frac{y-t-c_n z\epsilon}{c_n d_{nkj}}\right)$ and we have used (2.57) in the last inequality. Similarly, for $1 \leq k \leq n$,

$$\mathrm{E}\left(|Y_{nk}(z)| + |Y_{nk}(z)|^2 \right)$$

$$\leq 4K(c_n d_{nk0})^{-1} \int_{-\infty}^{\infty} \big[\widetilde{g}(y) + \widetilde{g}^2(y)\big] dy. \tag{2.72}$$

It follows from (2.71)-(2.72), together with the conditions (2.55) and (2.56), that, for all $z \in \mathrm{R}$,

$$\sum_{k=1}^{n} \big[\mathrm{E}\,|Y_{nk}(z)| + \mathrm{E}\,Y_{nk}^2(z)\big] \leq Cc_n^{-1} \sum_{k=1}^{n} (d_{nk0})^{-1} \leq C\,n/c_n, \tag{2.73}$$

and for any $|z| \leq \epsilon^{-1/2}$ (letting $\sum_{k=i}^{j} = 0$ if $i > j$),

$$\sum_{k=j+1}^{j+n_0} \big| \mathrm{E}\left\{Y_{nk}(z)\,Y_{nj}(z)\right\} \big| \leq 2\Big[\sum_{k=j+1}^{j+n_0} \mathrm{E}\,Y_{nk}^2(z) + n_0\,\mathrm{E}\,Y_{nj}^2(z)\Big],$$

$$\sum_{k=j+n_0+1}^{n} \big| \mathrm{E}\left\{Y_{nk}(z)\,Y_{nj}(z)\right\} \big|$$

$$\leq \mathrm{E}\,|Y_{nj}(z)|\Big(\sum_{k=j+n_0+1}^{\epsilon n} + \sum_{k=j+\epsilon n}^{n} \Big)\big| \mathrm{E}\left(Y_{nk}(z) \mid \mathcal{F}_{nj}\right) \big|$$

$$\leq C\,(c_n^2 d_{nj0})^{-1}\Big[\sum_{k=j+1}^{\epsilon n} (d_{nkj})^{-1} + \epsilon^{(1-\alpha)/2} \sum_{k=j+\epsilon n}^{n} (d_{nkj})^{-1}\Big]$$

$$\leq C\epsilon^{\min\{\beta, (1-\alpha)/2\}} n\,(c_n^2 d_{nj0})^{-1}. \tag{2.74}$$

Consequently, for any $|z| \leq \epsilon^{-1/2}$, we have

$$
\Lambda_n(\epsilon) \equiv \frac{c_n^2}{n^2} \mathrm{E} \Big[\sum_{k=1}^{n} Y_{nk}(z) \Big]^2
$$

$$
\leq \frac{c_n^2}{n^2} \sum_{k=1}^{n} \mathrm{E} Y_{nk}^2(z) + \frac{2\,c_n^2}{n^2} \sum_{j=1}^{n} \sum_{k=j+1}^{n} \big| \mathrm{E} \{ Y_{nk}(z) \, Y_{nj}(z) \} \big|
$$

$$
\leq C c_n/n + C \epsilon^{\min\{\beta,(1-\alpha)/2\}} \frac{1}{n} \sum_{j=1}^{n} (d_{nj0})^{-1}
$$

$$
\leq C c_n/n + C \epsilon^{\min\{\beta,(1-\alpha)/2\}} \tag{2.75}
$$

We are now ready to prove (2.69). Since $\int_{-\infty}^{\infty} \phi(x)dx = 1$, it is readily seen from (2.72) and (2.75) that

$$
\mathrm{E} \, |L_n - L_{n,\epsilon}| \leq \int_{-\infty}^{\infty} \frac{c_n}{n} \mathrm{E} \Big| \sum_{k=1}^{n} Y_{nk}(z) \Big| \phi(z)dz
$$

$$
\leq \int_{|z| \geq \epsilon^{-1/2}} \frac{c_n}{n} \sum_{k=1}^{n} \mathrm{E} \, |Y_{nk}(z)| \phi(z)dz + \int_{|z| \leq \epsilon^{-1/2}} \Lambda_n^{1/2}(\epsilon) \phi(z)dz
$$

$$
\leq C \int_{|z| \geq \epsilon^{-1/2}} \phi(z)dz + C \sqrt{c_n/n + \epsilon^{\min\{\beta,(1-\alpha)/2\}}}
$$

$$
\to 0,
$$

as $n \to \infty$ first and then $\epsilon \to 0$, which yields (2.69). $\qquad \square$

Theorem 2.19 can be strengthened to functional convergence on $D[0,1]$, which is formally stated in the following theorem.

Theorem 2.20. *Under the conditions of Theorem 2.19 on X_{nk} and $g(s,x)$, for any fixed $x \in \mathrm{R}$, we have*

$$
\frac{c_n}{n} \sum_{k=1}^{[nt]} g\big[\tfrac{k}{n}, c_n \, (X_{nk} + x c_n')\big] \Rightarrow \int_0^t G(s) \, \widetilde{L}_X(ds, x), \tag{2.76}
$$

on $D[0,1]$, for any $c_n \to \infty$ and $c_n/n \to 0$, where

$$
\widetilde{L}_X(r, x) = \begin{cases} L_X(r, 0), & \text{if } c_n' \to 0, \\ L_X(r, -x), & \text{if } c_n' = 1. \end{cases}
$$

Proof. First assume $x = 0$. The finite dimensional convergence comes easily from the Cramér-Wold device and the similar arguments as in the proof of Theorem 2.19.

Write $V_k = g\left(\frac{k}{n}, c_n X_{nk}\right)$. Note that (letting $\sum_{k=i}^{j} = 0$ if $i > j$)

$$\sum_{j=1}^{[nt]} V_j = \sum_{k=0}^{[nt]/n_0} \left(V_{1+kn_0} + V_{2+kn_0} + \ldots + V_{(k+1)n_0-1}\right).$$

To establish the tightness, by Theorem 4 of Billingsley (1974), it suffices to show that, for each $1 \le i \le n_0 - 1$,

$$\max_{0 \le k \le n/n_0} |V_{i+kn_0}| = o_P[n/c_n], \tag{2.77}$$

and, for $0 \le t_1 \le t_2 \le \ldots \le t_m \le t \le 1$, $t - t_m \le \delta$,

$$P\left[|S_n(t) - S_n(t_m)| \ge \epsilon \mid \mathcal{F}_{n,[nt_m]}\right] \le \alpha_n(\epsilon, \delta), \quad a.s., \tag{2.78}$$

where $S_n(t) = \frac{c_n}{n} \sum_{k=0}^{[nt]/n_0} V_{i+kn_0}$ and $\alpha_n(\epsilon, \delta)$ is a sequence of positive real numbers such that, for each $\epsilon > 0$,

$$\lim_{\delta \to 0} \lim_{n \to \infty} \sup \alpha_n(\epsilon, \delta) = 0.$$

Using Theorem 2.17, we have

$$E\, V_{i+kn_0}^2 \le C\left(c_n\, d_{n,i+kn_0,0}\right)^{-1},$$

$$E\left[|V_{i+kn_0}| \mid \mathcal{F}_{n,jn_0}\right] \le C\left(c_n\, d_{n,i+kn_0,jn_0}\right)^{-1}, \quad \text{for } j < k.$$

Now it follows from (2.55) and (2.56) that

$$\max_{0 \le k \le n/n_0} |V_{i+kn_0}| \le \Big[\sum_{k=0}^{n/n_0} V_{i+kn_0}^2 \Big]^{1/2} = O_P[(n/c_n)^{1/2}],$$

which yields (2.77) due to $c_n/n \to 0$, and whenever $t - t_m \le \delta$,

$$P\left[|S_n(t) - S_n(t_m)| \ge \epsilon \mid \mathcal{F}_{n,[nt_m]}\right]$$

$$\le \frac{c_n}{n\epsilon} \sum_{k=1+[nt_m]/n_0}^{[nt]/n_0} E\left[|V_{i+kn_0}| \mid \mathcal{F}_{n,[nt_m]}\right]$$

$$\le \frac{C}{n\epsilon} \sum_{k=1+[nt_m]/n_0}^{[nt]/n_0} (d_{n,i+kn_0,[nt_m]})^{-1} \le \frac{C}{n\epsilon} \sum_{k=1+[nt_m]}^{[nt]} (d_{n,i+k,[nt_m]})^{-1}$$

$$\le C\epsilon^{-1}\delta^{\min\{\beta, 1-\alpha/2\}} \to 0,$$

as $n \to \infty$ first and then $\delta \to 0$, i.e., (2.78) holds with

$$\alpha_n(\epsilon, \delta) = C\epsilon^{-1}\delta^{\min\{\beta, 1-\alpha/2\}}.$$

This proves (2.76) with $x = 0$.

For $x \ne 0$, it suffices to notice that, for any given $x \in \mathbb{R}$,

$$X_{n,[nt]} + xc_n' \Rightarrow \begin{cases} X_t, & \text{if } c_n' \to 0, \\ X_t + x, & \text{if } c_n' = 1; \end{cases}$$

The claim follows directly from (2.76) and the fact that $X + x$ has a local time $L_X(t, -x)$. $\qquad\square$

Remark 2.7. Let $Z_n(x) = \frac{c_n}{n} \sum_{k=1}^{n} g\left[\frac{k}{n}, c_n\left(X_{nk} + xc_n'\right)\right]$. Using similar arguments as in the proofs of Theorems 2.19 and 2.20, together with the Cramér-Wold theorem, we have

$$\left(Z_n(x_1), \cdots, Z_n(x_m)\right) \to_D \left(Z(x_1), \cdots, Z(x_m)\right),$$

where $Z(x) = \int_0^1 G(s)\widetilde{L}_X(ds, x)$. This provides finite dimensional distribution convergence for the process $Z_n(x), x \in \mathbb{R}$. Functional convergence of $Z_n(x)$ will be discussed in Section 2.5.

Remark 2.8. Riemann integrability of $G(t)$ in Theorems 2.19 and 2.20 cannot be removed. For instance, let $g(s, x) = g_1(s) g_2(x)$, where $|g_2(x)| \leq A/(1 + |x|^{1+b})$ for some $b > 0$, $g_1(s) = 1$ if s is irrational and $g_1(s) = 0$ if s is rational. Then $G(s) = g_1(s) \int_{-\infty}^{\infty} g_2(x) dx$ is not Riemann integrable and it can be easily checked that

$$\frac{c_n}{n} \sum_{k=1}^{n} g\left(\frac{k}{n}, c_n X_{nk}\right) = 0 \text{ and } \int_0^1 G(t) L_X(dt, 0) = \int_{-\infty}^{\infty} g_2(x) dx\, L_X(1, 0).$$

Hence (2.76) cannot be true whenever $\int_{-\infty}^{\infty} g_2(x) dx = 0$.

Corollary 2.1. *If, for $\forall \delta > 0$, there exists a continuous function $g_1(s, y)$ on $[0, 1] \times \mathbb{R}$ such that $\int_{-\infty}^{\infty} \sup_{0 \leq s \leq 1} |g_1(s, y)| dy < \infty$ and*

$$\int_{-\infty}^{\infty} \sup_{0 \leq s \leq 1} |g(s, y) - g_1(s, y)| dy \leq \delta. \tag{2.79}$$

Then (2.76) remains true. In particular, if $g(s, y) = f(y)$ and $\int_{-\infty}^{\infty} |f(y)| dy < \infty$, then

$$\frac{c_n}{n} \sum_{k=1}^{[nt]} f\left[c_n\left(X_{nk} + c_n' x_0\right)\right] \Rightarrow \tau \widetilde{L}_X(t, x_0), \tag{2.80}$$

on $D[0, 1]$, for any fixed x_0, where $\tau = \int_{-\infty}^{\infty} f(t) dt$ and $\widetilde{L}_X(t, x_0)$ is defined as in Theorem 2.20.

Proof. Let $g_{1N}(s, y) = g_1(s, y)\mathrm{I}(|y| \leq N)$. Due to the continuity of $g_1(s, y)$, (2.76) holds for any $N \geq 1$ if $g(s, y)$ is replaced by $g_{1N}(s, y)$. Now the claim follows from some routine calculations by using (2.38). □

We remark that (2.79) is quite weak in terms of Lusin's theorem, which states that a Lebesgue integrable function can be approximated by a continuous function. Indeed, when applied to $g(s, y) = f(y)$, the Lebesgue integrability of $f(y)$ is necessary to establish (2.80).

Corollary 2.2. *Suppose the conditions of Corollary 2.1 hold and let $\sigma(x)$ be a real function satisfying that $\sup_x |\sigma(x)| < \infty$ and $\sigma(x)$ is continuous at 0. Then, for any $d_n > 0$ satisfying $d_n/c_n \to 0$, we have*

$$\frac{c_n}{n} \sum_{k=1}^{[nt]} g\left(\frac{k}{n}, c_n X_{nk}\right) \sigma(d_n X_{nk}) \to_D \sigma(0) \int_0^t G(s) L_X(ds, 0), \quad (2.81)$$

on $D[0, 1]$.

Proof.   For any N and $\delta > 0$, there exists an n_0 such that, for all $n \geq n_0$,

$$\left| g\left(\frac{k}{n}, c_n X_{nk}\right) \sigma(d_n X_{nk}) - g\left(\frac{k}{n}, c_n x_{nk}\right) \sigma(0) \right|$$

$$\leq C \sup_{0 \leq s \leq 1} |g(s, c_n X_{nk})| \mathrm{I}(|c_n X_{nk}| \geq N) + \delta \left| g\left(\frac{k}{n}, c_n X_{nk}\right) \right|,$$

due to $d_n/c_n \to 0$ and the continuity of $\sigma(x)$ at 0. Claim follows from Corollary 2.1 and routine calculations by using (2.38).   □

Remark 2.9. If $\sigma(x) \to \infty$, as $x \to \infty$ or as $x \to x_0$, similar result to (2.81) can be achieved, but heavily depending on the structure of $\sigma(x)$. For instance, if $\sigma(x) = x^2$ and in addition $\int_{-\infty}^{\infty} y^2 \sup_{0 \leq s \leq 1} g(s, y) dy < \infty$, then

$$\frac{c_n}{n} \sum_{k=1}^{n} g\left(\frac{k}{n}, c_n X_{nk}\right) \sigma(d_n X_{nk}) = o_P(1),$$

but we have

$$\frac{c_n^2}{d_n^2} \frac{c_n}{n} \sum_{k=1}^{n} g\left(\frac{k}{n}, c_n X_{nk}\right) \sigma(d_n X_{nk}) \to_D \int_0^1 G_1(t) L_X(dt, 0),$$

where $G_1(t) = \int_{-\infty}^{\infty} g(t, y) y^2 dy$. We refer to Phillips (2009) for further discussions. When there exists certain structure such as X_{nk} is a standardized partial sum of linear processes, further extensions of Theorem 2.20 are possible. See next section for more details.

2.3.3   *Example: linear processes*

Let $\xi_j, j \geq 1$, be a linear process defined in Example 2.12. Set $S_k = \sum_{j=1}^{k} \xi_j$, $d_n^2 = E S_n^2$ and

$$W(t) = \begin{cases} B_{3/2-u}(t), & \text{under } \mathbf{LM}, \\ B_{1/2}(t), & \text{under } \mathbf{SM}, \end{cases}$$

where $B_H = \{B_H(t)\}_{t\geq 0}$ is a fractional Brownian motion defined in Example 2.1. Wang, et al. (2003) proved (2.62) and

$$\Big(\frac{1}{\sqrt{n}}\sum_{j=1}^{[nt]}\epsilon_j,\ \frac{1}{\sqrt{n}}\sum_{j=1}^{[nt]}\epsilon_{-j},\ \frac{1}{d_n}S_{[nt]}\Big) \Rightarrow \big(B_{1t},\ B_{2t},\ W(t)\big),$$

where $B = (B_{1t}, B_{2t})_{t\geq 0}$ is a standard 2-dimensional Brwonian motion and $W = \{W(t)\}_{t\geq 0}$ is a functional of B.

Let $Z_{nk} = y_{nk}/d_n$ be defined as in (2.61), i.e., $y_{n0} = 0$ and

$$y_{nk} = \gamma\, y_{n,k-1} + \xi_k,$$

where $\tau \geq 0$ is a constant and $\gamma = 1 - \tau/n$. Write, for $t \geq 0$,

$$Z_t = W(t) + \tau \int_0^t e^{-\tau(t-s)}W(s)ds. \tag{2.82}$$

As indicated in Example 2.2, Z_t is a fractional Ornstein-Uhlenbeck process, having a continuous local time $L_Z(t, x)$.

Theorem 2.21. *On $D_{\mathbf{R}^3}[0, \infty)$, we have*

$$\Big(\frac{1}{\sqrt{n}}\sum_{j=1}^{[nt]}\epsilon_j,\ \frac{1}{\sqrt{n}}\sum_{j=1}^{[nt]}\epsilon_{-j},\ Z_{n,[nt]}\Big) \Rightarrow \big(B_{1t},\ B_{2t},\ Z_t\big). \tag{2.83}$$

If in addition $\lim_{|t|\to\infty}|t|^\eta|\mathrm{E}\,e^{it\epsilon_0}| < \infty$ *for some* $\eta > 0$, *then, for any real function* $g(s, x)$ *such that* $G(s) = \int_{-\infty}^{\infty} g(s, x)dx$ *is Riemann integrable on* $[0, 1]$ *and* $\int_{-\infty}^{\infty}[\widetilde{g}(x) + \widetilde{g}^2(x)] < \infty$, *where* $\widetilde{g}(x) = \sup_{0\leq s\leq 1}|g(s, x)|$,

$$\Big(\frac{1}{\sqrt{n}}\sum_{j=1}^{[nt]}\epsilon_j,\ \frac{1}{\sqrt{n}}\sum_{j=1}^{[nt]}\epsilon_{-j},\ \frac{c_n}{n}\sum_{k=k_0}^{n}g\big[\frac{k}{n}, c_n\,(Z_{nk} + xc_n')\big]\Big)$$

$$\Rightarrow \Big(B_{1t}, B_{2t}, \int_0^1 G(s)\,\widetilde{L}_Z(ds, x)\Big), \tag{2.84}$$

on $D_{\mathbf{R}^3}[0, \infty)$, where x is fixed, $c_n \to \infty$, $c_n/n \to 0$, and

$$\widetilde{L}_Z(r, x) = \begin{cases} L_Z(r, 0), & \text{if } c_n' \to 0, \\ L_Z(r, -x), & \text{if } c_n' = 1. \end{cases}$$

Proof. The proof of (2.83) is standard. See, e.g., Buchmann and Chan (2007) with a minor modification. Since $\{Z_{nk}\}_{k\geq 1, n\geq 1}$ is a strong smooth array by Theorem 2.18 and $Z_{n,[nt]} \Rightarrow Z_t$ by (2.83), result (2.84) is an immediate consequence of Theorem 2.19. $\qquad\square$

The remaining part of this section is to establish certain variations of (2.84). To this end, let $g(s, x_0, x_1, \cdots, x_d)$, $\Gamma(x_1, \cdots, x_m)$, etc, be real functions of their components. Further, in addition to the notation in Example 2.12, let $\eta_j = (\epsilon_j, \nu_j)'$ be a sequence of i.i.d. random vectors with $E\,\eta_0 = 0$ and $E\,||\eta_0||^2 < \infty$. Set $\mathcal{F}_s = \sigma(\eta_s, \eta_{s-1,...})$ and we always assume $\lim_{|t| \to \infty} |t|^\eta |E\,e^{it\epsilon_0}| < \infty$ for some $\eta > 0$.

We have the following corollaries. Their proofs will be given in next subsection.

Corollary 2.3. *Suppose* $\widetilde{g}(x) = \sup_{0 \le s \le 1} |g(s, x)|$ *is bounded and integrable, and* $G(s) = \int_{-\infty}^{\infty} g(s, x)dx$ *is Riemann integrable on* $[0, 1]$. *Then, for any* $c_n \to \infty$ *and* $c_n/n \to 0$, *the following claims hold.*

(i) If $E\Gamma^2(\eta_1, ..., \eta_m) < \infty$ *and* $\Gamma_0 = E\Gamma(\eta_1, \cdots, \eta_m) \ne 0$, *then*

$$\Gamma_n(t) := \frac{c_n}{n} \sum_{k=1}^{[nt]} g\left[\frac{k}{n}, c_n \left(Z_{nk} + xc_n'\right)\right] \Gamma(\eta_k, ..., \eta_{k-m})$$

$$\Rightarrow \Gamma_0 \int_0^t G(s)\,\widetilde{L}_Z(ds, x), \quad on\ D[0, 1]. \tag{2.85}$$

(ii) If $u_k = \sum_{j=0}^{\infty} \psi_j\,\eta_{k-j}'$, *where* $E\,||\eta_0||^4 < \infty$ *and the coefficient vector* $\psi_k = (\psi_{k1}, \psi_{k2})$ *satisfies* $\sum_{k=0}^{\infty}(|\psi_{1k}| + |\psi_{2k}|) < \infty$, *then*

$$\Gamma_{1n}(t) := \frac{c_n}{n} \sum_{k=1}^{[nt]} g\left[\frac{k}{n}, c_n \left(Z_{nk} + xc_n'\right)\right] u_k^2$$

$$\Rightarrow E\,u_0^2 \int_0^t G(s)\,\widetilde{L}_Z(ds, x), \quad on\ D[0, 1]. \tag{2.86}$$

Corollary 2.4. *Suppose that*

(i) there exist $0 < \gamma \le 1$ *and* $\beta \ge 1$ *such that, for all* $x, y_1, \cdots, y_d \in \mathbb{R}$ *and* $h_1, \cdots, h_d$ *sufficiently small,*

$$\sup_{0 \le s \le 1} |g(s, x, x + h_1 y_1, \cdots, x + h_d y_d) - g(s, x, x, \cdots, x)|$$

$$\le \max_{0 \le j \le d} |h_j|^\gamma\, f(x)(1 + \max_{0 \le j \le d} |y_j|^\beta), \tag{2.87}$$

and $f(x)$ *is integrable;*

(ii) $\widetilde{g}(x) = \sup_{0 \le s \le 1} |g(s, x, \cdots, x)|$ *is bounded and integrable,* $G_1(s) \equiv \int_{-\infty}^{\infty} g(s, x, \cdots, x)dx$ *is Riemann integrable on* $[0, 1]$;

(iii) $E\,|\epsilon_0|^{[\beta]+1} < \infty$ *and* ξ_k *is a short memory linear process, i.e.,* $\sum_{j=0}^{\infty} |\phi_j| < \infty$ *and* $\sum_{j=0}^{\infty} \phi_j \ne 0$.

Then, for any $c_n \to \infty$ and $c_n/\sqrt{n} \to 0$,

$$\frac{c_n}{n} \sum_{k=1}^{[nt]} g\left(\frac{k}{n}, c_n Z_{nk}, c_n Z_{n,k+1}, \cdots, c_n Z_{n,k+d}\right)$$

$$\Rightarrow \int_0^t G_1(s) L_Z(ds, 0), \qquad (2.88)$$

on $D[0,1]$.

Remark 2.10. Result (2.85) remains true if we replace $\Gamma(\eta_k, \cdots, \eta_{k-m})$ by $\Gamma(\eta_k, \cdots, \eta_{k-m}, \zeta_k, \zeta_{k-1}, ...)$, where ζ_k is an arbitrary random sequence independent of ϵ_k. If $E\Gamma(\eta_k, \cdots, \eta_{k-m}) = 0$, convergence to a mixture of normal distributions is obtained. See Section 3.5 for details.

Remark 2.11. Condition (2.87) is weak, e.g., which is satisfied by

$$g(s, x, y_1, ..., y_m) = (1 + |x|^k)^{-1} |y_1|^{\alpha_1} \cdots |y_m|^{\alpha_m},$$

where $\alpha_1, \cdots, \alpha_m \geq 0$ and $1 + \alpha_1 + \cdots + \alpha_m < k$. The proof of Corollary 2.4 is simple, but it fails to work for a long memory linear process.

Remark 2.12. If $c_n/\sqrt{n} \to \infty$, (2.88) fails to be true. In fact, for any $d \geq 1$,

$$\frac{c_n}{n} \sum_{k=1}^{n} g\left(\frac{k}{n}, c_n Z_{nk}, c_n Z_{n,k+1}, \cdots, c_n Z_{n,k+d}\right) \to \infty, a.s.$$

If $S_k = \sum_{i=1}^{k} \epsilon_i$ and $Z_{nk} = S_k/\sqrt{n}$, Borodin and Ibragimov (1995) established

$$\frac{1}{\sqrt{n}} \sum_{k=1}^{n} g\left(S_k, S_{k+1}, \cdots, S_{k+d}\right) \to_D \tau L_B(1, 0),$$

where $\tau = \int_{-\infty}^{\infty} E\, g(x, x + S_1, ..., x + S_d)dx$. This provides a result for (2.88) when $c_n = \sqrt{n}$. For other related results, we refer to an unpublished manuscript Jeganathan (2008).

2.3.4 Proofs of Corollaries 2.3 and 2.4

This section provides the proofs of Corollaries 2.3 and 2.4, starting with two lemmas. Notation will be the same as previous sections except explicitly mentioned.

Lemma 2.1. Let $\Lambda = \sigma(\eta_{t_1}, ..., \eta_{t_{m_0}})$, where $t_1, t_2, ..., t_{m_0} \in \mathbb{Z}$. There exists an $A_0 > 0$ such that the following results hold.

(i) *For any $h > 0$ and $t \geq A_0$, we have*

$$\mathrm{E}\left\{|\Lambda|\,|p(y_{nt}/h)|\right\} \leq \frac{C\,h\,\mathrm{E}\,|\Lambda|}{d_t}\int_{-\infty}^{\infty}|p(x)|dx. \qquad (2.89)$$

(ii) *For any $h > 0$ and $t - s \geq A_0$, we have*

$$\mathrm{E}\left\{|\Lambda|\,|p(y_{nt}/h)|\,\big|\,\mathcal{F}_s\right\} \leq \frac{Ch\mathrm{E}\,|\Lambda|}{d_{t-s}}\int_{-\infty}^{\infty}|p(x)|dx, \qquad (2.90)$$

whenever $s + 1 \leq t_1, ..., t_{m_0} \leq t$. If in addition $\mathrm{E}\,\Lambda = 0$, then

$$\left|\mathrm{E}\left\{\Lambda\,p(y_{nt}/h)\,\big|\,\mathcal{F}_s\right\}\right| \leq \frac{C\beta\,h}{d_{t-s}^2}\int_{-\infty}^{\infty}|p(x)|dx, \qquad (2.91)$$

where $\beta = m_0\,(\mathrm{E}\,\Lambda^2)^{1/2}\sum_{k=0}^{t-\min\{t_1,...,t_{m_0}\}}|\phi_k|$.

(iii) *For any $h > 0$ and $t - s \geq 1$, result (2.90) remains true if Λ is a constant and $\phi_0 \neq 0$.*

Proof.   We only prove (2.91). The other derivations are similar and the details are omitted. Let $\Lambda_{m_0} = \sum_{j=1}^{m_0}\epsilon_{t_j}a_{t-t_j}$ and $y_{ts}^* = y_{nts}^{(1)} - \Lambda_{m_0}$, where a_j and $y_{nkj}^{(1)}$ are defined as in (2.65). Recalling (2.63), there exists an $A_0 > 0$ such that, whenever $t - s \geq A_0$, $\mathrm{E}\,(y_{ts}^*)^2 \asymp d_{t-s}^2$. As a consequence, similar arguments as in the proof of Theorem 2.21 [part (ii), see, e.g., the fact $\mathbb{F}$ and (2.66)] yields that

whenever $t - s \geq A_0$, y_{ts}^*/d_{t-s} has a density function $\nu_{st}(x)$, which is uniformly bounded over x by a constant C and

$$\sup_x|\nu_{st}(x + u) - \nu_{st}(x)| \leq C\min\{|u|, 1\}. \qquad (2.92)$$

This, together with (2.65) and the independence of ϵ_i, implies that

$$\mathrm{E}\left\{\Lambda\,p(y_{nt}/h)\,\big|\,\mathcal{F}_s\right\} = \mathrm{E}\left\{\Lambda\,p\big[(x_{st} + \Lambda_{m_0} + y_{ts}^*)/h\big]\,\big|\,\mathcal{F}_s\right\}$$

$$= \mathrm{E}\left\{\Lambda\int_{-\infty}^{\infty}p\big[(x_{st} + \Lambda_{m_0} + d_{t-s}y)/h\big]\nu_{st}(y)dy\,\big|\,\mathcal{F}_s\right\}$$

$$= \frac{h}{d_{t-s}}\int_{-\infty}^{\infty}p(y)\,\nu_{st}^*(y)dy, \qquad (2.93)$$

where $x_{st} = y_{nt} + y_{nts}^{(2)}$ depending only on $\epsilon_s, \epsilon_{s-1}, ...,$ and

$$\nu_{st}^*(y) = \mathrm{E}\left\{\Lambda\,\nu_{st}\big(\frac{-x_{st} - \Lambda_{m_0} + hy}{d_{t-s}}\big)\,\big|\,\mathcal{F}_s\right\}.$$

Note that $\mathrm{E}\left\{\Lambda\,\nu_{st}\big(\frac{-x_{st}+hy}{d_{t-s}}\big)\,\big|\,\mathcal{F}_s\right\} = 0$. By (2.3.4), we have

$$|\nu_{s,t}^*(y)| \leq C\mathrm{E}\left[|\Lambda|\,\min\{|\Lambda_{m_0}|/d_{t-s}, 1\}\right].$$

Taking this estimate into (2.93), simple calculations yield (2.91). $\square$

Lemma 2.2. *Let* $\widetilde{g}(x) = \sup_{0 \le s \le 1} g(s, x)$ *be a bounded real function with* $\int_{-\infty}^{\infty} \widetilde{g}(x)dx < \infty$ *and assume* m_0 *is a fixed integer.*

(i) For any $h > 0$, *if* $\mathrm{E}\,|\sigma(\eta_1, ..., \eta_{m_0})| < \infty$, *then*

$$\sup_x \sum_{k=1}^{n} \mathrm{E}\left\{ |g[k/n, \ (y_{nk} + x)/h]\, \sigma(\eta_{i_1}, \cdots, \eta_{i_m})| \right\}$$

$$\le C(nh/d_n)\,\mathrm{E}\,|\sigma(\eta_1, ..., \eta_{m_0})|, \tag{2.94}$$

where $t_1, t_2, ..., t_{m_0} \in \mathbb{Z}$.

(ii) For any $h > 0$ *and* $nh/d_n \to \infty$, *if* $\mathrm{E}\sigma^2(\eta_1, ..., \eta_{m_0}) < \infty$ *and* $\mathrm{E}\sigma(\eta_1, ..., \eta_{m_0}) = 0$, *then*

$$\sup_x \mathrm{E}\,\Big| \sum_{k=1}^{[nt]} g[k/n, \ (y_{nk} + x)/h]\, \sigma(\eta_k, \cdots, \eta_{k-m_0}) \Big|^2$$

$$\le \frac{Cnh\,\mathrm{E}\sigma^2(\eta_1, ..., \eta_{m_0})}{d_n} \begin{cases} 1 + h, & under\ \mathbf{LM}, \\ 1 + h \log n, & under\ \mathbf{SM}, \end{cases} \tag{2.95}$$

uniformly for $0 \le t \le 1$.

(iii) For any $h > 0$, $nh/d_n \to \infty$ *and* $j \ge 0$, *if* $\mathrm{E}\sigma^2(\eta_1) < \infty$ *and* $\mathrm{E}\sigma(\eta_1) = 0$, *then*

$$\sup_x \mathrm{E}\,\Big| \sum_{k=1}^{[nt]} g[k/n, \ (y_{nk} + x)/h]\, \sigma(\eta_{k-j}) \Big|^2$$

$$\le \frac{C\,nh\,\mathrm{E}\,\sigma^2(\eta_1)}{d_n} \begin{cases} 1 + hj^{1/2}, & under\ \mathbf{LM}, \\ 1 + hj^{1/2} + h \log n, & under\ \mathbf{SM}, \end{cases} \tag{2.96}$$

uniformly for $0 \le t \le 1$.

(iv) Let $u_{k,m_0} = \sum_{j=m_0}^{\infty} \psi_j\, \lambda'_{k-j}$, *where* $\psi_j = (\psi_{1j}, \psi_{2j})$ *and* $\lambda_i = (\sigma_1(\eta_i), \sigma_2(\eta_i))$ *with* $\mathrm{E}\,\lambda_1 = 0$. *For all* $h \to 0$ *(*$h \log n \to 0$ *under* $\mathbf{SM}$*) and* $nh/d_n \to \infty$, *we have*

$$\sup_x \mathrm{E}\,\Big| \sum_{k=1}^{[nt]} u_{k,m_0}\, g[k/n, \ (y_{nk} + x)/h] \Big|^2$$

$$\le C\,\mathrm{E}\,||\lambda_1||^2\,(nh/d_n)\,\Big[\sum_{j=m_0}^{\infty} j^{1/4}(|\psi_{1j}| + |\psi_{2j}|) \Big]^2, \tag{2.97}$$

uniformly for $0 \le t \le 1$.

Proof.   (2.94) follows from an application of (2.89). We next prove (2.96). Let $\sum_{j=k}^{l} = 0$ for $k > l$ and write $g_k = g\big[k/n, \, (y_{nk} + x)/h\big]$. We have

$$\Delta_n \equiv \Big| \sum_{k=1}^{[nt]} g_k \, \sigma(\eta_{k-j}) \Big|^2$$

$$\leq 2\Big| \sum_{k=A_0}^{[nt]} g_k \, \sigma(\eta_{k-j}) \Big|^2 + C \, \Big(\sum_{k=1}^{A_0} |\sigma(\eta_{k-j})| \, \Big)^2$$

$$= 2\Big(\sum_{k=A_0}^{[nt]} \sum_{\substack{l=1 \\ |k-l|<A_0}}^{[nt]} + 2 \sum_{k=A_0}^{[nt]-1} \sum_{l=k+A_0}^{[nt]} \Big) \, \sigma(\eta_{k-j}) \, \sigma(\eta_{l-j}) \, g_k \, g_l$$

$$+ C \, \Big(\sum_{k=1}^{A_0} |\sigma(\eta_{k-j})| \, \Big)^2$$

$$= \Delta_{1n}(t) + \Delta_{2n}(t) + \Delta_{3n}, \quad say, \tag{2.98}$$

where A_0 is chosen as in Lemma 2.1. Using $|g(x)| \leq C$ and (2.89) of Lemma 2.1 with $p(y) = g(k/n, y - x/h)$, we have

$$\big| \mathrm{E} \, \{ \sigma(\eta_{k-j}) \, \sigma(\eta_{l-j}) \, g_k \, g_l \} \big| \leq C \, \mathrm{E} \, \big| \sigma(\eta_{k-j}) \, \sigma(\eta_{l-j}) \, g_k \, \big|$$

$$\leq C \, \mathrm{E} \, \sigma^2(\eta_1) \, h/d_k,$$

for $k \geq A_0$ and $|k - l| < A_0$. Similarly, by (2.90) and (2.91),

$$\big| \mathrm{E} \, \{ \sigma(\eta_{k-j}) \, \sigma(\eta_{l-j}) \, g_k \, g_l \} \big|$$

$$\leq \begin{cases} \mathrm{E} \, \big| \sigma(\eta_{k-j}) \, \sigma(\eta_{l-j}) \, g_k \, \mathrm{E} \, \{ g_l | \mathcal{F}_k \} \big| & \text{if } l - j \leq k, \\ \mathrm{E} \, \big| g_k \, \sigma(\eta_{k-j}) \, \mathrm{E} \, \{ \sigma(\eta_{l-j}) \, g_l | \mathcal{F}_k \} \big| & \text{if } l - j > k, \end{cases}$$

$$\leq C \, \mathrm{E} \, \sigma^2(\eta_1) \, h^2 \, d_k^{-1} \begin{cases} d_{l-k}^{-1} & \text{if } l - j \leq k, \\ \sum_{k=0}^{j} |\phi_k| \, d_{l-k}^{-2} & \text{if } l - j > k, \end{cases}$$

for $k \geq A_0$ and $l - k \geq A_0$. It follows from these facts that, uniformly for $0 \leq t \leq 1$,

$$\sup_x \mathrm{E} \, |\Delta_{1n}(t)| \leq C \mathrm{E} \, \sigma^2(\eta_1) \, h \sum_{k=1}^{n} \sum_{\substack{l=1 \\ |k-l|<A_0}}^{n} 1/d_k$$

$$\leq C_1 \, \mathrm{E} \, \sigma^2(\eta_1) \, nh/d_n,$$

$$\sup_x \mathrm{E}\,|\Delta_{2n}(t)|$$

$$\leq CE\,\sigma^2(\eta_1)\,h^2 \sum_{k=A_0}^{n-1} d_k^{-1} \Big(\sum_{l=k+A_0}^{n\wedge(k+j)} d_{l-k}^{-1} + \sum_{k=0}^{j} |\phi_k| \sum_{l=k+j}^{n} d_{l-k}^{-2} \Big)$$

$$\leq C\,\mathrm{E}\,\sigma^2(\eta_1)\,(nh^2/d_n) \begin{cases} j/d_j + \sum_{k=0}^{j} |\phi_k|, & \text{under } \mathbf{LM}, \\ j/d_j + \sum_{k=0}^{j} |\phi_k| \log n, & \text{under } \mathbf{SM}, \end{cases}$$

$$\leq C\,\mathrm{E}\,\sigma^2(\eta_1)\,(nh^2/d_n) \begin{cases} j^{1/2}, & \text{under } \mathbf{LM}, \\ j^{1/2} + \log n, & \text{under } \mathbf{SM}. \end{cases}$$

On the other hand, it is readily seen $\sup_x \mathrm{E}\,|\Delta_{3n}| \leq C\,\mathrm{E}\,\sigma^2(\eta_1)$. Taking these estimates into (2.98) and noting that $nh/d_n \to \infty$, (2.96) is proved.

The proof of (2.95) is similar to that of (2.96) and hence the details are omitted. By virtue of the Hölder's inequality, (2.97) follows from

$$\mathrm{E}\,\Big| \sum_{k=1}^{[nt]} u_{k,m_0}\, g_k \Big|^2 = \mathbb{E}\,\Big| \sum_{j=m_0}^{\infty} \sum_{k=1}^{[nt]} \psi_j\, \lambda'_{k-j}\, g_k \Big|^2$$

$$\leq \sum_{j=m_0}^{\infty} j^{1/4}(|\psi_{1j}| + |\psi_{2j}|) \sum_{j=m_0}^{\infty} j^{-1/4}(|\psi_{1j}| + |\psi_{2j}|)^{-1} \mathrm{E}\,\Big| \sum_{k=1}^{[nt]} \psi_j\, \lambda'_{k-j}\, g_k \Big|^2$$

$$\leq 2 \sum_{j=m_0}^{\infty} j^{1/4}(|\psi_{1j}| + |\psi_{2j}|) \sum_{j=m_0}^{\infty} j^{-1/4}(|\psi_{1j}| + |\psi_{2j}|)$$

$$\Big(\mathrm{E}\,\Big| \sum_{k=1}^{[nt]} \sigma_1(\eta_{k-j})\, g_k \Big|^2 + \mathrm{E}\,\Big| \sum_{k=1}^{[nt]} \sigma_2(\eta_{k-j})\, g_k \Big|^2 \Big)$$

$$\leq C\,\mathrm{E}\,||\lambda_1||^2\,(nh/d_n) \Big[\sum_{j=m_0}^{\infty} j^{1/4}(|\psi_{1j}| + |\psi_{2j}|) \Big]^2,$$

uniformly for $0 \leq t \leq 1$, where we employ $h \to 0$ ($h \log n \to 0$ under $\mathbf{SM}$) and (2.96) with $\sigma(.) = \sigma_1(.)$ and $\sigma_2(.)$, respectively. $\qquad\square$

Proof of Corollary 2.3. We first prove (2.85). Without loss of generality, assume $x = 0$. It follows from (2.95) with $h = d_n/c_n$ that

$$\Big(\frac{c_n}{n} \Big)^2 \sup_{0 \leq t \leq 1} \mathrm{E}\,\Big| \sum_{k=1}^{[nt]} g\Big(\frac{k}{n}, c_n Z_{nk} \Big) \big[\Gamma(\eta_k, ..., \eta_{k-m}) - \Gamma_0 \big] \Big|^2 = o(1).$$

On the other hand, by recalling parts (i) and (ii) of Theorem 2.21, Theorem 2.20 implies

$$\frac{c_n}{n} \sum_{k=1}^{[nt]} g\Big(\frac{k}{n}, c_n Z_{nk} \Big) \Rightarrow \int_0^t G(s)\, L_Z(ds, x), \quad \text{on } D[0,1].$$

Hence the finite dimensional distributions of $\Gamma_n(t)$ converge to those of $\Gamma_0 \int_0^t G(s) L_Z(ds,x)$. The tightness of $\{\Gamma_n(t)\}_{n\geq 1}$ on $D[0,1]$ can be similarly proved as that in Theorem 2.20 by virtue of Lemma 2.1. We omit the details.

Result (2.86) follows from an application of (2.85) and Lemma 2.2. To see this, let $\mu_{k,m_0} = \sum_{j=m_0+1}^{\infty} \psi_j \, \eta'_{k-j}$ and $v_{k,m_0} = \sum_{j=0}^{m_0} \psi_j \, \eta'_{k-j}$. It follows from (2.85) that, for any $m_0 \geq 1$,

$$\Gamma_{2n}(t) := \frac{c_n}{n} \sum_{k=1}^{[nt]} g\Big[\frac{k}{n}, c_n \left(Z_{nk} + xc'_n\right)\Big] \, v_{k,m_0}^2$$

$$\Rightarrow E v_{0,m_0}^2 \int_0^t G(s) \, \widetilde{L}_Z(ds,x), \quad \text{on } D[0,1]. \tag{2.99}$$

On the other hand, by letting $A(m_0) = \sum_{j=m_0+1}^{\infty} (|\psi_{1k}| + |\psi_{2k}|)$ and noting that

$$\mu_{k,m_0}^2 \leq A(m_0) \sum_{j=m_0+1}^{\infty} (|\psi_{1j}| + |\psi_{2j}|) \|\eta_{k-j}\|^2$$

due to the Hölder's inequality, routine calculations by (2.94) yield

$$\Gamma_{3n} := \frac{c_n}{n} \sum_{k=1}^{n} |g\Big[\frac{k}{n}, c_n \left(Z_{nk} + xc'_n\right)\Big]| \, \mu_{k,m_0}^2 = o_P(1),$$

as $n \to \infty$ first and then $m_0 \to \infty$. Result (2.86) now follows from (2.99), $E v_{0,m_0}^2 \to E\, u_0^2$ as $m_0 \to \infty$ and

$$\sup_{0\leq t\leq 1} |\Gamma_{2n}(t) - \Gamma_{1n}(t)|$$

$$\leq \frac{c_n}{n} \sum_{k=1}^{n} |g\Big[\frac{k}{n}, c_n \left(Z_{nk} + xc'_n\right)\Big]| \, (\mu_{k,m_0}^2 + 2|\mu_{k,m_0}||v_{k,m_0}|)$$

$$\leq \Gamma_{3n} + 2\Gamma_{3n}^{1/2}\Gamma_{2n}^{1/2}(1) = o_P(1),$$

as $n \to \infty$ first and then $m_0 \to \infty$. $\qquad\square$

Proof of Corollary 2.4. Due to Theorem 2.21 and the boundedness of $\widetilde{g}(x)$, we have

$$\frac{c_n}{n} \sum_{k=1}^{[nt]} g\Big(\frac{k}{n}, c_n Z_{nk}, c_n Z_{nk}, \cdots, c_n Z_{nk}\Big) \Rightarrow \int_0^t G_1(t) \, L_Z(dt, 0),$$

on $D[0,1]$. On the other hand, by noting that $|Z_{n,k+i} - Z_{nk}| \leq Cn^{-1/2} \sum_{j=1}^{i} |\xi_{k+j}|$, it follows from (2.87) that

$$\Delta_n \equiv \frac{c_n}{n} \sum_{k=1}^{n} |g\left(\frac{k}{n}, c_n Z_{nk}, c_n Z_{n,k+1}, \cdots, c_n Z_{n,k+d}\right)$$

$$-g\left(\frac{k}{n}, c_n Z_{nk}, c_n Z_{nk}, \cdots, Z_{nk}\right)|$$

$$\leq \left(\frac{C c_n}{\sqrt{n}}\right)^{\gamma} \frac{c_n}{n} \sum_{k=1}^{n} f\left(c_n Z_{nk}\right) \left(1 + \sum_{j=1}^{d} |\xi_{k+j}|^{\beta}\right).$$

The result (2.88) follows if $\Delta_n = o_P(1)$, which implies that we only need to prove

$$\Delta_{1n} \equiv \frac{c_n}{n} \sum_{k=1}^{n} E\left\{f\left(c_n Z_{nk}\right)\left(1 + \sum_{j=1}^{d} |\xi_{k+j}|^{\beta}\right)\right\} = O(1), \quad (2.100)$$

due to $c_n/\sqrt{n} \to 0$. To prove (2.100), set $a_0 \equiv \sum_{i=0}^{\infty} |\phi_i|$. Hölder's inequality yields

$$|\xi_{k+j}|^{2^m} \leq \left(a_0 \sum_{i=0}^{\infty} |\phi_i| \epsilon_{k+j-i}^2\right)^{2^{m-1}} \leq \dots \leq a_0^{2^m} \sum_{i=0}^{\infty} |\phi_i| |\epsilon_{k+j-i}|^{2^m}.$$

Consequently, by (2.94) of Lemma 2.1, we have

$$\Delta_{1n} \leq \frac{C c_n}{n} \sum_{i=0}^{\infty} |\phi_i| \sum_{k=1}^{n} E\left\{f\left(c_n Z_{nk}\right)\left(1 + |\epsilon_{k+j-i}|^{[\beta]+1}\right)\right\}$$

$$\leq \frac{C}{\sqrt{n}} \sum_{k=1}^{n} k^{-1/2} \leq C,$$

as required. $\qquad\qquad\qquad\qquad\qquad\qquad\qquad\qquad\qquad\qquad\qquad\qquad\square$

2.3.5 *Convergence to local time: framework II*

Using the concept of strong smooth array, Section 2.3.2 established a framework on the convergence to local time. When Theorem 2.19 is quite general, the condition imposed in Definition 2.5 that,

> conditional on $\mathcal{F}_{nj} = \sigma(X_{n1}, ..., X_{nj})$, $(X_{nk} - X_{nj})/d_{nkj}$ has a density $h_{nkj}(x)$ which is uniformly bounded by a constant K,

can hardly be checked, particularly in the situation that X_{nk} is a standardized partial sum of mixing random variables.

By using characteristic functions, this section provides an alternative condition that enables us to investigate the convergence to local time. This new framework has advantages when it is applied to stationary variables. Examples in relation to this framework will be given in Section 2.3.6.

Definition 2.6. A random array $\{X_{nk}\}_{k\geq 1, n\geq 1}$, is said to satisfy *smooth CF (characteristic function) condition* with $c_n > 0$ if, for any $M > 0$,

$$\frac{c_n}{n^2} \sum_{k=1}^{n} \int_{|s|\leq Mc_n} \left| \mathrm{E}e^{is\,X_{nk}} \right| ds \to 0, \qquad (2.101)$$

$$\frac{1}{n^2} \sum_{k=1}^{n-1} \sum_{j=k+1}^{n} \int_{|s|\leq 2Mc_n} \int_{A<|t|\leq Mc_n}$$
$$\left| \mathrm{E}e^{is\,X_{nk}-it(X_{nj}-X_{nk})} \right| ds\,dt \to 0, \qquad (2.102)$$

as $n \to \infty$ first and then $A \to \infty$.

We assume, throughout this section, that X_{nk} satisfies smooth CF condition with $c_n > 0$ such that $c_n \to \infty$ and $c_n/n \to 0$, and

$$X_{n,[nt]} \Rightarrow X_t \quad \text{on } D[0,\infty),$$

where $X = \{X_t\}_{t\geq 0}$ has a local time $L_X = \{L_X(t,x)\}_{t\geq 0, x\in\mathbb{R}}$ so that $L_X(t,x)$ is continuous in t for fixed x and $L_X(t,x)$ can be expressed as

$$L_X(t,x) = \frac{1}{2\pi} \int_{-\infty}^{\infty} e^{-i\,ux} \int_0^t e^{iuX_s}\,ds\,du. \qquad (2.103)$$

As in (2.8), the right hand side of (2.103) is understood as a limit of $L^N(t,x)$ in L^2-theory, where

$$L^N(t,x) = \frac{1}{2\pi} \int_{-N}^{N} e^{-i\,ux} \int_0^t e^{iuX_s}\,ds\,du.$$

Under this convention, for any bounded Borel function $f(t)$ on $[0,1]$, $\int_0^1 f(u)L_X(du,x)$ is well defined and we have

$$\int_0^1 f(u)L^N(du,x) = \frac{1}{2\pi} \int_{-N}^{N} e^{-i\,tx} \int_0^1 f(u)e^{itX_u}\,du\,dt$$
$$\to \int_0^1 f(u)L_X(du,x), \qquad (2.104)$$

in probability, as $N \to \infty$. The conditions on X such that $L_X(t,x)$ is continuous in t and has an expression (2.103) can be found in Section 2.1, including Gaussian and Lévy processes.

Theorem 2.22. *Suppose that $g(s,x)$ is Riemann integrable on $[0,1] \times \mathbb{R}$ and $\sup_{0 \le s \le 1} |g(s,x)| \le C/(1 + |x|^{1+b})$ for some $b > 0$ and $C > 0$. Then, on $D_{\mathbb{R}^2}[0,\infty)$, we have*

$$\left(X_{n,[nt]}, \; \frac{c_n}{n} \sum_{k=1}^{n} g\Big[\frac{k}{n}, c_n\left(X_{nk} + c'_n x\right)\Big] \right)$$

$$\Rightarrow \left(X_t, \; \int_0^1 G(t)\widetilde{L}_X(dt, x) \right), \tag{2.105}$$

where $G(t) = \int_{-\infty}^{\infty} g(t,x)dx$, x is fixed and

$$\widetilde{L}_X(r,x) = \begin{cases} L_X(r,0), & \text{if } c'_n \to 0, \\ L_X(r,-x), & \text{if } c'_n = 1. \end{cases}$$

Proof. Only consider $x = 0$ and $g(s,x) \ge 0$. Extension to $x \ne 0$ and general $g(s,x)$ is standard. Hence the details are omitted. Let $\widetilde{g}(x) = \sup_{0 \le s \le 1} g(s,x)$. First assume:

Con: $g(s,x)$ is continuous with $\int_{-\infty}^{\infty} \widetilde{g}(x)dx < \infty$ and there exists an $N_0 > 0$ such that $\widehat{g}(s,x) = 0$ for all $0 \le s \le 1$ and $|x| \ge N_0$, where $\widehat{g}(s,x) = \int_{-\infty}^{\infty} e^{ixt} g(s,t)dt$. \hfill (2.106)

Since $g(s,x) = \frac{1}{2\pi} \int_{-\infty}^{\infty} e^{itx} \widehat{g}(s,-t)dt$ due to (2.106), we have

$$\frac{c_n}{n} \sum_{k=1}^{n} g\Big(\frac{k}{n}, c_n X_{nk}\Big) = \frac{1}{2\pi n} \sum_{k=1}^{n} \int_{-\infty}^{\infty} \widehat{g}\Big(\frac{k}{n}, -\frac{t}{c_n}\Big) e^{it X_{nk}} dt$$

$$= R_{1n} + R_{2n}, \tag{2.107}$$

where, for some $A > 0$,

$$R_{1n} = \frac{1}{2\pi n} \sum_{k=1}^{n} \int_{|t| \le A} \widehat{g}\Big(\frac{k}{n}, -\frac{t}{c_n}\Big) e^{it X_{nk}} dt,$$

$$R_{2n} = \frac{1}{2\pi n} \sum_{k=1}^{n} \int_{|t| > A} \widehat{g}\Big(\frac{k}{n}, -\frac{t}{c_n}\Big) e^{it X_{nk}} dt.$$

Furthermore, R_{1n} can be written as

$$R_{1n} = \frac{1}{2\pi} \int_{|t| \le A} \int_0^1 \widehat{g}\Big(\frac{[nu]}{n}, -\frac{t}{c_n}\Big) e^{it X_{n,[nu]}} du\, dt + o_P(1).$$

Recall $c_n \to \infty$ and $\int_{-\infty}^{\infty} \widetilde{g}(x)dx < \infty$. It is readily seen that

$$\sup_{0 \le s \le 1} \sup_{|t| \le A} |\widehat{g}(s, -t/c_n) - \widehat{g}(s,0)| \to 0$$

for any $A > 0$. Hence, by Theorem 2.13, we have

$$\left(X_{n,[nt]}, R_{1n}\right) \Rightarrow \left(X_t, \frac{1}{2\pi}\int_{|t|\le A}\int_0^1 \widehat{g}(u,0)e^{itX_u}du\,dt\right), \quad (2.108)$$

on $D_{\mathbb{R}^2}[0,\infty)$, for any $A > 0$, as $n \to \infty$. As $\widehat{g}(u,0) = \int_{-\infty}^{\infty} g(u,x)dx$, by recalling (2.104), (2.105) will follow if we prove

$$E|R_{2n}|^2 \to 0, \quad (2.109)$$

as $n \to \infty$ first and then $A \to \infty$. In fact, by noting that $\sup_{0\le s\le 1}|\widehat{g}(s,x)| = 0$ for $|x| \ge N_0$ and

$$\sup_{0\le s\le 1}|\widehat{g}(s,x)| \le \int_{-\infty}^{\infty}\widetilde{g}(x)dx < \infty,$$

due to (2.106), it follows from smooth CF condition with $c_n > 0$ satisfying $c_n \to \infty$ and $c_n/n \to 0$ that

$$E|R_{2n}|^2 \le \frac{C}{n^2}\sum_{k=1}^n\sum_{l=1}^n\int_{|s|>A}\int_{|t|>A}|\widehat{g}(\frac{k}{n},-\frac{s}{c_n})|\,|\widehat{g}(\frac{l}{n},\frac{t}{c_n})|$$
$$\left|Ee^{isX_{nk}-itX_{nl}}\right|dsdt$$

$$\le \frac{C}{n^2}\sum_{k=1}^n\int_{|s|>A}\int_{|t|>A}|\widehat{g}(\frac{k}{n},-\frac{s}{c_n})|\,|\widehat{g}(\frac{k}{n},\frac{t}{c_n})|\left|Ee^{i(s-t)X_{nk}}\right|dsdt$$

$$+\frac{2C}{n^2}\sum_{1\le k<l\le n}\int_{|s|>A}\int_{|t|>A}|\widehat{g}(\frac{k}{n},-\frac{s}{c_n})|\,|\widehat{g}(\frac{l}{n},\frac{t}{c_n})|$$
$$\left|Ee^{i(s-t)X_{nk}-it(X_{nl}-X_{nk})}\right|dsdt$$

$$\le \frac{Cc_n}{n^2}\sum_{k=1}^n\int_{|s|\le 2N_0c_n}\left|Ee^{isX_{nk}}\right|ds$$

$$+\frac{2C}{n^2}\sum_{1\le k<l\le n}\int_{|s|\le 2N_0c_n}\int_{A<|t|\le N_0c_n}\left|Ee^{isX_{nk}-it(X_{nl}-X_{nk})}\right|dsdt$$

$$\to 0,$$

as $n \to \infty$ first and then $A \to \infty$. This proves (2.109) and also completes the proof of (2.105) under the additional condition (2.106).

It remains to remove (2.106). To this end, it suffices to show that,

for any $\epsilon > 0$, there exist $g_1(s,x)$ and $g_2(s,x)$ such that $g_1(s,x) \le g(s,x) \le g_2(s,x)$, both $g_1(s,x)$ and $g_2(s,x)$ satisfy (2.106), and

$$\int_{-\infty}^{\infty}\sup_{0\le s\le 1}[g_2(s,x) - g_1(s,x)]dx < \epsilon. \quad (2.110)$$

Indeed it follows from these facts that, by letting $G_i(t) = \int_{-\infty}^{\infty} g_i(t, x) dx$,

$$\frac{c_n}{n} \sum_{k=1}^{n} g\left(\frac{k}{n}, c_n X_{nk}\right) \leq \frac{c_n}{n} \sum_{k=1}^{n} g_2\left(\frac{k}{n}, c_n X_{nk}\right)$$

$$\to_D \int_0^1 G_2(t) L_X(dt, 0) \leq \int_0^1 [G(t) + \epsilon] L_X(dt, 0),$$

$$\frac{c_n}{n} \sum_{k=1}^{n} g\left(\frac{k}{n}, c_n X_{nk}\right) \geq \frac{c_n}{n} \sum_{k=1}^{n} g_1\left(\frac{k}{n}, c_n X_{nk}\right)$$

$$\to_D \int_0^1 G_1(t) L_X(dt, 0) \geq \int_0^1 [G(t) - \epsilon] L_X(dt, 0).$$

This yields (2.105) due to the arbitrarity of ϵ.

We next construct $g_1(s, x)$ and $g_2(s, x)$ that satisfy (2.110). To start with, by recalling that $g(s, x)$ is Riemann integrable, for $\forall \epsilon > 0$, there exist two continuous functions $g_{1\epsilon}(s, x)$ and $g_{2\epsilon}(s, x)$ such that,

$$g_{1\epsilon}(s, x) \leq g(s, x) \leq g_{2\epsilon}(s, x) \quad \text{and} \quad g_{2\epsilon}(s, x) - g_{1\epsilon}(s, x) \leq \epsilon. \quad (2.111)$$

Here $g_{i\epsilon}(s, x), i = 1, 2$, can be chosen such that

$$\sup_{0 \leq s \leq 1} |g_{i\epsilon}(s, x)| \leq C/(1 + |x|^{1+b})$$

and

$$\int_{-\infty}^{\infty} \sup_{0 \leq s \leq 1} [g_{2\epsilon}(s, x) - g_{1\epsilon}(s, x)] dx < \epsilon. \quad (2.112)$$

Define, for $i = 1, 2$,

$$f_{i\delta}(s, x) = \frac{\delta}{\pi} \int_{-\infty}^{\infty} \frac{\sin^2[(x - y)/\delta]}{(x - y)^2} g_{i\epsilon}(s, y) dy$$

$(\sin(y)/y \equiv 1$ if $y = 0)$ and write

$$f(x) = \sum_{n=1}^{\infty} n^{-1-b/2} \frac{\sin^2(x - n)}{(x - n)^2}.$$

Simple calculation shows that

$$c_1 (1 + |x|^{-1-b/2}) \leq f(x) \leq c_2 (1 + |x|^{-1-b/2}), \quad (2.113)$$

where $c_1 > 0$ and $c_2 > 0$ are constants. On the other hand, by noting that $\int_{-\infty}^{\infty} \frac{\sin^2(x)}{x^2} dx = \pi$ and

$$f_{i\delta}(s, x) = \frac{1}{\pi} \int_{-\infty}^{\infty} \frac{\sin^2(y)}{y^2} g_{i\epsilon}(s, x + \delta y) dy,$$

it follows from (2.112) that, for any $\delta > 0$,

$$\int_{-\infty}^{\infty} \sup_{0 \le s \le 1} |f_{2\delta}(s,x) - f_{1\delta}(s,x)| dx$$

$$\le \frac{1}{\pi} \int_{-\infty}^{\infty} \frac{\sin^2(y)}{y^2} \int_{-\infty}^{\infty} \sup_{0 \le s \le 1} |g_{2\epsilon}(s,x) - g_{1\epsilon}(s,x)| dx dy$$

$$\le \epsilon. \tag{2.114}$$

It will be proved below that, for any $\epsilon > 0$, there exists a δ_0 such that for all $0 < \delta \le \delta_0$,

$$\sup_{0 \le s \le 1} |f_{i\delta}(s,x) - g_{i\epsilon}(s,x)| \le \epsilon f(x). \tag{2.115}$$

Now the required $g_1(s,x)$ and $g_2(s,x)$ can be defined by

$$g_2(s,x) = f_{2\delta_0}(s,x) + \epsilon f(x) \quad \text{and} \quad g_1(s,x) = f_{1\delta_0}(s,x) - \epsilon f(x).$$

Indeed, by noting

$$\int_{-\infty}^{\infty} \frac{\sin^2(x)}{x^2} e^{itx} dx = \begin{cases} \pi(1 - |t|/2), & \text{if } |t| < 2 \\ 0, & \text{otherwise,} \end{cases}$$

it is readily seen that, $g_1(s,x)$ is continuous, $\int_{-\infty}^{\infty} \sup_{0 \le s \le 1} g_1(s,x) dx < \infty$ and for all $0 \le s \le 1$ and $\min\{t, t\delta_0\} \ge 2$,

$$\widehat{g}_1(s,t) = \int_{-\infty}^{\infty} e^{itx} g_1(s,x) dx,$$

$$= \frac{1}{\pi} \int_{-\infty}^{\infty} g_{1\epsilon}(s,y) e^{ity} dy \int_{-\infty}^{\infty} \frac{\sin^2(x)}{x^2} e^{it\delta_0 x} dx$$

$$- \epsilon \sum_{n=1}^{\infty} n^{-1-b/2} e^{itn} \int_{-\infty}^{\infty} \frac{\sin^2(x)}{x^2} e^{it x} dx$$

$$= 0.$$

That is, $g_1(s,x)$ satisfies (2.106). Similarly, $g_2(s,x)$ satisfies (2.106). Furthermore, by (2.111) and (2.115),

$$g_1(s,x) \le g_{1\epsilon}(s,x) \le g(s,x) \le g_{2\epsilon}(s,x) \le g_2(s,x),$$

and by (2.113)–(2.114),

$$\int_{-\infty}^{\infty} \sup_{0 \le s \le 1} [g_2(s,x) - g_1(s,x)] dx$$

$$\le \int_{-\infty}^{\infty} \sup_{0 \le s \le 1} [f_{2\delta_0}(s,x) - f_{1\delta_0}(s,x)] dx + 2\epsilon \int_{-\infty}^{\infty} f(x) dx$$

$$\le C \epsilon,$$

where C is a constant. Collecting all these facts, (2.110) is verified due to the arbitrarity of ϵ.

Finally, we only need to prove (2.115). We have

$$
\sup_{0\leq s\leq 1} \left| f_{1\delta}(s,x) - g_{1\epsilon}(s,x) \right|
$$

$$
\leq \frac{1}{\pi} \int_{-\infty}^{\infty} \frac{\sin^2(y)}{y^2} \sup_{0\leq s\leq 1} \left| g_{1\epsilon}(s, x+\delta y) - g_{1\epsilon}(s,x) \right| dy
$$

$$
\leq \frac{1}{\pi} \left(\int_{|x|\leq A} + \int_{\substack{|x|>A \\ |\delta y|\geq |x|/2}} + \int_{\substack{|x|>A \\ |\delta y|<|x|/2}} \right) \frac{\sin^2(y)}{y^2} \left| g_{1\epsilon}(s, x+\delta y) - g_{1\epsilon}(s,x) \right| dy
$$

$$
:= R_{1A} + R_{2A} + R_{3A}, \tag{2.116}
$$

where A will be chosen later. Recall (2.113) and $\sup_{0\leq s\leq 1}|g_{1\epsilon}(s,x)| \leq C/(1+|x|^{1+b})$ for some $b > 0$. For any $\epsilon > 0$, there exists an $A_0 > 0$ such that, whenever $A \geq A_0$,

$$
R_{3A} \leq 2C\,(A/2)^{-b/2}\,[1 + (|x|/2)^{-1-b/2}]\mathrm{I}(|x| \geq A)
$$

$$
\leq \epsilon\, f(x)\mathrm{I}(|x| \geq A)/3.
$$

For this A_0, routine calculations yield that there exists a δ_0 such that, for all $0 < \delta \leq \delta_0$,

$$
R_{1A_0} \leq \epsilon\, f(x)\mathrm{I}(|x| \leq A_0)/3,
$$

due to the continuity of $g_{1\epsilon}(s,x)$, and

$$
R_{2A_0} \leq \frac{C\delta}{\pi} \int_{|y|\geq |x|/2} y^{-2} \left[|x|^{-1-b} + (1+|x+y|^{1+b})^{-1} \right] dy\, \mathrm{I}(|x| \geq A_0)
$$

$$
\leq \epsilon\, f(x)\mathrm{I}(|x| \geq A_0)/3.
$$

Taking $A = A_0$ in (2.116), it follows from these estimates that, for any $\epsilon > 0$, there exists a δ_0 such that for all $0 < \delta \leq \delta_0$,

$$
\sup_{0\leq s\leq 1} \left| f_{1\delta}(s,x) - g_{1\epsilon}(s,x) \right| \leq \epsilon\, f(x).
$$

Similarly, $\sup_{0\leq s\leq 1}\left| f_{2\delta}(s,x) - g_{2\epsilon}(s,x) \right| \leq \epsilon\, f(x)$, and hence (2.115) is proved. $\qquad\square$

Remark 2.13. If in addition that X_{nk} is a smooth array (see Definition 2.2), conclusions of Corollaries 2.1 and 2.2 still hold. We mention that if there exists a sequence of $\sigma_{nk} > 0$ such that $\sup_{n\geq 1} \frac{1}{n}\sum_{k=1}^{n} \sigma_{nk}^{-1} < \infty$ and $\mathrm{E}\, e^{itX_{nk}/\sigma_{nk}}, k \geq 1, n \geq 1$, is uniformly integrable, i.e.,

$$
\lim_{A\to\infty} \sup_{n,k} \int_{|t|\geq A} \left| \mathrm{E}\, e^{itX_{nk}/\sigma_{nk}} \right| = 0,
$$

then X_{nk} is a smooth array and we also have (2.101).

Smooth CF condition is "high level" for a framework, which is similar to the conditions that enable X_{nk} to be a strong smooth array. However, the smooth CF condition is easy to verify for certain stationary time series. The following is an example.

Let $\epsilon_1, \epsilon_2, \ldots$ be a sequence of stationary random variables with mean zero on a probability space $(\Omega, \mathcal{F}, \mathrm{P})$. Let $\mathcal{D}_j, j = 0, \pm 1, \pm 2, \ldots$, be a sequence of sub σ-fields of $\mathcal{F}$. Set

$$X_{nk} = \frac{1}{\sqrt{n}} \sum_{j=1}^{k} \epsilon_j$$

and $\mathcal{D}_p^q = \sigma(\mathcal{D}_j, p \leq j \leq q)$. We introduce restrictions on ϵ_i and $\mathcal{D}_p^q$:

C1. $\mathrm{E}|\epsilon_1|^4 < \infty$ and $\lim_{n \to \infty} var(X_{nn}) = \sigma^2 > 0$.

C2. There exist $d > 0$ and a $\mathcal{D}_{n-m}^{n+m}$ measurable random variable $\epsilon_{n,m}^*$ such that for $n, m = 1, 2, \ldots$ with $m > d^{-1}$,

$$\mathrm{E}|\epsilon_n - \epsilon_{n,m}^*| \leq d^{-1} e^{-dm}.$$

C3. There exists $d > 0$ such that for $n, m = 1, 2, \ldots$, $A \in \mathcal{D}_{-\infty}^n, B \in \mathcal{D}_{n+m}^\infty$,

$$|\mathrm{P}(A \cap B) - \mathrm{P}(A)\mathrm{P}(B)| \leq d^{-1} e^{-dm}.$$

C4. There exists $d > 0$ such that for $n, m = 1, 2, \ldots$ with $n > m > d^{-1}$ and all $t \in \mathrm{R}$ with $|t| \geq d$,

$$\mathrm{E}|\mathrm{E}\{\exp[it(\epsilon_{n-m} + \ldots + \epsilon_{n+m})]|\mathcal{D}_j : j \neq n\}| \leq e^{-d}.$$

C5. There exists $d > 0$ such that for $n, m, p = 1, 2, \ldots$, $A \in \mathcal{D}_{n-p}^{n+p}$,

$$\mathrm{E}|\mathrm{P}(A|\mathcal{D}_j : j \neq n) - \mathrm{P}(A|\mathcal{D}_j : 0 < |n - j| \leq m + p)| \leq d^{-1} e^{-dm}.$$

This condition set is the same as that in Götze and Hipp (1983), where the authors developed asymptotic expansions for dependent random variables. Due to the flexible choice of D_i, the condition set is suitable for many interesting models. Examples are given in Section 2.3.6.

Theorem 2.23. *Suppose* **C1**–**C5** *hold. Then X_{nk} satisfies smooth CF condition with $c_n > 0$ such that $c_n \to \infty$ and $c_n \log^{5/2} n/n \to 0$.*

Proof.   We refer to Wang (2015).   □

2.3.6 Example: stationary processes

This section provides several examples of stationary processes such that X_{nk} satisfies smooth CF condition with some $c_n > 0$ and $X_{n,[nt]} \Rightarrow X_t$ on $D[0, \infty)$.

Example 2.13. Let $\eta_0, \eta_1, \ldots$ be a sequence of i.i.d. random variables. Let $m \geq 1$ and $h : \mathbb{R}^m \to \mathbb{R}$ be continuously differentiable, and define

$$\epsilon_i = h(\eta_{i+1}, \ldots, \eta_{i+m}), \quad i = 1, 2, \ldots$$

$\epsilon_k, k \geq 1$, is a sequence of stationary m-dependent shift.

Suppose that $E\epsilon_1 = 0$, $E\epsilon_1^2 < \infty$ and $\frac{1}{n} var\left(\sum_{k=1}^n \epsilon_k\right) \to \sigma^2 > 0$. It is well-known (see, e.g., Billingsley (1968)) that

$$X_{n,[nt]} := \frac{1}{\sqrt{n}\sigma} \sum_{k=1}^{[nt]} \epsilon_k \Rightarrow B_t, \quad \text{on } D[0, \infty),$$

where $B = \{B_t\}_{t \geq 0}$ is a Brownian motion.

Furthermore, if in addition that η_0 has a density function $p(x)$, $E\epsilon_0^4 < \infty$ and there exist $y_1, \ldots, y_{2m-1} \in \mathbb{R}$ and an open set $U \supset \{y_1, \ldots, y_{2m-1}\}$ such that $p > 0$ on U and

$$\sum_{j=1}^m \frac{\partial}{\partial x_j} h(x_1, \ldots, x_m) \mid_{(x_1, \ldots, x_m) = y_j, \ldots, y_{m+j-1}} \neq 0,$$

then X_{nk} satisfies smooth CF condition with all $c_n > 0$ such that $c_n \to \infty$ and $c_n \log^{5/2} n/n \to 0$. Indeed, Example 1.11 of Götze and Hipp (1983) shows that ϵ_k satisfies **C1–C5** and then claim follows from Theorem 2.23.

Example 2.14. Let $\eta_0, \eta_1, \ldots$ be a strictly Markov chain, and the transition kernel $P(x, A)$ of the Markov chain satisfies

$$\sup |P(x, A) - P(x', A)| < 1$$

where the sup is taken over all x, x' and A. The last condition is satisfied whenever there exists a positive measure μ such that, for all x and A, $P(x, A) \geq \mu(A)$. Let f be a measurable function on the state space I of the chain and define

$$\epsilon_j = f(\eta_j), \quad j = 1, 2, \ldots$$

Suppose that $E\epsilon_1 = 0$, $E |\epsilon_1|^3 < \infty$ and $\frac{1}{n} var\left(\sum_{k=1}^n \epsilon_k\right) \to \sigma^2 > 0$. It follows from Formanov(1975) that

$$X_{n,[nt]} := \frac{1}{\sqrt{n}\sigma} \sum_{k=1}^{[nt]} \epsilon_k \Rightarrow B_t, \quad \text{on } D[0, \infty),$$

where $B = \{B_t\}_{t \geq 0}$ is a Brownian motion.

Furthermore, if in addition that ϵ_1 satisfies Cramér's condition, i.e., $\lim_{|t| \to \infty} |Ee^{it\epsilon_0}| < 1$, and $E\epsilon_1^4 < \infty$, then X_{nk} satisfies smooth CF condition with all $c_n > 0$ such that $c_n \to \infty$ and $c_n \log^{5/2} n/n \to 0$. Indeed, Example 1.13 of Götze and Hipp (1983) shows that ϵ_k satisfies **C1–C5** and then claim follows from Theorem 2.23.

Example 2.15. Let $\{\epsilon_j, j \geq 1\}$ be a stationary sequence of Gaussian random variables with $E\epsilon_1 = 0$ and covariances $\gamma(j - i) = E\epsilon_i\epsilon_j$ satisfying the following condition, for some $0 < \alpha < 2$,

$$d_n^2 \equiv \sum_{1 \leq i, j \leq n} \gamma(j - i) \sim n^\alpha h(n),$$

as $n \to \infty$, where $h(n)$ is a slowly varying function at ∞. Note that $d_n^2 = var\left(\sum_{k=1}^n \epsilon_k\right)$. It follows from Lemma 5.1 of Taqqu (1975) that

$$X_{n,[nt]} := \frac{1}{d_n} \sum_{k=1}^{[nt]} \epsilon_k \Rightarrow B_{\alpha/2}(t), \quad \text{on } D[0, \infty),$$

where $B_{\alpha/2} = \{B_{\alpha/2}(t)\}_{t \geq 0}$ is a fractional Brownian motion. Furthermore, routine calculations show that X_{nk} satisfies smooth CF condition with all $c_n > 0$ such that $c_n \to \infty$ and $c_n/n \to 0$.

Example 2.16. Let $\tau \geq 0$ be a constant, $\gamma = 1 - \tau/n$, $y_{n0} = 0$,

$$y_{nk} = \gamma y_{n,k-1} + \xi_k \quad \text{and} \quad Z_{nk} = y_{nk}/d_n,$$

where ξ_k is a linear process discussed in Example 2.12 and $d_n^2 = var(\sum_{j=1}^n \xi_j)$. It is proved in Theorem 2.21 that $Z_{n,[nt]} \Rightarrow Z_t$ on $D[0, \infty)$, where $Z = \{Z_t\}_{t \geq 0}$ is an Ornstein-Uhlenbeck process defined by (2.82). Furthermore, if $\lim_{|t| \to \infty} |t|^\eta |E e^{it\epsilon_0}| < \infty$ for some $\eta > 0$, then there exists a $k_0 \geq 1$ such that $Z_{nk}, k \geq k_0, n \geq 1$, is a strong smooth array (see Theorem 2.18). The following theorem indicates that Z_{nk} satisfies smooth CF condition under less restriction on ϵ_0.

Theorem 2.24. *If* $\limsup_{|t| \to \infty} |Ee^{it\epsilon_0}| < 1$, *i.e.,* ϵ_0 *satisfies the Cramér's condition, then* Z_{nk} *satisfies smooth CF condition with all* $c_n > 0$ *such that* $c_n \to \infty$ *and* $c_n/n \to 0$.

Proof. We start with some preliminaries. First notice that

$$y_{nk} = \sum_{j=1}^k \gamma^{k-j} \xi_j = \sum_{j=1}^k \gamma^{k-j} \left(\sum_{q=1}^j + \sum_{q=-\infty}^0 \right) \epsilon_q \phi_{j-q}$$

$$= \sum_{q=1}^k \epsilon_q a_{k,q} + \sum_{q=-\infty}^0 \epsilon_q a'_{k,q}, \tag{2.117}$$

where $a_{k,q} = \sum_{j=q}^{k} \gamma^{k-j} \phi_{j-q} = \sum_{j=0}^{k-q} \gamma^{k-j-q} \phi_j$ and $a'_{k,q} = \sum_{j=1}^{k} \gamma^{k-j} \phi_{j-q}$. Obviously, $a_{k,q} = a_{k-q,0}$. Furthermore, by recalling $\gamma = 1 - \tau/n$, for $1 \leq q \leq k/2$, $n \geq k$ and k sufficiently large, there exist $0 < c_1 < c_2 < \infty$ such that $c_1/\sqrt{k} \leq |a_{k,q}|/d_k \leq c_2/\sqrt{k}$ and

$$c_1 d_k^2 \leq \sum_{q=1}^{k/2} a_{k,q}^2 \leq \sum_{i=1}^{k} a_{k,i}^2 + \sum_{i=-\infty}^{0} a_{k,i}'^2 \leq c_2 d_k^2, \qquad (2.118)$$

where d_k is given in (2.62).

Due to (2.117), it follows from the independence of ϵ_q that, for all $k < j$,

$$I_{kj} := \int_{|s| \leq 2Mc_n} \int_{A < |t| \leq Mc_n} \left| E e^{is Z_{nk} - it(Z_{nj} - Z_{nk})} \right| ds\, dt$$

$$\leq \int_{|s| \leq 3Mc_n} \int_{A \leq |t| \leq Mc_n} \left| E\, e^{is Z_{nk} + it Z_{nj}} \right| ds\, dt$$

$$\leq \int_{A \leq |t| \leq Mc_n} \left| E\, e^{it z^{(2)}} \right| \Lambda(t, k)\, dt, \qquad (2.119)$$

where $\Lambda(t, k) = \int_{|s| \leq 3Mc_n} \left| E\, e^{iz_k^{(1)}(t,s)} \right| ds$,

$$z_k^{(1)}(t, s) = \sum_{q=1}^{k} \epsilon_q \left(t\, a_{j,q} + s\, a_{k,q} \right)/d_n, \quad z^{(2)} = \sum_{q=k+1}^{j} \epsilon_q a_{j,q}/d_n.$$

In order to estimate (2.119), write $\Omega_1 \equiv \Omega_1(t, s)$ (Ω_2, respectively) for the set of $1 \leq q \leq k/2$ such that $|t\, a_{j,q} + s\, a_{k,q}| \geq d_n$ ($|t\, a_{j,q} + s\, a_{k,q}| < d_n$, respectively), and

$$B_{1k} = \sum_{q \in \Omega_2} a_{k,q}^2, \quad B_2 = \sum_{q \in \Omega_2} a_{j,q} a_{k,q} \quad \text{and} \quad B_3 = \sum_{q \in \Omega_2} a_{j,q}^2.$$

It will be shown later that, for k sufficiently large and all t, there exist constants $\gamma_1 > 0$ and $\gamma_2 > 0$ such that

$$\left| E\, e^{iz_k^{(1)}(t,s)} \right| \leq \exp \left\{ -\gamma_1 \#(\Omega_1) - \gamma_2 B_{1k} \left(s + t B_2/B_{1k} \right)^2/d_n^2 \right\}, \qquad (2.120)$$

where $\#(A)$ denotes the number of elements in A.

Note that $\Omega_1 + \Omega_2 = k/2$. By recalling (2.118), we have $B_{1k} \geq C d_k^2$ whenever $\#(\Omega_1) \leq \sqrt{k}$ and k sufficiently large (say, $k \geq k_0$). This, together with (2.120), implies that, for all $t \in \mathbb{R}$ and $k \geq k_0$,

$$\Lambda(t, k) \leq \int_{\#(\Omega_1) \geq \sqrt{k}} e^{-\gamma_1 \#(\Omega_1)}\, I(|s| \leq 3Mc_n)\, ds$$

$$+ \int_{\#(\Omega_1) \leq \sqrt{k}} e^{-\gamma_2 B_{1k} (s + t B_2/B_{1k})^2/d_n^2}\, ds$$

$$\leq C\, k^{-2} c_n + \int_{-\infty}^{\infty} e^{-C d_k^2 s^2/d_n^2}\, ds$$

$$\leq C\, (k^{-2} c_n + d_n/d_k). \qquad (2.121)$$

Similarly, by noting that $sz^{(2)} =_D z^{(1)}_{j-k}(0, s)$, it follows from (2.120) that, for $j - k \geq k_0$ and for all $A > 0$,

$$
\int_{A \leq |t| \leq Mc_n} \left| \mathrm{E}\, e^{itz^{(2)}} \right| dt \leq \int_{\#(\Omega_1) \geq \sqrt{j-k}} e^{-\gamma_1 \#(\Omega_1)} \mathrm{I}(|t| \leq 3Mc_n) dt
$$

$$
+ \int_{\#(\Omega_1) \leq \sqrt{j-k}} e^{-\gamma_2 B_{1,j-k} t^2/d_n^2} \mathrm{I}(|t| \geq A) dt
$$

$$
\leq C\, (j-k)^{-2} c_n + \int_{|t| \geq A} e^{-C d_{j-k}^2 t^2/d_n^2} dt. \quad (2.122)
$$

Note that both (2.121) and (2.122) hold for $k, j - k \leq k_0$. Taking these estimates into (2.119), routine calculations show that

$$
\frac{1}{n^2} \sum_{k=1}^{n-1} \sum_{j=k+1}^{n} I_{kj}
$$

$$
\leq Cc_n/n + \frac{1}{n^2} \sum_{k=1}^{n-1} d_n/d_k \sum_{j=k+1}^{n} \int_{|t| \geq A} e^{-C d_{j-k}^2 t^2/d_n^2} dt
$$

$$
\leq Cc_n/n + \frac{C}{n} \sum_{k=1}^{n} (d_n/d_k) \mathrm{I}(d_k/d_n \leq A^{-1/4}) + CA^{1/4} \int_{|t| \geq \sqrt{A}} e^{-C t^2} dt
$$

$$
\to 0,
$$

as $n \to \infty$ first and then $A \to \infty$, due to $c_n/n \to 0$. This proves (2.102). Similarly, by noting that

$$
\int_{|s| \leq Mc_n} \left| \mathrm{E} e^{is\, Z_{nk}} \right| ds \leq \Lambda(0, k),
$$

it follows from (2.121) that

$$
\frac{c_n}{n^2} \sum_{k=1}^{n} \int_{|s| \leq Mc_n} \left| \mathrm{E} e^{is\, Z_{nk}} \right| ds \leq \frac{Cc_n^2}{n^2} + \frac{Cc_n}{n} \frac{d_n}{n} \sum_{k=1}^{n} d_k^{-1} \to 0,
$$

due to $c_n/n \to 0$. That is, (2.101) holds.

It remains to prove (2.120). To see this, first notice that there exist constants $\gamma_1 > 0$ and $\gamma_2 > 0$ such that

$$
\left| \mathrm{E}\, e^{i \epsilon_1 t} \right| \leq \begin{cases} e^{-\gamma_1} & \text{if } |t| \geq 1, \\ e^{-\gamma_2 t^2} & \text{if } |t| \leq 1, \end{cases} \quad (2.123)
$$

since $E\epsilon_1 = 0$, $E\epsilon_1^2 = 1$ and ϵ_1 satisfies the Cramer's condition. See, e.g., Chapter 1 of Petrov (1995). On the other hand, by noting that $B_2^2 \leq B_{1k} B_3$

due to Hölder's inequality, we have

$$\sum_{q \in \Omega_2} (t\, a_{j,q} + s a_{k,q})^2 = t^2 B_3 + 2t\, s\, B_2 + s^2\, B_{1k}$$

$$= B_{1k}(s + tB_2/B_{1k})^2 + t^2(B_3 - B_2^2/B_{1k})$$

$$\geq B_{1k}(s + tB_2/B_{1k})^2.$$

By virtue of these facts, it follows from the independence of ϵ_t that, for k sufficiently large and all t,

$$\left| \mathrm{E}\, e^{iz_k^{(1)}(t,s)} \right| \leq \prod_{q=1}^{k/2} \left| \mathrm{E}\, e^{i\epsilon_1(t\, a_{j,q} + s\, a_{k,q})} \right|$$

$$\leq \exp\left\{ -\gamma_1 \#(\Omega_1) - \gamma_2 \sum_{q \in \Omega_2} (t\, a_{j,q} + s a_{k,q})^2/d_n^2 \right\}$$

$$\leq \exp\left\{ -\gamma_1 \#(\Omega_1) - \gamma_2\, B_{1k}\, (s + tB_2/B_{1k})^2/d_n^2 \right\},$$

as required. $\qquad\square$

Example 2.17. Let $\{\xi_j, j \geq 1\}$ be a fractional process defined by

$$(1 - B)^d \xi_j = \epsilon_j,$$

where $0 \leq d < 1/2$, B is a backshift operator and $\{\epsilon_j\}_{j \in \mathbb{Z}}$ is a sequence of i.i.d. random variables with $\mathrm{E}\epsilon_0 = 0$ and $\mathrm{E}\epsilon_0^2 = 1$. The fractional difference operator $(1 - B)^\gamma$ is defined by its Maclaurin series (by its binomial expansion, if γ is an integer):

$$(1 - B)^\gamma = \sum_{j=0}^\infty \frac{\Gamma(-\gamma + j)}{\Gamma(-\gamma)\Gamma(j+1)} B^j \quad \text{where} \quad \Gamma(z) = \begin{cases} \int_0^\infty s^{z-1} e^{-s} ds & \text{if } z > 0 \\ \infty & \text{if } z = 0. \end{cases}$$

If $z < 0$, $\Gamma(z)$ is defined by the recursion formula $z\Gamma(z) = \Gamma(z+1)$.

Let $\tau \geq 0$ be a constant, $\gamma = 1 - \tau/n$, $y_{n0} = 0$,

$$y_{k,n} = \gamma\, y_{k-1,n} + \xi_k \quad \text{and} \quad Z_{nk} = y_{nk}/d_n,$$

where $d_n^2 = c_d\, n^{1+2d}$ with

$$c_d = \frac{1}{\Gamma^2(d)d(1 + 2d)} \int_0^\infty x^{d-1}(x+1)^{d-1} dx.$$

Note that $\xi_k = \sum_{j=0}^\infty a(j)\, \epsilon_{k-j}$ with

$$a(j) = \frac{\Gamma(j + d)}{\Gamma(j+1)\,\Gamma(d)} \sim \frac{1}{\Gamma(d)} j^{d-1}.$$

It follows from Example 2.16 that, on $D[0,1]$,

$$Z_{n,[nt]} \Rightarrow W_{d+1/2}(t) + \tau \int_0^t e^{\tau.(t-s)} W_{d+1/2}(s) ds,$$

a fractional Ornstein-Uhlenbeck process. Furthermore, if ϵ_0 satisfies the Cramér's condition, then Z_{nk} satisfies smooth CF condition with all $c_n > 0$ such that $c_n \to \infty$ and $c_n/n \to 0$.

2.4   Convergence to self-intersection local time

Let $g(x)$ be a real function on R. Let $\tau \geq 0$ be a constant, $\gamma = 1 - \tau/n$, $y_{n0} = 0$,

$$y_{nk} = \gamma \, y_{n,k-1} + \xi_k \quad \text{and} \quad Z_{nk} = y_{nk}/d_n,$$

where ξ_k is a linear process given in Example 2.12 and $d_n^2 = var(\sum_{j=1}^n \xi_j)$. Theorem 2.21 yields $Z_{n,[nt]} \Rightarrow Z_t$ on $D[0,1]$, where $Z = \{Z_t\}_{t \geq 0}$ is a Gaussian process given in (2.82). Using Remark 2.2, routine calculations show that Z has a self-intersection local time V_t that can be expressed as

$$V_t = \frac{1}{2\pi} \lim_{N \to \infty} \int_{-N}^{N} \int_0^t \int_0^t e^{iu(Z_s - Z_v)} ds dv \, du, \qquad (2.124)$$

where the limit is understood in the framework of L^2-theory. The following theorem shows that V_1 is a limit of S_n defined by

$$S_n = \frac{c_n}{n^2} \sum_{k,j=1}^{n} g\big[c_n \, (Z_{nk} - Z_{nj})\big],$$

where c_n is a sequence of positive constants.

Theorem 2.25. *Suppose that* $\int_{-\infty}^{\infty} |g(x)| dx < \infty$, $\int_{-\infty}^{\infty} g(x) dx \neq 0$ *and* $\lim_{|t| \to \infty} |t|^\eta |\mathrm{E} \, e^{it\epsilon_0}| < \infty$ *for some* $\eta > 0$. *Then, for any* $c_n \to \infty$ *and* $n/c_n \to \infty$,

$$\big(X_{n,[nt]}, \, S_n\big) \Rightarrow \Big(Z_t, \int_{-\infty}^{\infty} g(x) dx \, V_1\Big). \qquad (2.125)$$

Proof. We only provide an outline, as the proof is similar to that of Theorem 2.22 and Theorem 2.24. We first prove (2.125) under an additional condition:

Con: $g(x)$ is continuous and $\widehat{g}(t)$ has a compact support,

where $\widehat{g}(x) = \int_{-\infty}^{\infty} e^{ixt} g(t) dt.$ \hfill (2.126)

To start with, noting that $g(x) = \frac{1}{2\pi} \int_{-\infty}^{\infty} e^{itx} \widehat{g}(-t) dt$, we have

$$\frac{c_n}{n^2} \sum_{k,j=1}^{n} g\big[c_n \, (Z_{nk} - Z_{nj})\big]$$

$$= \frac{1}{2\pi n^2} \sum_{k,j=1}^{n} \int_{-\infty}^{\infty} \widehat{g}(-s/c_n) \, e^{is \, (Z_{nk} - Z_{nj})} ds$$

$$= R_{1n} + R_{2n}, \qquad (2.127)$$

where, for some $A > 0$,

$$R_{1n} = \frac{1}{2\pi n^2} \sum_{k,j=1}^{n} \int_{|s| \leq A} \widehat{g}(-s/c_n) \, e^{is \, (Z_{nk} - Z_{nj})} ds,$$

$$R_{2n} = \frac{1}{2\pi n^2} \sum_{k,j=1}^{n} \int_{|s| > A} \widehat{g}(-s/c_n) \, e^{is \, (Z_{nk} - Z_{nj})} ds.$$

Recall that $c_n \to \infty$. It follows that $\sup_{|s| \leq A} |\widehat{g}(-s/c_n) - \widehat{g}(0)| \to 0$ for any fixed $A > 0$. This, together with Theorem 2.21 (i), yields that, for any $A > 0$,

$$\left(X_{n,[nt]}, \; R_{1n} \right) \Rightarrow \left(Z_t, \; \frac{\widehat{g}(0)}{2\pi} \int_{|s| \leq A} \int_0^1 \int_0^1 e^{is[Z_u - Z_v]} du \, dv \, ds \right), \tag{2.128}$$

on $D_{\mathrm{R}^2}[0,\infty)$, as $n \to \infty$. By (2.124) and (2.128), noting that $\widehat{g}(0) = \int_{-\infty}^{\infty} g(x) dx$, (2.125) will follow under the additional condition (2.126), if we prove

$$\mathrm{E} \, |R_{2n}|^2 \to 0, \tag{2.129}$$

as $n \to \infty$ first and then $A \to \infty$.

In order to prove (2.129), write $\eta_r' = \sum_{q=1}^{r} \epsilon_q \phi_{r-q}$ and

$$y_{\lambda,n}' = \sum_{r=1}^{\lambda} \gamma^{\lambda-r} \eta_r' = \sum_{q=1}^{\lambda} \epsilon_q \sum_{r=0}^{\lambda-q} \gamma^{\lambda-q-r} \phi_r = \sum_{q=1}^{\lambda} \epsilon_q \, a(\lambda - q),$$

where $a(v) = \sum_{r=0}^{v} \gamma^{v-r} \phi_r$. We have, whenever $k \geq j \geq l \geq m$,

$$s \left(y_{k,n}' - y_{j,n}' \right) - t \left(y_{l,n}' - y_{m,n}' \right)$$

$$= s \sum_{q=j+1}^{k} \epsilon_q \, a(k-q) + s \sum_{q=l+1}^{j} \epsilon_q \, [a(k-q) - a(j-q)]$$

$$+ \sum_{q=m+1}^{l} \epsilon_q \left\{ s \, [a(k-q) - a(j-q)] - t \, a(l-q) \right\}$$

$$+ \sum_{q=1}^{m} \epsilon_q \left\{ s \, [a(k-q) - a(j-q)] - t \, [a(l-q) - a(m-q)] \right\}.$$

Due to the independence of ϵ_q, it follows that, whenever $k \geq j \geq l \geq m$,

$$\left| \mathrm{E} \exp \left\{ \left[is \left(y_{k,n}' - y_{j,n}' \right) - it \left(y_{l,n}' - y_{m,n}' \right) \right] \right\} \right| \leq I_1(s) \, I_2(s,t),$$

where

$$I_1(s) = \left| E \exp\left\{ is \sum_{q=j+1}^{k} \epsilon_q\, a(k-q) \right\} \right|,$$

$$I_2(s,t) = \left| E \exp\left(i \sum_{q=m+1}^{l} \epsilon_q \left\{ s\left[a(k-q) - a(j-q) \right] - t\, a(l-q) \right\} \right) \right|.$$

Similar arguments as in the proof of Theorem 2.24 yield

$$\int_{|s|\geq A} I_1\!\left(\frac{s}{d_n}\right) |\widehat{g}(-s/c_n)|\, ds$$

$$\leq C\, c_n\, (k-j)^{-2} + \int_{|s|\geq A} e^{-C\, d_{k-j}^2\, s^2/d_n^2}\, ds, \tag{2.130}$$

and for each $s \in \mathbb{R}$,

$$\int_{|t|\geq A} I_2\!\left(\frac{s}{d_n},\frac{t}{d_n}\right) |\widehat{g}(-t/c_n)|\, dt \leq C\left[c_n\,(l-m)^{-2} + d_n/d_{l-m} \right]. \tag{2.131}$$

Hence, by noting that

$$s\,(Z_{nk} - Z_{nj}) - t\,(Z_{nl} - Z_{nm})$$
$$= s\,(y'_{k,n} - y'_{j,n}) - t\,(y'_{l,n} - y'_{m,n}) + F(\epsilon_0,\epsilon_{-1},...),$$

where $F(\epsilon_0,\epsilon_{-1},...)$ depends only on $\epsilon_j, j \leq 0$, it follows from (2.130) and (2.131) that

$$\mathrm{E}\,|R_{2n}|^2 \leq \frac{C}{n^4} \sum_{k,j,l,m=1}^{n} \int_{|s|\geq A} \int_{|t|\geq A} |\widehat{g}(-s/c_n)|\, |\widehat{g}(-t/c_n)|$$

$$\left| E \exp\left\{ \left[is\,(y'_{k,n} - y'_{j,n}) - it\,(y'_{l,n} - y'_{m,n}) \right] \right\} \right|\, ds\, dt$$

$$\leq \frac{C}{n^4} \sum_{k>j\geq l>m} \int_{|s|\geq A} I_1\!\left(\frac{s}{d_n}\right) |\widehat{g}(-s/c_n)|\, ds$$

$$\int_{|t|\geq A} I_2\!\left(\frac{s}{d_n},\frac{t}{d_n}\right) |\widehat{g}(-t/c_n)|\, dt + o(1)$$

$$\leq C\left[\frac{c_n}{n} + \frac{1}{n^2} \sum_{k>j} \int_{|s|\geq A} e^{-C\, d_{k-j}^2\, s^2/d_n^2}\, ds \right]$$

$$\times \left[\frac{c_n}{n} + \frac{1}{n^2} \sum_{l>m} d_n/d_{l-m} \right]$$

$$\to 0,$$

when $n \to \infty$ first and then $A \to \infty$. This proves (2.129), and hence the result (2.125) under the additional condition (2.126).

Removal of (2.126) is similar to that of Theorem 2.22 and hence the details are omitted. $\qquad\square$

Remark 2.14. The integrability condition on the characteristic function of ϵ_0 can be weakened if we place further restrictions on $g(x)$. Indeed Theorem 2.25 still holds if $\lim_{|t|\to\infty} |t|^\eta |\mathrm{E}\,e^{it\epsilon_0}| < \infty$ for some $\eta > 0$ is replaced by the Cramér condition, i.e., $\limsup_{|t|\to\infty} |\mathrm{E}\,e^{it\epsilon_0}| < 1$, and in addition to the stated conditions already on $g(x)$ we have $|g(x)| \leq M/(1 + |x|^{1+b})$ for some $b > 0$, where M is a constant. See Theorem 2.22.

Asymptotic behavior of S_n when $c_n = 1$ is quite different, as seen in the following theorem. We remark that, in this situation, further smoothness condition on ϵ_0 is not necessary.

Theorem 2.26. *Suppose that $g(x)$ is Borel measurable function satisfying*

$$\lim_{h\to 0} \int_{-K}^{K} |x|^{\alpha-1} \sup_{|u|\leq h} |g(x+u) - g(x)|dx = 0, \qquad (2.132)$$

for all $K > 0$ and some $0 < \alpha \leq 1$. Then,

$$\frac{1}{n^2} \sum_{k,j=1}^{n} g\big(Z_{nk} - Z_{nj}\big) \to_D \int_0^1 \int_0^1 g[Z_u - Z_v]dudv.$$

Since (2.132) holds if and only if $g(x)$ is locally Riemann integrable, Theorem 2.26 can be proved by the similar arguments as in part (ii) of Theorem 2.10. As noticed in Berkes and Horváth (2006), without further smoothness condition on ϵ_0, condition (2.132) cannot be replaced by

$$\lim_{h\to 0} \int_{-K}^{K} |x|^{\alpha-1}|g(x+u) - g(x)|dx = 0,$$

for all $K > 0$ and some $0 < \alpha \leq 1$.

2.5 Uniform approximation to local time

Let $g(x)$ be a real function on R and $\{X_{nk}\}_{k\geq 1, n\geq 1}$ be a triangular array. Write, for $x \in$ R,

$$S_n(x) = \frac{c_n}{n} \sum_{k=1}^{n} g[c_n (X_{nk} + x)],$$

where $c_n \to \infty$ and $c_n/n \to 0$. Unlike Sections 2.3.2-2.3.6 in which pointwise asymptotics of $S_n(x)$ were established, this section is concerned with uniform approximation of $S_n(x)$ to local time and uniform upper and lower bounds of $S_n(x)$. We further consider uniform approximation of $S_{1n}(t)$ defined by

$$S_{1n}(t) = \frac{c_n}{n} \sum_{k=1}^{n} g[c_n (X_{nk} - X_{n,[nt]})], \quad t \in [0, 1].$$

The following assumptions are used in this section.

Assumption 2.1. X_{nk} is a strong smooth array (see Definition 2.5) with $\sup_{n,1 \leq k \leq n} EX_{nk}^2 < \infty$ and $d_{nkj} = [(k - j)/n]^d$, where $0 < d < 1$.

Assumption 2.2. On a rich probability space, there exist a process $G = \{G(t)\}_{t \geq 0}$ that has a local time $L_G(1, x)$ satisfying

$$|L_G(1, x) - L_G(1, y)| \leq C|x - y|^\eta, \quad a.s., \tag{2.133}$$

for some $\eta > 0$ and a sequence of stochastic processes $G_n(t)$ such that $\{G_n(t); 0 \leq t \leq 1\} =_D \{G(t); 0 \leq t \leq 1\}$ for each $n \geq 1$ and

$$\sup_{0 \leq t \leq 1} |X_{n,[nt]} - G_n(t)| = o_{a.s.}(n^{-\delta}), \tag{2.134}$$

for some $0 < \delta < 1/2$.

Assumption 2.3. (i) There is an $\epsilon_0 > 0$ such that $n^{-\epsilon_0} c_n \to \infty$ and $n^{-1+\epsilon_0} c_n \to 0$; (ii) $\int_{-\infty}^{\infty} (1 + |x|) |g(x)| dx < \infty$ and $|g(x) - g(y)| \leq C|x - y|$ whenever $|x - y|$ is sufficiently small on R.

Assumptions 2.1 and 2.3 are natural and standard for uniform convergence problem. Assumption 2.2 is a strong approximation version of the result $X_{n,[nt]} \Rightarrow G(t)$ on $D[0, 1]$, and it is widely obtainable for many random sequences. A typical example in statistics and econmetrics is provided in Theorem 2.31 (in Section 2.5.4) where we establish Assumption 2.2 for general linear processes. Note that, $G_n(x)$ cannot be replaced by one single process $G(x)$ which is independent of n. Explanation in this regards can be found in Csörgö and Révész (1981). As for the Lipschitz type condition (2.133), it is satisfied by the classical Gaussian processes, Lévy processes and many semimartingales. See Section 2.1 for examples.

2.5.1   *An invariance principle*

Let $C[-M, M]$, $M > 0$, denote the continuous functional space on $[-M, M]$ with uniform topology. $S_n(x)$ is a random element of $C[-M, M]$ only if $g(x)$

is a continuous function. Remark 2.7 establishes finite dimensional convergence of $S_n(x)$. The following theorem provides an invariance principle for $S_n(x)$ on $C[-M, M]$.

Theorem 2.27. *In addition to Assumptions 2.1 and 2.3, there exists a stochastic process $X = \{X_t\}_{t\geq 0}$ having a continuous local time $L_X(t, s)$ such that $X_{n,[nt]} \Rightarrow X_t$ on $D[0, \infty)$. Then, for any $\alpha > 0$ and any $M > 0$,*

$$S_n(n^{-\alpha}x) \Rightarrow \int_{-\infty}^{\infty} g(t)dt \, L_X(1, 0), \qquad (2.135)$$

on $C[-M, M]$.

We will prove Theorem 2.27 in Section 2.5.5. Extension to $\alpha = 0$ requires more restrictive Assumption 2.2, which will be discussed in Section 2.5.2.

Let $K(x)$ be a real function having compact support and $|K(x) - K(y)| \leq C|x - y|$ whenever $|x - y|$ is sufficiently small. Let $m(x)$ and $\pi(x)$ be real functions satisfying:

(i) there exist $m_1(x)$ and $\gamma \in (0, 1]$ such that, for any y sufficiently small,

$$|m(x + y) - m(x)| \leq C \, |y|^{\gamma} \, m_1(x); \qquad (2.136)$$

(ii) $\int_{-\infty}^{\infty} [1 + m^2(x) + m_1^2(x)] \, \pi(x) \, dx < \infty$.

The following corollary is an application of Theorem 2.27.

Corollary 2.5. *Suppose the conditions of Theorem 2.27 hold for $X_{nk} = y_{nk}/d_n$, where $d_n \asymp n^d$ for some $0 < d < 1$. Then, for any $M > 0$, $\beta < d$ and any bounded h satisfying $n^{1-\epsilon_0}h/d_n \to \infty$, where ϵ_0 can be chosen as small as required,*

$$\frac{d_n}{nh} \sum_{k=1}^{n} K[(y_{nk} - n^{\beta}x)/h] \Rightarrow \int_{-\infty}^{\infty} K(t)dt \, L_X(1, 0), \qquad (2.137)$$

on $C[-M, M]$. If $h \to 0$, we further have

$$(\frac{d_n}{nh})^2 \int_{-\infty}^{\infty} \left\{ \sum_{k=1}^{n} K[(y_{nk} - x)/h] \, m(y_{nk}) \right\}^2 \pi(x)dx$$

$$\to_D \tau_1 \, L_X^2(1, 0), \qquad (2.138)$$

where $\tau_1 = \left(\int_{-\infty}^{\infty} K(x)dx \right)^2 \int_{-\infty}^{\infty} m^2(x)\pi(x)dx$.

Proof. Note that, by letting $c_n = d_n/h$,

$$\frac{d_n}{nh} \sum_{k=1}^{n} K[(y_{nk} - n^\beta x)/h] = \frac{c_n}{n} \sum_{k=1}^{n} K[c_n(X_{nk} - xn^\beta/d_n)].$$

Since $n^\beta/d_n \leq Cn^{-(d-\beta)}$, result (2.137) follows immediately from (2.135) with $\alpha = d - \beta$. To prove (2.138), we may write

$$\int_{-\infty}^{\infty} \left\{ \sum_{k=1}^{n} K[(y_{nk} - x)/h] \, m(y_{nk}) \right\}^2 \pi(x) dx$$
$$= V_{1n} + V_{2n} + 2V_{3n}, \tag{2.139}$$

where $V_{3n} \leq V_{1n}^{1/2} V_{2n}^{1/2}$,

$$V_{1n} = \int_{-\infty}^{\infty} \left\{ \sum_{k=1}^{n} K[(y_{nk} - x)/h] \right\}^2 m^2(x)\pi(x) dx,$$

$$V_{2n} = \int_{-\infty}^{\infty} \left\{ \sum_{k=1}^{n} K[(y_{nk} - x)/h] \, [m(y_{nk}) - m(x)] \right\}^2 \pi(x) dx.$$

It follows from (2.137) and the continuous mapping theorem that

$$\left(\frac{d_n}{nh}\right)^2 \int_{-M}^{M} \left\{ \sum_{k=1}^{n} K[(y_{nk} - x)/h] \right\}^2 m^2(x)\pi(x) dx$$
$$\to_D \int_{-M}^{M} m^2(x)\pi(x) dx \left(\int_{-\infty}^{\infty} K(x) dx \right)^2 L_X^2(1,0),$$

for any $M > 0$. Since, by Lemma 2.5 below with $g_n(y) = K[(d_n y - x)/h]$,

$$\sup_x \mathrm{E}\left\{ \sum_{k=1}^{n} K[(y_{nk} - x)/h] \right\}^2 \leq C \, (nh/d_n)^2,$$

routine calculations yield

$$\left(\frac{d_n}{nh}\right)^2 V_{1n} \to_D \tau_1 L_G^2(1,0).$$

On the other hand, by recalling (2.136) and since $K(x)$ has compact support,

$$V_{2n} \leq Ch^\gamma \int_{-\infty}^{\infty} \left\{ \sum_{k=1}^{n} K[(y_{nk} - x)/h] \right\}^2 m_1^2(x)\pi(x) dx$$
$$= o_P[(d_n/nh)^2]$$

and $V_{3n} \leq V_{1n}^{1/2} V_{2n}^{1/2} = o_P[(d_n/nh)^2]$, due to $h \to 0$. Taking these estimates into (2.139), (2.138) is proved. $\qquad \square$

2.5.2 Uniform approximation to local time

If $X_{n,[nt]} \Rightarrow X_t$ required in Theorem 2.27 is replaced by more restrictive Assumption 2.2, a more general uniform approximation result can be achieved. As a consequence, result (2.135) can be improved.

Theorem 2.28. *Suppose Assumptions 2.1–2.3 hold. On the same probability space as in Assumption 2.2, for any $\beta > 0$, we have*

$$\sup_{x \in \mathbb{R}} |S_n(x) - \tau L_{G_n}(1, -x)| = o_\mathrm{P}(\log^{-\beta} n) \qquad (2.140)$$

where $\tau = \int_{-\infty}^{\infty} g(t)dt \neq 0$. Consequently, for any $M > 0$,

$$S_n(c'_n x) \Rightarrow \tau \widetilde{L}_G(1, -x), \qquad (2.141)$$

on $C[-M, M]$, where

$$\widetilde{L}_G(r, x) = \begin{cases} L_G(r, 0), & \text{if } c'_n \to 0, \\ L_G(r, -x), & \text{if } c'_n = 1. \end{cases}$$

The proof of Theorem 2.28 will be given in Section 2.5.5. An application of Theorem 2.28 is to establish uniform upper and lower bounds for $S_n(x)$.

Corollary 2.6. *Under Assumptions 2.1–2.3, we have*

$$\sup_{x \in \mathbb{R}} |S_n(x)| = O_\mathrm{P}(1). \qquad (2.142)$$

Furthermore, if $\int_{-\infty}^{\infty} g(x)dx \neq 0$ and $\lim_{n \to \infty} \mathrm{P}(\inf_{x \in \Omega_n} L_G(1, -x) = 0) = 0$ where Ω_n is a subset of $\mathbb{R}$, then

$$\left[\inf_{x \in \Omega_n} |S_n(x)| \right]^{-1} = O_\mathrm{P}(1). \qquad (2.143)$$

Remark 2.15. Let $G(x)$ be a standard Wiener process. Then $\mathrm{P}(L_G(1,0) = 0) = 0$, implying $\lim_{n \to \infty} \mathrm{P}(\inf_{|x| \le r_n} L_G(1, -x) = 0) = 0$ for any $0 < r_n \to 0$. Corollary 2.6 yields

$$\left[\inf_{|x| \le r_n} |S_n(x)| \right]^{-1} = O_\mathrm{P}(1), \qquad (2.144)$$

for any $0 < r_n \to 0$. By noting $\mathrm{P}(L_G(1,x) = 0) > 0$ for any fixed $x \neq 0$ (see, for instance, Takacs (1995)), the range $|x| \le r_n$ in (2.144) is optimal in the sense that it cannot be improved to $|x| \le b$ for any constant $b > 0$.

More applications of Theorem 2.28 are given as follows.

Corollary 2.7. *Suppose Assumptions 2.1–2.3 hold. Then, for any continuous function $m(x)$ on $\mathbb{R}$ and any real function $\pi(x)$ satisfying $\int_{-\infty}^{\infty} |\pi(x)|dx < \infty$, we have*

$$\int_{-\infty}^{\infty} m[S_n(x)]\pi(x)dx \to_D \int_{-\infty}^{\infty} m[\tau L_G(1, -x)]\pi(x)dx. \qquad (2.145)$$

Proof. It follows from (2.141) and the continuous mapping theorem that

$$\int_{-M}^{M} m[S_n(x)]\pi(x)dx \to_D \int_{-M}^{M} m[\tau\, L_G(1,-x)]\pi(x)dx, \qquad (2.146)$$

for any $M > 0$. Note that $\sup_{x\in\mathbb{R}}|m(S_n(x))| = O_P(1)$ and $\sup_{x\in\mathbb{R}}|m[\tau L_G(1,-x)]| = O_P(1)$ by Corollary 2.6 and the continuity of $m(x)$. Result (2.145) follows from (2.146) and some routine calculations. $\qquad\square$

There is a similar result to Theorem 2.28 for $S_{1n}(t)$, as shown in the following theorem.

Theorem 2.29. *Suppose Assumptions 2.1–2.3 hold. On the same probability space as in Assumption 2.2, for any $\beta > 0$, we have*

$$\sup_{0\le t\le 1} |S_{1n}(t) - \tau\, L_{nt}| = o_P(\log^{-\beta} n), \qquad (2.147)$$

where $\tau = \int_{-\infty}^{\infty} g(t)dt \ne 0$ and $L_{nt} = \lim_{\epsilon\to 0}\frac{1}{2\epsilon}\int_0^1 \mathrm{I}(|G_n(s)-G_n(t)| \le \epsilon)ds$. Consequently, if $\mathrm{P}(\inf_{0\le t\le 1} L_t = 0) = 0$, where

$$L_t = \lim_{\epsilon\to 0}\frac{1}{2\epsilon}\int_0^1 \mathrm{I}(|G(s)-G(t)| \le \epsilon)ds,$$

then $\left[\min_{j=1,2,\dots,n}\left|\sum_{k=1}^{n} g[c_n\,(X_{nk}-X_{nj})]\right|\right]^{-1} = O_P(c_n/n)$.

The proof of Theorem 2.29 is similar to that of Theorem 2.28, where the details can be found in Linton and Wang (2014).

2.5.3   *Uniform approximation to local time: random bandwidth*

As in Corollary 2.5, supposing that the conditions of Theorem 2.28 hold for $X_{nk} = y_{nk}/d_n$, where $d_n \asymp n^d$ for some $0 < d < 1$, we have

$$\sup_{x\in\mathbb{R}}\left|\frac{d_n}{nh}\sum_{k=1}^{n} g\Big[\frac{y_{nk}+c_n' n^d x}{h}\Big] - \tau\,\widetilde{L}_{G_n}(1,-x)\right| = o_P(\log^{-\beta} n), \qquad (2.148)$$

for any $\beta > 0$ and any bounded constant sequence h satisfying $n^{1-\epsilon_0}h/d_n \to \infty$, where

$$\widetilde{L}_{G_n}(r,x) = \begin{cases} L_{G_n}(r,0), & \text{if } c_n' \to 0, \\ L_{G_n}(r,-x), & \text{if } c_n' = 1. \end{cases}$$

In many applications, it is useful to allow for the bandwidth sequence h in (2.148) to be data-dependent (random). Toward this extension, the following uniform bound for zero energy functional of X_{nk} is crucial. Write, for some constant $A_0 > 0$,

$$\Omega = \Big\{ f : \int_{-\infty}^{\infty} f(y)dy = 0, \quad \int_{-\infty}^{\infty} |f(y)|dy \leq A_0 \Big\}.$$

Theorem 2.30. *Let Ψ be a subset of Ω such that $\#(\Psi) \leq n^m$ for some $m > 0$. Suppose Assumptions 2.1 holds. Suppose, uniformly for $f \in \Psi$,*

$$\inf_{t \in R} \int_{-\infty}^{\infty} |f(y-t)| \, |y| \, dy \leq C \, n^{d-\delta_0}, \quad \sup_y |f(y)| \leq C \, n^{1-d-\delta_0}, \quad (2.149)$$

for some $0 < \delta_0 < \min\{d, 1-d\}$, where $0 < d < 1$ is defined in Assumption 2.1. Then there exists a $\gamma > 0$ such that

$$\max_{f \in \Psi} \Big| \sum_{k=1}^{n} f(n^d X_{nk}) \Big| = O(n^{1-d-\gamma\delta_0}), \quad a.s. \quad (2.150)$$

We will prove Theorem 2.30 in Section 2.5.5. An application of Theorem 2.30 yields the following corollary.

Corollary 2.8. *Suppose Assumption 2.1 holds for $X_{nk} = y_{nk}/d_n$, where $d_n \asymp n^d$ for some $0 < d < 1$. Suppose Assumption 2.3 (ii) holds. Then, for any $0 < \delta_0 < \min\{d, 1-d\}$, there exists a $\gamma > 0$ such that*

$$\sup_{(h,x) \in \mathcal{L}_n} \Big| \sum_{k=1}^{n} \big[g_h(y_{nk} + x) - g(y_{nk} + x) \big] \Big| = O_{a.s.}(n^{1-d-\gamma\delta_0}), \quad (2.151)$$

where $g_h(y) = h^{-1} g(y/h)$ and, for $0 < t_1 < t_2 < \infty$,

$$\mathcal{L}_n = \{(t,s) : t_1 n^{d-1+\delta_0} \leq t \leq t_2, \ |s| \in R\}.$$

Proof. Set $t_1 n^{d-1+\delta_0} = h_1 < \ldots < h_{q_{n1}} = t_2$ and $-Mn^5 = x_1 < \ldots < x_{q_{n2}} = Mn^5$, with $h_i - h_{i-1} \sim n^{-7}$ and $x_i - x_{i-1} \sim n^{-10}$. By the Borel-Cantelli lemma, $n^d X_{nk} = o_{a.s.}(n^5)$ due to $\sup_{n, 1 \leq k \leq n} EX_{nk}^2 < \infty$. Recall Assumption 2.3 (ii). To prove (2.151), it suffices to show:

$$\max_{1 \leq i \leq q_{n1}} \max_{1 \leq j \leq q_{n2}} \Big| \sum_{k=1}^{n} f_{h_i, x_j}(n^d X_{nk}) \Big| = O_{a.s.}(n^{1-d-\gamma\delta_0}), \quad (2.152)$$

where $f_{h_i, x_j}(y) = g_{h_i}(d_n y/n^d + x_j) - g(d_n y/n^d + x_j)$.

The proof of (2.152) is simple by using Theorem 2.30. Indeed, by letting

$$\Psi = \{ f_{h_i, x_j} : 1 \leq i \leq q_{n1}, \ 1 \leq j \leq q_{n2} \},$$

(2.152) follows from Theorem 2.30, due to $f_{h_i, x_j}(y) \in \Omega$, $\sup_y |f_{h_i, x_j}(y)| \leq Ch_i^{-1} \leq C\, n^{1-d-\delta_0}$ and

$$\inf_{t \in R} \int_{-\infty}^{\infty} |f_{h_i, x_j}(y-t)| \, |y| \, dy \leq C \int_{-\infty}^{\infty} |g_{h_i}(y) - g(y)| |y| \, dy \leq C,$$

uniformly on Ψ. $\qquad\square$

Let h_n be a random sequence satisfying, as $n \to \infty$,

$$P\big(h_n \in [t_1 n^{d-1+\delta_0}, \ t_2]\big) \to 1,$$

where $0 < \delta_0 < \min\{d, 1-d\}$ can be chosen as small as required. Corollary 2.8, together with Theorem 2.27 (result (2.148), respectively), implies the following results, which provide weak (uniform, respectively) convergence to local time with random bandwidths.

Corollary 2.9. *Suppose Assumption 2.1 holds for $X_{nk} = y_{nk}/d_n$, where $d_n \asymp n^d$ for some $0 < d < 1$, and there exists a stochastic process $X = \{X_t\}_{t \geq 0}$ having a continuous local time $L_X(t, s)$ such that $X_{n,[nt]} \Rightarrow X_t$ on $D[0, \infty)$. Suppose Assumption 2.3 (ii) holds. Then, for any $\beta \leq d - \delta_0$,*

$$\frac{d_n}{n h_n} \sum_{k=1}^{n} g[(y_{nk} - n^\beta x)/h_n] \Rightarrow \int_{-\infty}^{\infty} g(t)dt \, L_X(1, 0), \qquad (2.153)$$

on $C[-M, M]$.

Corollary 2.10. *Suppose Assumptions 2.1–2.3 hold for $X_{nk} = y_{nk}/d_n$, where $d_n \asymp n^d$ for some $0 < d < 1$. Then, on the same probability space as in Assumption 2.2,*

$$\sup_{x \in \mathbb{R}} \left| \frac{d_n}{n h_n} \sum_{k=1}^{n} g\Big(\frac{y_{nk} + c'_n d_n x}{h_n}\Big) - \int_{-\infty}^{\infty} g(t)dt \, \widetilde{L}_{G_n}(1, -x) \right|$$

$$= o_P(\log^{-\beta} n), \qquad (2.154)$$

for any $\beta > 0$.

We end this section with a further corollary of Theorem 2.30 that is used in the proof of Theorem 2.28. Its proof is similar to that of Corollary 2.8 and hence the details are omitted.

Corollary 2.11. *Suppose Assumptions 2.1 and 2.3 hold. Then, for any $\beta > 0$, we have*

$$\sup_{t \in \mathbb{R}} \sup_{s:|s-t| \leq c_n n^{-\delta_0}} \left| \frac{c_n}{n} \sum_{k=1}^{n} \big[g(c_n X_{nk} + t) - g(c_n X_{nk} + s)\big] \right|$$

$$= O_{a.s.}(\log^{-\beta} n) \qquad (2.155)$$

where $\delta_0 > 0$ can be chosen as small as required.

2.5.4 Example: linear processes

Let $\{\xi_j\}_{j\geq 1}$ be a linear process defined by

$$\xi_j = \sum_{k=0}^{\infty} \phi_k \, \epsilon_{j-k},$$

where $\{\epsilon_j\}_{j\in\mathbb{Z}}$ is a sequence of i.i.d. random variables with $E\epsilon_0 = 0$, $E\epsilon_0^2 = 1$, $E|\epsilon_0|^r < \infty$ for some $r > 2$ and $\lim_{|t|\to\infty} |t|^\eta |E\, e^{it\epsilon_0}| < \infty$ for some $\eta > 0$. Furthermore, the coefficients ϕ_k, $k \geq 0$ are assumed to satisfy one of the following conditions:

C1. $\phi_k \sim k^{-\mu}$, where $1/2 < \mu < 1$.
C2. $\sum_{k=0}^{\infty} k|\phi_k| < \infty$ and $\phi \equiv \sum_{k=0}^{\infty} \phi_k \neq 0$.

Theorem 2.31. *On a rich probability space, there exists a fractional Brownian motion $B_H = \{B_H(t)\}_{t\geq 0}$ such that, under condition* **C1**,

$$\sup_{0\leq t\leq 1} \left| c_\mu^{-1/2} \sum_{k=1}^{[nt]} \xi_k - B_{3/2-\mu}(nt) \right| = o_{a.s.}[n^{(r+1)/r-\mu}], \qquad (2.156)$$

where $c_\mu = \frac{1}{(1-\mu)(3-2\mu)} \int_0^\infty x^{-\mu}(x+1)^{-\mu}dx$; under condition **C2**,

$$\sup_{0\leq t\leq 1} \left| \phi^{-1} \sum_{k=1}^{[nt]} \xi_k - B_{1/2}(nt) \right| = o_{a.s.}(n^{1/r}). \qquad (2.157)$$

The proof of (2.156) is given in Wang et al. (2003). For a proof of (2.157) we refer to Csörgő and Horvath (1993, page 18) and Linton and Wang (2014).

Let $x_k = \sum_{j=1}^{k} \xi_j$ and $d_n^2 = \begin{cases} c_\mu \, n^{3-2\mu}, & \text{under } \mathbf{C1}, \\ \phi^2 \, n, & \text{under } \mathbf{C2}. \end{cases}$ As in Section 2.3.3, x_k/d_n satisfies Assumption 2.1. By using Theorem 2.31 and noting that fractional process has a continuous local time satisfying (2.133), x_k/d_n satisfies Assumption 2.2. As a consequence of Theorems 2.28 and 2.29, we have the following result.

Corollary 2.12. *Let $g(x)$ satisfy Assumption 2.3 (ii). Then, for any $h = O(1)$ and $n^{1-\epsilon_0}h/d_n \to \infty$ for some $\epsilon_0 > 0$,*

$$\frac{d_n}{nh} \sum_{k=1}^{n} g\big[(x_k + x\,d_n)/h\big] \Rightarrow L_Y(1, -x),$$

$$\frac{d_n}{nh} \sum_{k=1}^{n} g\big[(x_k + x\,c_n)/h\big] \Rightarrow L_Y(1, 0)$$

on $C[-M, M]$ for any $M > 0$, where $c_n = o(d_n)$ and

$$Y(t) = \begin{cases} B_{3/2-\mu}(t), & under \ \mathbf{C1}, \\ B_{1/2}(t), & under \ \mathbf{C2}. \end{cases}$$

We further have

$$\sup_{x \in \mathbb{R}} |\sum_{k=1}^{n} g[h^{-1}(x_k + x)]| = O_P(nh/d_n);$$

and if $\int_{-\infty}^{\infty} g(x)dx = 0$, then

$$\sup_{x \in \mathbb{R}} |\sum_{k=1}^{n} g[h^{-1}(x_k + x)]| = O_P\big[(nh/d_n)\log^{-\beta} n\big];$$

for any $\beta > 0$; and if $\int_{-\infty}^{\infty} g(x)dx \neq 0$, then

$$\Big[\inf_{|x| \leq r_n d_n} |\sum_{k=1}^{n} g[h^{-1}(x_k + x)]|\Big]^{-1} = O_P(d_n/(nh)),$$

for any $0 < r_n \to 0$, and

$$\Big[\min_{j=1,2,\ldots,n} |\sum_{k=1}^{n} g[(x_k - x_j)/h]|\Big]^{-1} = O_P(d_n/nh).$$

2.5.5 Proofs of main results

We start with several lemmas. Recall $\mathcal{F}_{nk} = \sigma(X_{n1}, \ldots, X_{nk}), k \geq 1, \mathcal{F}_{n0} = \sigma(\phi, \Omega)$ and K, C, ϵ_0 and n_0 are constants given in Assumption 2.3 and Definition 2.5.

Lemma 2.3. *Let $Y_{n1}, Y_{n2}, \ldots, n \geq 1$, be an arbitrary random array. If $\sup_{k \geq 1} \mathrm{E}\, |Y_{nk}|^p \leq H_0^p p!$ for $p = \log n$, where $H_0 > 0$ is a constant, then*

$$\max_{1 \leq k \leq n^m} |Y_{nk}| = O(\log n), \quad a.s, \tag{2.158}$$

for each fixed $m \geq 1$.

Proof. Since, by the Markov's inequality,

$$\mathrm{P}\big(\max_{1 \leq k \leq n^m} |Y_{nk}| \geq e^{m+2} H_0 \log n\big)$$

$$\leq e^{-(m+2)p} \log^{-p} n H_0^{-p} \sum_{k=1}^{n^m} \mathrm{E}\, |Y_{nk}|^p$$

$$\leq e^{-(m+2)\log n} n^m (\log n)^{-\log n} (\log n)! \leq n^{-2},$$

the claim follows from the Borel-Cantelli lemma. $\qquad\square$

Lemma 2.4. *For each $k \geq 1$, let $\{Z_{nj}(k), \mathcal{F}_{nj}\}_{j \geq 1, n \geq 1}$ be a martingale difference such that*

$$\sup_{k \geq 1} \sum_{j=1}^{n} \mathrm{E} \, |Z_{nj}(k)|^{\log n} \leq A_n^{\log n}, \tag{2.159}$$

$$\sup_{k \geq 1} \sum_{j=1}^{n} \mathrm{E} \left[Z_{nj}^2(k) \mid \mathcal{F}_{n,j-1} \right] \leq A_n^2, \quad a.s., \tag{2.160}$$

where A_n is a sequence of positive numbers. Then, for each fixed $m \geq 1$, we have

$$\max_{1 \leq k \leq n^m} \left| \sum_{j=1}^{n} Z_{nj}(k) \right| = O(A_n \log n), \quad a.s. \tag{2.161}$$

Proof. Let $Y_{nk} = A_n^{-1} \sum_{j=1}^{n} Z_{nj}(k)$. By the Rosenthal's inequality for martingale [see, e.g., Theorem 2.12 of Hall ad Heyde (1980)], we have

$$\sup_{k \geq 1} \mathrm{E} \, |Y_{nk}|^p \leq 2^p \, p!$$

for $p = \log n$. Result follows immediately from Lemma 2.3. $\qquad\square$

Lemma 2.5. *Let $g_n(y), n \geq 1$, be real functions on R. Suppose Assumption 2.1 holds. Then there exists a constant $H_0 > 0$ depending only on K and C such that, for all $0 \leq t_1 \leq t_2 - n_0$ and $t_3 - t_2 \geq n_0 \, m$,*

$$\mathrm{E} \left(\left| \sum_{k=t_2}^{t_3} g_n(X_{nk}) \right|^m \mid \mathcal{F}_{n,t_1} \right)$$

$$\leq H_0^m \, m! \, n^d \, (t_3 - t_1)^{1-d} \, \delta_n^{m-1} \int_{-\infty}^{\infty} |g_n(y)| dy, \tag{2.162}$$

where $\delta_n = \sup_y |g_n(y)| + (t_3 - t_2)^{1-d} n^d \int_{-\infty}^{\infty} |g_n(y)| dy$. In particular, for any $\omega \geq 0$ and $b_n \geq n_0 m$, we have

$$\mathrm{E} \left| \sum_{k=\omega b_n + 1}^{(\omega+1) b_n} g_n(X_{nk}) \right|^m \leq H_0^m \, m! \, \delta_{1n}^m, \tag{2.163}$$

where $\delta_{1n} = \sup_y |g_n(y)| + b_n (n/b_n)^d \int_{-\infty}^{\infty} |g_n(y)| dy$.

Proof. Note that, by using Theorem 2.17,

$$I_k := \sum_{j=k+n_0}^{t_3} \mathrm{E} \left[|g_n(X_{nj})| \mid \mathcal{F}_{nk} \right] \leq K n^d \int_{-\infty}^{\infty} |g_n(y)| dy \sum_{j=k+n_0}^{t_3} (j-k)^{-d}$$

$$\leq H_0 n^d (t_3 - k)^{1-d} \int_{-\infty}^{\infty} |g_n(y)| dy,$$

for any $k \geq 0$. Result (2.162) follows from

$$
\mathrm{E}\left(\left|\sum_{k=t_2}^{t_3} g_n(X_{nk})\right|^m \mid \mathcal{F}_{n,t_1}\right)
$$

$$
\leq m! \sum_{t_2 \leq k_1 \leq k_2 \leq \ldots \leq k_{m-1} \leq t_3} \left[n_0 \sup_y |g_n(y)| + I_{k_{m-1}}\right]
$$

$$
\mathrm{E}\left\{\left|g_n(X_{nk_1}) \cdots g_n(X_{nk_{m-1}})\right| \mid \mathcal{F}_{n,t_1}\right\}
$$

$$
\leq \cdots \cdots \leq
$$

$$
\leq m! \, (H_0 \delta_n)^{m-1} \sum_{t_2 \leq k_1 \leq t_3} \mathrm{E}\left\{\left|g_n(X_{nk_1})\right| \mid \mathcal{F}_{n,t_1}\right\}
$$

$$
\leq H_0^m \, m! \, n^d \, (t_3 - t_1)^{1-d} \, \delta_n^{m-1} \int_{-\infty}^{\infty} |g_n(y)| dy.
$$

□

We are now ready to prove main results in previous sections.

Proof of Theorem 2.30. We start with some preliminaries. Let

$$
b_n = [n^{1-v}], \quad \text{where } v = (2d)^{-1} \delta_0 \min\{1/4, (1-d)/d\},
$$

and take γ sufficiently small such that

$$
0 < \gamma < \frac{1}{4} \min\{1/4, (1-d)/d\} \quad \text{and} \quad (1-d)v \geq 2\gamma\delta_0.
$$

Recalling (2.149), simple calculations show that $2dv < (1-\gamma)\delta_0$,

$$
n^{dv}[\sup_y |f(y)| + n^{1-d-\eta}] \leq n^{1-d-2\gamma\delta_0} \log^{-2} n, \qquad (2.164)
$$

for any $\eta \geq \delta_0 \min\{1/4, (1-d)/d\}$, and

$$
\sup_y |f(y)| + b_n^{1-d} \leq 2n^{1-d-2\gamma\delta_0}, \qquad (2.165)
$$

for n sufficiently large. We further mention that, when $g_n(x) = f(n^d x)$,

$$
n^d \int_{-\infty}^{\infty} |g_n(x)| dx = \int_{-\infty}^{\infty} |f(x)| dx \leq A_0 < \infty.
$$

We now turn back to the proof of Theorem 2.30. Let T_n be the largest integer s such that $sb_n \leq n$. Set, for $w \geq 0$,

$$
\Delta_{nw} = \sum_{k=wb_n+1}^{(w+1)b_n} f(n^d X_{nk}), \quad \Delta_{nw}^* = \Delta_{nw} - \mathrm{E}\left[\Delta_{nw} \mid \mathcal{F}_{n,(w-1)b_n}\right].
$$

We may write

$$\sum_{k=1}^{n} f(n^d X_{nk})$$

$$= \sum_{w=0}^{T_n-1} \mathrm{E}\left[\Delta_{nw} \mid \mathcal{F}_{n,(w-1)b_n}\right] + \sum_{w=0}^{T_n-1} \Delta_{nw}^* + \sum_{w=T_n b_n+1}^{n} f(n^d X_{nk})$$

$$= R_{1n} + R_{2n} + R_{3n}, \quad \text{say.} \tag{2.166}$$

Note that $n - T n b_n \leq b_n$. It follows from (2.163) with $g_n(x) = f(n^d x)$ that, for any $p \geq 2$,

$$\max_{f \in \Psi} \mathrm{E}\,|R_{3n}|^p \leq H_0^p\, p!\, \max_{f \in \Psi}\left[\sup_y |f(y)| + b_n^{1-d}\right]^p$$

$$\leq H_0^p\, p!\, \left(n^{1-d-2\gamma\delta_0}\right)^p,$$

due to (2.165). This yields $\max_{f \in \Psi} |R_{3n}| = O_{a.s.}(n^{1-d-\gamma\delta_0})$ by Lemma 2.3.

We next estimate R_{1n} and R_{2n}. Similar to the proof of Theorem 2.17, it follows from (2.57) and $\int_{-\infty}^{\infty} f(y)dy = 0$ that

$$I_{kj} := \left|\mathrm{E}\left[f(n^d X_{nk}) \mid \mathcal{F}_{nj}\right]\right|$$

$$\leq C\,(k-j)^{-d}\,\min\left\{(k-j)^{-d}\,\inf_{t \in \mathrm{R}}\int_{-\infty}^{\infty} |f(y-t)|\,|y|\,dy,\ 1\right\},$$

for $k - j \geq n_0$. This, together with (2.149) and (2.165), yields that

$$\max_{f \in \Psi} |R_{1n}| \leq \sum_{w=0}^{T_n-1} \sum_{k=wb_n+1}^{(w+1)b_n} \max_{f \in \Psi} |I_{k,(w-1)b_n}|$$

$$\leq C\,n\,b_n^{-2d}\,\inf_{t \in \mathrm{R}}\int_{-\infty}^{\infty} |f(y-t)|\,|y|\,dy$$

$$\leq C n^{1-d+(2dv-\delta_0)} \leq C n^{1-d-\gamma\delta_0}.$$

Similarly, by letting $\zeta = \delta_0/d$, we have

$$\sum_{k=j+n_0}^{(w+1)b_n} \max_{f \in \Psi} |I_{kj}| \leq C \sum_{k=1}^{b_n} k^{-d}\,\min\{n^{d-\delta_0}k^{-d}, 1\}$$

$$\leq \sum_{k=1}^{n^{1-\zeta}} k^{-d} + n^{d-\delta_0} \sum_{k=n^{1-\zeta}+1}^{b_n} k^{-2d} \leq C\,n^{1-d-\eta_0},$$

for $j \geq wb_n$, where

$$\eta_0 = \begin{cases} \delta_0 + \nu(1-2d), & if \quad 0 < d < 1/2, \\ \delta_0/4, & if \quad d = 1/2, \\ \left(\frac{1-d}{d}\right)\delta_0, & if \quad 1/2 < d < 1. \end{cases}$$

Hence, uniformly for $f \in \Psi$, we have

$$E\left[\Delta_{nw}^2 \mid \mathcal{F}_{n,(w-1)b_n}\right]$$

$$\leq n_0 \sum_{j=wb_n+1}^{(w+1)b_n} E\left(|f(n^d X_{nj})| \mid \mathcal{F}_{n,(w-1)b_n}\right) \sup_y |f(y)|$$

$$+2 \sum_{j=wb_n+1}^{(w+1)b_n} E\left(|f(n^d X_{nj})| \mid \mathcal{F}_{n,(w-1)b_n}\right) \sum_{k=j+n_0}^{(w+1)b_n} \max_{f\in\Psi} |I_{kj}|$$

$$\leq H_0\, b_n^{1-d}\left(\sup_y |f(y)| + n^{1-d-\eta_0}\right),$$

by (2.162) with $g_n(y) = f(n^d y)$ and $m = 1$. As a consequence, uniformly for $f \in \Psi$,

$$\sum_{w=0}^{T_n-1} E\left[\Delta_{nw}^{*2} | \mathcal{F}_{n,(w-1)b_n}\right] \leq 2H_0\, nb_n^{-d}\left(\sup_y |f(y)| + n^{1-d-\eta_0}\right)$$

$$\leq H_0 n^{2(1-d)-2\gamma\,\delta_0} \log^{-2} n,$$

due to (2.164), for n sufficiently large. Furthermore, by (2.163) with $g_n(x) = f(n^d x)$ and (2.165), we have

$$\sum_{w=0}^{T_n-1} E\left|\Delta_{nw}^*\right|^{\log n} \leq H_0^{\log n} (\log n)!\, nb_n^{-1} \left[\sup_y |f(y)| + b_n^{1-d}\right]^{\log n}$$

$$\leq (2H_0)^{\log n} \left[n^{(1-d-2\gamma\,\delta_0)}\, n^{v/\log n} \log n\right]^{\log n}$$

$$\leq (2H_0)^{\log n} \left[n^{1-d-\gamma\,\delta_0} \log^{-1} n\right]^{\log n},$$

uniformly for $f \in \Psi$ and n sufficiently large. Now, by noting that $\{\Delta_{nw}^*, \mathcal{F}_{n,wb_n}\}_{w\geq 0}$ forms a martingale difference, an application of Lemma 2.4 yields

$$\max_{f\in\Psi} |R_{2n}| = O(n^{1-d-\gamma\delta_0}), \quad a.s.$$

Taking these estimates into (2.166), we prove (2.150). □

Proof of Theorem 2.27. Finite dimensional convergence comes from Remark 2.7 and the tightness of process $S_n(n^{-\alpha}x)$ follows from Corollary 2.11. □

Proof of Theorem 2.28. Let $\psi(x) = (1 - |x|)I(|x| \leq 1)$ and

$$\psi_\epsilon(x) = \epsilon^{-1}\psi(\epsilon^{-1}x), \quad \epsilon = n^{-\alpha}$$

with $0 < \alpha < \min\{\delta/2, \epsilon_0/4\}$, where δ and ϵ_0 are defined in Assumptions 2.2 and 2.3. Write,

$$T_n(x) = \frac{\tau}{n} \sum_{k=1}^{n} \psi_\epsilon(X_{nk} + x),$$

$$U_n(x) = \frac{1}{n} \sum_{k=1}^{n} \int_{-\infty}^{\infty} g(y)\psi_\epsilon(X_{nk} + x - y/c_n)dy.$$

Since $\sup_{x\in\mathrm{R}} \left| S_n(x) - \tau L_{G_n}(1, -x) \right| \le \Phi_{1n} + \Phi_{2n} + \Phi_{3n}$, where

$$\Phi_{1n} = \sup_{x\in\mathrm{R}} \left| S_n(x) - U_n(x) \right|,$$

$$\Phi_{2n} = \sup_{x\in\mathrm{R}} \left| U_n(x) - T_n(x) \right|,$$

$$\Phi_{3n} = \sup_{x\in\mathrm{R}} \left| T_n(x) - \tau L_{G_n}(1, -x) \right|,$$

it suffices to show, for any $\beta > 0$,

$$\Phi_{in} = o_\mathrm{P}(\log^{-\beta} n), \quad i = 1, 2, 3. \tag{2.167}$$

Note that $\int_{-\infty}^{\infty} \psi_\epsilon(x)dx = 1$ and

$$U_n(x) = \frac{c_n}{n} \sum_{k=1}^{n} \int_{-\infty}^{\infty} g[c_n(X_{nj} + x - y)]\psi_\epsilon(y)dy.$$

It follows from Corollary 2.11 that, for any $\beta > 0$,

$$\Phi_{1n} \le \sup_{x} \int_{-\infty}^{\infty} |S_n(x - y) - S_n(x)|\psi_\epsilon(y)dy$$

$$\le \sup_{t} \sup_{s:|t-s|\le n^{-\alpha}} |S_n(t) - S_n(s)| = o_{a.s}(\log^{-\beta} n).$$

Similarly, by recalling $n^{-\epsilon_0} c_n \to \infty$ and noting that $\psi(x)$ satisfies the Lipschitz condition, we have

$$\Phi_{2n} \le \sup_{x} \frac{1}{n} \sum_{k=1}^{n} \int_{-\infty}^{\infty} |g(y)| \, |\psi_\epsilon(X_{nk} + x - y/c_n) - \psi_\epsilon(X_{nk} + x)|dy$$

$$\le C\epsilon^{-2} c_n^{-1} \le Cn^{2\alpha-\epsilon_0} = o_{a.s}(\log^{-\beta} n),$$

for any $\beta > 0$.

We next prove $\Phi_{3n} = o_\mathrm{P}(\log^{-\beta} n)$. Recalling the definition of $\psi_\epsilon(y)$ and $\int_{-\infty}^{\infty} \psi_\epsilon(y)dy = 1$, it follows from (2.133) that

$$\left| \int_0^1 \psi_\epsilon(G(t) + x)dt - L_G(1, -x) \right|$$

$$= \left| \int_{-\infty}^{\infty} \psi_\epsilon(y + x)L_G(1, y)dy - L_G(1, -x) \right|$$

$$\leq \int_{-\infty}^{\infty} \psi_\epsilon(y)|L_G(1, y - x) - L_G(1, -x)|dy = O_{a.s.}(\epsilon^\eta),$$

uniformly for all $x \in \mathbb{R}$. This, together with $\{G_n(t); 0 \leq t \leq 1\} =_D \{G(t); 0 \leq t \leq 1\}$ for all $n \geq 1$ due to Assumption 2.2, implies that

$$\sup_x \left| \int_0^1 \psi_\epsilon[G_n(t) + x]dt - L_{G_n}(1, -x) \right| = O_P(\epsilon^\eta).$$

Hence it follows from (2.134) in Assumption 2.2 that

$$\Phi_{3n} \leq \tau \left| \frac{1}{n} \sum_{j=1}^{n} \psi_\epsilon(X_{nj} + x) - L_{G_n}(1, -x) \right|$$

$$\leq \left| \int_0^1 \psi_\epsilon(X_{n,[nt]} + x)dt - \int_0^1 \psi_\epsilon[G_n(t) + x]dt \right| + 2/(\varepsilon n)$$

$$+ \left| \int_0^1 \psi_\epsilon[G_n(t) + x]dt - L_{G_n}(1, -x) \right|$$

$$= O_P\left[\varepsilon^{-2}n^{-\delta} + 2/(\varepsilon n) + \varepsilon^\eta\right] = o_P(\log^{-\beta} n)$$

uniformly for all $x \in \mathbb{R}$, due to $\varepsilon = n^{-\alpha}$ and $\alpha < \delta/2$, as required.

Combining all estimates above, we establish (2.167) and also complete the proof of Theorem 2.28. □

2.6　Bibliographical Notes

Section 2.1. Local time theory has a long history. In combining with the purpose of this book, we consider existence of local time using Fourier analysis. The main results are taken from Berman (1969) and Geman and Horwicz (1976, 1980). Different notion of local time is used in Section 2.1.3 for semimartingale. The materials in this part are mainly from Protter (2005). We also refer to Revuz and Yor (2003). We do not collect the most sharp results on the existence of local time, except those that are closely related to asymptotics in the remaining parts of this chapter. The current literature on random fields can be found in Dozzi (2003), Xiao (2006) and the references therein.

Section 2.2. If $g(x)$ is a continuous real-valued function, (2.33) can be easily established by the continuous mapping theorem at least as long as the limit process $X = \{X_t\}_{t \geq 0}$ has continuous paths. Under certain conditions on X_{nk}, several articles have contributed to establish (2.33) under the minimum condition that $g(x)$ is locally integrable (in the Lebesgue sense). We refer to Park and Phillips (1999, 2001), de Jong (2004), Pötscher (2004), Berkes and Horváth (2006), and Christopeit (2009). This section further extends these previous works, allowing for various different settings that appear in many applications.

Section 2.3. On the convergence to local time, this section provides two frameworks and various applications. The major idea to establish first framework is from Wang and Phillips (2009a), where conditional density arguments are used. The related work can be found in Jeganathan (2004), who investigated the asymptotics that are similar to (2.67) when X_{nk} is the partial sum of a linear process. For the particular situation where $c_n X_{nk}$ is a partial sum of independent and identically distributed random variables, some other related results can be found in Borodin and Ibragimov (1995), Akonom (1993), and Phillips and Park (1999). We establish second framework using characteristic functions, developed in Wang (2015). This framework has advantages when it is applied to certain stationary variables such as m-dependent shift, Markov process and linear process. It is expected that this framework will provide a platform for further studies on the convergence to local time.

Section 2.4. The main results are taken from Wang and Phillips (2012), providing certain extensions to near $I(1)$ processes. For related works on partial sums of i.i.d. random variables, we refer to Aldous (1986), van der Hofstad, et al. (1997), van der Hofstad and Wolfgang (2001), and van der Hofstad, et al. (2003).

Section 2.5. The main results are taken from Liu, Chan and Wang (2015), and Linton and Wang (2015). For other related results, we refer to Wang and Wang (2013), Chan and Wang (2014) and Duffy (2014 a, b).

Chapter 3

Convergence to a mixture of normal distributions

Let $(S_n, Z_n)_{n \geq 1}$ be a sequence of random vectors. It is an important issue to consider the joint convergence:

$$(S_n, Z_n) \to_D (S, Z). \tag{3.1}$$

In many situations, we may prove $S_n \to_D VN$, where $V > 0$ a.s. is a random variable independent of $N \sim N(0,1)$, that is, S_n converges to a mixture of normal distributions. This is a typical limit theorem in the framework of semimartingales. Usually, the distribution of V is unknown, making the limit theorem useless for statistical purpose. However, if we can prove (3.1) with $Z = V$, the continuous mapping theorem show that

$$S_n / Z_n \to_D N(0, 1), \tag{3.2}$$

having a big advantage in statistical applications. This chapter investigates the convergence to a mixture of normal distributions that enables (3.2).

3.1 Convergence on product space

Let $S = S_1 \times S_2$ be the product of two separable metric spaces S_1 and S_2. Let (X, Y) and (X_n, Y_n) be random elements of S, defined on the same probability space $(\Omega, \mathcal{F}, \mathrm{P})$.

We say (X_n, Y_n) converges to (X, Y), denoted by $(X_n, Y_n) \Rightarrow (X, Y)$, if

$$\mathrm{P}(X_n \in A, Y_n \in B) \to \mathrm{P}(X \in A, Y \in B),$$

for all X-continuous sets A and all Y-continuous sets B. If $(X_n, Y_n) \Rightarrow (X, Y)$, then two marginal distributions converge, i.e.,

$$X_n \Rightarrow X \quad \text{and} \quad Y_n \Rightarrow Y.$$

In converse, if $X_n \to_P X$ and $Y_n \to_P X$, then $(X_n, Y_n) \to_P (X, Y)$, implying that $(X_n, Y_n) \Rightarrow (X, Y)$. However $(X_n, Y_n) \Rightarrow (X, Y)$ may fail if $X_n \Rightarrow X$ and $Y_n \to_P X$.

Example 3.1. Let $X_n = -X$ and $Y_n = X \sim N(0, 1)$. Then $X_n \to_D X$ due to $-X =_D X$, but $X_n + Y_n = 0 \not\to_D 2X = X + Y$, i.e., $(X_n, Y_n) \not\Rightarrow (X, Y)$.

Example 3.1 indicates that it is not enough to establish the joint convergence of (X_n, Y_n) if only $X_n \Rightarrow X$ and $Y_n \to_P Y$ hold. The condition $X_n \Rightarrow X$ has to be strengthened, requiring the concept of stable (mixing) convergence.

Let $X, X_n, n \geq 1$, be random elements of a separable metric space S_1.

Definition 3.1. Suppose $X_n \Rightarrow X$. The convergence is said to be *stable*, written as

$$X_n \to_D X \quad \text{(stably)},$$

if, for all $B \in \mathcal{F}$ and all X-continuous sets A,

$$\lim_{n \to \infty} P(X_n \in A, \ B), \quad \text{exists.} \tag{3.3}$$

The convergence is said to be *mixing*, written as

$$X_n \to_D X \quad \text{(mixing)},$$

if, for all $B \in \mathcal{F}$ and all X-continuous sets A,

$$\lim_{n \to \infty} P(X_n \in A, \ B) = P(X \in A) P(B), \tag{3.4}$$

i.e., X can be taken to be independent of $\mathcal{F}$.

Theorem 3.1. *The following statements are equivalent:*

(i) $X_n \to_D X$ *(stably);*
(ii) $(X_n, Y) \Rightarrow (X, Y)$, *for any* $Y \in \mathcal{F}$;
(iii) $g(X_n, Y) \to_D g(X, Y)$ *(stably), for any* $Y \in \mathcal{F}$ *and continuous function* $g(x, y)$ *on* $S_1 \times R$;
(iv) For all fixed k *and* $B \in \sigma(X_1, ..., X_k), P(B) > 0$,

$$\lim_{n \to \infty} P(X_n \in A | B) \quad \text{exists},$$

for all X-*continuous sets* A.

Theorem 3.2. *The following statements are equivalent:*

(i) $X_n \to_D X$ *(mixing);*

(ii) $(X_n, Y) \Rightarrow (X^*, Y)$, for any $Y \in \mathcal{F}$, where $X^* =_D X$ is independent of $\mathcal{F}$;

(iii) $g(X_n, Y) \to_D g(X^*, Y)$ (mixing), for any $Y \in \mathcal{F}$ and continuous function $g(x, y)$ on $S_1 \times \mathrm{R}$, where $X^* =_D X$ is independent of $\mathcal{F}$;

(iv) For all fixed k and $B \in \sigma(X_1, ..., X_k), \mathrm{P}(B) > 0$,

$$\lim_{n \to \infty} \mathrm{P}(X_n \in A | B) = \mathrm{P}(X \in A),$$

for all X-continuous sets A.

Theorems 3.1 and 3.2 provide the properties of stable (mixing) convergence. For the proofs of Theorems 3.1 and 3.2, we refer to Aldous and Eagleson (1978). See also Jacod and Shiryaev (2003).

By the definition, we have

$$X_n \to_\mathrm{P} X \quad \text{implies} \quad X_n \to_D X \text{ (stably)} \quad \text{implies} \quad X_n \Rightarrow X,$$

but the converse fails, i.e., stable convergence is weaker than the convergence in probability, but stronger than the convergence in distribution.

Example 3.2. Let $X_{2k} = X$ and $X_{2k+1} = X'$, where X and X' are independent and have identical distributions. Then $\mathrm{P}(X_{2k} \leq a, X \leq b) \to \mathrm{P}(X \leq a \wedge b)$ and $\mathrm{P}(X_{2k+1} \leq a, X \leq b) \to \mathrm{P}(X \leq a)\mathrm{P}(X \leq b)$, so that $X_n \to_D X$ but $X_n \not\to_D X$ (stably).

If $X_n \to_D X$ (stably) and $X \in \mathcal{F}$, Theorem 3.1 yields

$$(X_n, X) \Rightarrow (X, X),$$

which implies $X_n \to_\mathrm{P} X$. On the other hand, if ϵ_j are i.i.d. random variables with $\mathrm{E}\epsilon_1 = 0$ and $\mathrm{E}\epsilon_1^2 = 1$, the classical central limit theorem states:

$$X_n \equiv \frac{1}{\sqrt{n}} \sum_{j=1}^{n} \epsilon_j \to_D X \quad \text{(mixing)}, \quad \text{but} \quad X_n \not\to_\mathrm{P} X, \tag{3.5}$$

where $X \sim N(0, 1)$. See, e.g., Rényi (1963). As a consequence, stable convergence is a property of the sequence of rv's X_n, but the limit X must be defined on an extension of the original $(\Omega, \mathcal{F}, \mathrm{P})$.

Theorem 3.3. *Suppose that* $Y_n \to_\mathrm{P} Y$.

(i) *If* $X_n \to_D X$ *(stably), then* $(X_n, Y_n) \Rightarrow (X, Y)$. *Furthermore,* $g(X_n, Y_n) \to_D g(X, Y)$ *(stably) for any continuous function* $g(x, y)$ *on* $S_1 \times \mathrm{R}$.

(ii) If $X_n \to_D X$ (mixing), then $(X_n, Y_n) \Rightarrow (X, Y)$, where X and Y are independent. Furthermore, $g(X_n, Y_n) \to_D g(X, Y)$ (mixing) for any continuous function $g(x, y)$ on $S_1 \times$ R.

Proof. Note that $Y \in \mathcal{F}$ and

$$(X_n, Y_n) = (X_n, Y) + (0, Y_n - Y) = (X_n, Y) + o_P(1),$$

due to $Y_n \to_P Y$. The claims (i) and (ii) follow from Theorems 3.1 and 3.2 respectively. □

Remark 3.1. Using the concept of stable (mixing) convergence, Theorem 3.3 provides sufficient conditions on the joint convergence of (X_n, Y_n). As in (3.5), many classical limit theorems are mixing (stable), but the convergence in probability does not hold. Theorem 3.4 below, which is from Eagleson (1976), provides a typical mixing (stable) convergence result.

Theorem 3.4. Let $\epsilon_k, k \geq 1$, be a sequence of random variables and $T_n = T_n(\epsilon_1, \epsilon_2, ...)$ be a statistic. Suppose that there exists a random variable T such that $T_n \to_D T$ and for each fixed k,

$$T_n(\epsilon_1, \epsilon_2, ...) - T_n(\epsilon_{k+1}, \epsilon_{k+2}, ...) \to_P 0,$$

as $n \to \infty$. If the invariant σ-field of ϵ_n is trivial, then

$$T_n \to_D T \quad (mixing).$$

Consequently, if η_n is a strictly stationary, ergodic sequence and $\frac{1}{B_n}(\sum_{j=1}^{n} \eta_j - A_n) \to_D S$, where $0 < B_n \to \infty$, then this limit theorem must be mixing.

Remark 3.2. The condition that $Y_n \to_P Y$ is not necessary to establish the joint convergence of (X_n, Y_n). Two illustration examples are given in the following theorems. More results on martingale limit theorems will be discussed in the following sections.

Theorem 3.5. Let $h, h_n, n \geq 1$, be measurable mappings from S_1 to R. Let E be the set of $x \in S_1$ such that $h_n(x_n) \not\to h(x)$ for some $x_n \to x$. If $(X_n, Y_n) \Rightarrow (X, Y)$ and $P(X \in E) = 0$, then

$$\big[h_n(X_n), Y_n\big] \Rightarrow \big[h(X), Y\big]. \tag{3.6}$$

The proof of (3.6) is straightforward. Indeed, by using the Skorokhod representation theorem, there exists an extended probability space such that $(X_n, Y_n) \to (X, Y), a.s.$ As $P(X \in E) = 0$, $X_n \to X, a.s.$ implies $h_n(X_n) \to h(X), a.s.$, and hence the claim follows.

Let D_h be the set of discontinuous points of h. Then $E = D_h$ if $h_n \equiv h$. Furthermore, $P(X \in E) = 0$ if $P(X \in D_h) = 0$ and for each compact set $A \subset S_1$,

$$\sup_{x \in A} |h_n(x) - h(x)| \to 0.$$

If $g(t)$ is a continuous real function on R, typical mappings h on $D_{\mathrm{R}^d}[0, \infty)$ satisfying $P(X \in D_h) = 0$ include

$$h(X) = \int_0^1 g[X(s)]ds, \quad \sup_{s \in [0,1]^d} g[X(s)] \quad \text{and} \quad g[X(1)].$$

Theorem 3.6. *Let $X_k = (X_{1k}, X_{2k}), k \geq 1$, be a sequence of i.i.d. random vectors. Let $X_\alpha = \{X_\alpha(t)\}_{t \geq 0}$ be a stable Lévy process with index $0 < \alpha < 2$ and $B = \{B_t\}_{t \geq 0}$ be a Brownian motion. Suppose there exist constant sequences $a_{1n} > 0, a_{2n} > 0$ and $b_{1n}, b_{2n} \in$ R such that, on $D_{\mathrm{R}}[0, \infty)$,*

$$\frac{1}{a_{1n}} \sum_{k=1}^{[nt]} (X_{1k} - b_{1k}) \Rightarrow X_\alpha(t), \quad \frac{1}{a_{2n}} \sum_{k=1}^{[nt]} (X_{2k} - b_{2k}) \Rightarrow B_t.$$

Then, on $D_{\mathrm{R}^2}[0, \infty)$,

$$S_n(t) := \Big(\frac{1}{a_{1n}} \sum_{k=1}^{[nt]} (X_{1k} - b_{1k}), \frac{1}{a_{2n}} \sum_{k=1}^{[nt]} (X_{2k} - b_{2k}) \Big)$$
$$\Rightarrow \big(X_\alpha(t), B_t \big), \tag{3.7}$$

and X_α is independent of B.

In order to prove (3.7), note that every convergent subsequence of $S_n = \{S_n(t)\}_{t \geq 0}$ has the limit of the form $S = W + J$ where W is a Wiener process, J is a Lévy process without Wiener component, and W and J are independent. On the other hand, we get $S = (X_\alpha, B)$ from the assumed marginal convergence, which indicates $W = (0, B)$ and $J = (X_\alpha, 0)$. This shows X_α and B are independent and the claim comes from the fact that every convergent subsequence has the same limit.

3.2   Convergence to a process with stationary independent increments

Let $\{X_{ni}\}_{n \geq 1, i \geq 1}$ be an array of random variables. For each $n \geq 1$, let $\{\mathcal{F}_{ni}\}_{i \geq 1}$ be a sequence of σ-fields such that

$$\mathcal{F}_{ni} \subseteq \mathcal{F}_{n,i+1}, \quad i \geq 1,$$

and $\{X_{ni}\}_{i\geq1}$ is adapted to $\{\mathcal{F}_{ni}\}_{i\geq1}$. Let $\tau_n(t), t \geq 0$, be a stopping time with respect to $\{\mathcal{F}_{ni}\}_{i\geq1}$ for each $n \geq 1$ and $t > 0$, and the sample paths of $\tau_n(.)$ are integer-valued, non-decreasing and right-continuous with $\tau_n(0) = 0$. This section is concerned with the convergence of $\sum_{i=1}^{\tau_n(t)} X_{ni}$ to a process with stationary independent increments.

3.2.1   Central limit theorem for dependent random variables

Let $\tau_n = \tau_n(t_0)$ for some $t_0 > 0$ and $S_n = \sum_{i=1}^{\tau_n} X_{ni}$.

Theorem 3.7. *Suppose that*

$$\max_{1\leq i\leq\tau_n} |X_{ni}| \to_P 0, \tag{3.8}$$

$$\sum_{i=1}^{\tau_n} \left| \mathrm{E}\left[X_{ni}\mathrm{I}(|X_{ni}| \leq 1) \mid \mathcal{F}_{n,i-1} \right] \right| \to_P 0, \tag{3.9}$$

$$\sum_{i=1}^{\tau_n} X_{ni}^2 \to_P 1. \tag{3.10}$$

Then $S_n \to_D N(0,1)$. If in addition $\mathcal{F}_{ni}$ is nested, i.e., $\mathcal{F}_{ni} \subseteq \mathcal{F}_{n+1,i}$ for $n \geq 1, i \geq 1$, then $S_n \to_D N(0,1)$ (mixing).

Proof.   First prove $S_n \to_D N(0,1)$. Write $S_{1n} = \sum_{j=1}^{\tau_n} Y_{nj}$, where

$$Y_{nj} = X_{nj}\mathrm{I}(|X_{nj}| \leq 1) - \mathrm{E}\left[X_{ni}\mathrm{I}(|X_{ni}| \leq 1) \mid \mathcal{F}_{n,i-1} \right].$$

Note that $\mathrm{P}(S_n \neq \sum_{i=1}^{\tau_n} X_{nj}\mathrm{I}(|X_{nj}| \leq 1)) \leq \mathrm{P}(\max_{1\leq i\leq\tau_n} |X_{ni}| \geq 1)$. By (3.8) and (3.9), it suffices to show that

$$S_{1n} \to_D N(0,1). \tag{3.11}$$

To this end, let $S_{2n} = \sum_{j=1}^{\tau_n} Z_{nj}$, where $Z_{nj} = Y_{nj}\mathrm{I}(\sum_{k=1}^{j-1} Y_{nk}^2 \leq 2)$. Define $r(x)$ by $e^{ix} = (1+ix)\exp\{-x^2/2+r(x)\}$, and note that $|r(x)| \leq |x|^3$ for $|x| \leq 1$. Let $I_n = e^{itS_{2n}}$, $T_n \equiv T_n(t) = \Pi_{j=1}^{\tau_n}(1 + itZ_{nj})$ and

$$W_n = \exp\left\{ -\frac{t^2}{2} \sum_{j=1}^{\tau_n} Z_{nj}^2 + \sum_{j=1}^{\tau_n} r(tZ_{nj}) \right\}$$

Then

$$I_n = T_n e^{-t^2/2} + T_n(W_n - e^{-t^2/2}). \tag{3.12}$$

Since Z_{nj} is a martingale difference, $ET_n = 1$. Furthermore, $\sum_{i=1}^{\tau_n} Y_{ni}^2 \to_P 1$ and $\sum_{i=1}^{\tau_n} Z_{ni}^2 \to_P 1$ by (3.9) and the fact that (3.8) is equivalent to

$\sum_{i=1}^{\tau_n} X_{ni}^2 I(|X_{ni}| \geq \epsilon) \to_P 0$ for any $\epsilon > 0$ [see (3.18) below]. Result (3.11) will follow if we prove

$$E\,|T_n(W_n - e^{-t^2/2})| \to 0. \tag{3.13}$$

Indeed (3.12) and (3.13) imply $EI_n \to e^{-t^2/2}$ for all $t \in \mathbb{R}$. We hence have $S_{2n} \to_D N(0, 1)$, which implies (3.11) due to

$$P(S_{1n} \neq S_{2n}) \leq P\Big(\sum_{k=1}^{\tau_n} Y_{nk}^2 > 2\Big) \to 0.$$

In order to prove (3.13), note that

$$|T_n|^2 = \Pi_{k=1}^{\tau_n}(1 + t^2 Z_{nk}^2) \leq \exp\Big\{t^2 \sum_{j=1}^{J_n-1} Y_{nj}^2\Big\}(1 + t^2 Y_{nJ_n})$$

$$\leq (1 + 2t^2)e^{2t^2}, \tag{3.14}$$

where J_n is defined by

$$J_n = \begin{cases} \min\{j \leq \tau_n : \sum_{i=1}^{j} Y_{ni}^2 > 2\}, & \text{if } \sum_{i=1}^{\tau_n} Y_{ni}^2 > 2, \\ \tau_n, & \text{otherwise.} \end{cases}$$

Result (3.14) and $|I_n| \leq 1$ imply

$$T_n(W_n - e^{-t^2/2}) = I_n - T_n e^{-t^2/2}$$

is uniformly integrable. On the other hand, by recalling

$$\max_{1 \leq i \leq \tau_n} |Z_{ni}| \to_P 0 \quad \text{and} \quad \sum_{i=1}^{\tau_n} Z_{ni}^2 \to_P 1,$$

we have

$$\Big|\sum_{i=1}^{\tau_n} r(t Z_{nj})\Big| \leq |t|^3 \sum_{i=1}^{\tau_n} |Z_{nj}|^3 \leq |t|^3 \max_{1 \leq i \leq \tau_n} |Z_{ni}| \sum_{i=1}^{\tau_n} Z_{nj}^2 \to_P 0,$$

which implies $|T_n(W_n - e^{-t^2/2})| \leq |T_n|\,|W_n - e^{-t^2/2})| \to_P 0$. In view of the convergence in probability and the uniform integrability, (3.13) holds and the proof for $S_n \to_D N(0, 1)$ is completed.

We next prove $S_n \to_D N(0, 1)$ (mixing). By (3.12) and (3.13), it suffices to show $E\big[T_n I(B)\big] \to P(B)$ for any $B \in \mathcal{F}$; or equivalently,

$$E\,|T_n(t) - 1| = o(1), \quad \text{as } n \to \infty, \tag{3.15}$$

for all real t. Note that $E \max_{1 \leq i \leq \tau_n} |Z_{ni}|^2 \leq \sum_{j=1}^{J_n} E\,Y_{nj}^2 \leq 4$. The proof of (3.15) is the same as the verification of (3.14) in Hall and Heyde (1980), given in their proof of Theorem 3.2. We omit the details. $\square$

Theorem 3.7 provides a basic central limit theorem for dependent random variables. There are other alternative conditions to establish similar results, depending on the following lemmas.

Lemma 3.1. *Suppose $\{Z_{ni}\}_{n\geq 1, i\geq 1}$ is a positive random array adapted to $\{\mathcal{F}_{ni}\}_{i\geq 1}$ for each $n \geq 1$. Then,*

$$\sum_{i=1}^{\tau_n} \mathrm{E}\left(Z_{ni} \mid \mathcal{F}_{n,i-1}\right) \to_{\mathrm{P}} 0 \quad \text{implies} \quad \sum_{i=1}^{\tau_n} Z_{ni} \to_{\mathrm{P}} 0. \tag{3.16}$$

If in addition $\{\max_{1\leq i\leq \tau_n} |Z_{ni}|\}_{n\geq 1}$ is uniformly integrable, then

$$\sum_{i=1}^{\tau_n} Z_{ni} \to_{\mathrm{P}} 0 \quad \text{implies} \quad \sum_{i=1}^{\tau_n} \mathrm{E}\left(Z_{ni} \mid \mathcal{F}_{n,i-1}\right) \to_{\mathrm{P}} 0. \tag{3.17}$$

Proof. To prove (3.16), let $\nu_n = \inf\{k \geq 1 : \sum_{i=1}^{k} \mathrm{E}\left(Z_{ni} \mid \mathcal{F}_{n,i-1}\right) \geq 1\}$, $k_n = (\nu_n - 1) \wedge \tau_n$ and $A_n = \sum_{i=1}^{k_n} \mathrm{E}\left(Z_{ni} \mid \mathcal{F}_{n,i-1}\right)$. It is readily seen that

$$\mathrm{P}(k_n \neq \tau_n) \leq \mathrm{P}(\nu_n \geq \tau_n + 1) \to 0$$

and, due to $0 \leq A_n \leq 1$ and $A_n \leq \sum_{i=1}^{\tau_n} \mathrm{E}\left(Z_{ni} \mid \mathcal{F}_{n,i-1}\right) \to_{\mathrm{P}} 0$,

$$\mathrm{E}\sum_{i=1}^{k_n} Z_{ni} = E\, A_n \to_{\mathrm{P}} 0.$$

Hence, for any $\epsilon > 0$,

$$\mathrm{P}\big(\sum_{i=1}^{\tau_n} Z_{ni} \geq \epsilon\big) \leq \mathrm{P}\left(k_n \neq \tau_n\right) + \mathrm{P}\big(\sum_{i=1}^{k_n} Z_{ni} \geq \epsilon\big) \to 0,$$

which yields (3.16).

Similarly, to prove (3.17), define in this case $\nu_n = \inf\{k \geq 1 : \sum_{i=1}^{k} Z_{ni} > 1\}$ and $k_n = \nu_n \wedge \tau_n$. Note that

$$0 \leq \sum_{i=1}^{k_n} Z_{ni} \leq 1 + \max_{1\leq i\leq \tau_n} |Z_{ni}|,$$

implying $\mathrm{E}\sum_{i=1}^{k_n} Z_{ni} \to 0$ under given conditions. It follows that, for any $\epsilon > 0$,

$$\mathrm{P}\big(\sum_{i=1}^{\tau_n} \mathrm{E}\left(Z_{ni} \mid \mathcal{F}_{n,i-1}\right) \geq \epsilon\big) \leq \mathrm{P}(k_n \neq \tau_n) + \epsilon^{-1}\mathrm{E}\sum_{i=1}^{k_n} Z_{ni} \to 0.$$

Hence (3.17) holds.   □

Lemma 3.2. *(i)* $\max_{1 \leq i \leq \tau_n} |X_{ni}| \to_P 0$ *if and only if*

$$\sum_{i=1}^{\tau_n} |X_{ni}|^m I(|X_{ni}| \geq \epsilon) \to_P 0, \quad \text{for all } \epsilon > 0, \tag{3.18}$$

where $m \geq 1$ is an integer, and if and only if

$$\sum_{i=1}^{\tau_n} P(|X_{ni}| \geq \epsilon \mid \mathcal{F}_{n,i-1}) \to_P 0, \quad \text{for all } \epsilon > 0, \tag{3.19}$$

(ii) If $E \max_{1 \leq i \leq \tau_n} |X_{ni}|^m \to 0$, where $m \geq 1$ is an integer, then

$$\sum_{i=1}^{\tau_n} E\left[|X_{ni}|^m I(|X_{ni}| \geq \epsilon) \mid \mathcal{F}_{n,i-1}\right] \to_P 0, \quad \text{for all } \epsilon > 0, \tag{3.20}$$

(iii) If (3.20) holds with $m = 2$, then $\max_{1 \leq i \leq \tau_n} |X_{ni}| \to_P 0$ and (3.10) is equivalent to

$$\sum_{i=1}^{\tau_n} E[X_{ni}^2 \mid \mathcal{F}_{n,i-1}] \to_P 1. \tag{3.21}$$

Proof. (i) Note that $\max_{1 \leq i \leq \tau_n} |X_{ni}| \geq \epsilon$ if and only if

$$\sum_{i=1}^{\tau_n} |X_{ni}|^m I(|X_{ni}| \geq \epsilon) \geq \epsilon^m,$$

and if and only if $\sum_{i=1}^{\tau_n} I(|X_{ni}| \geq \epsilon) \neq 0$. The claims follow from Lemma 3.1 and some routine calculations.

(ii) $E \max_{1 \leq i \leq \tau_n} |X_{ni}|^m \to 0$ implies that $\max_{1 \leq i \leq \tau_n} |X_{ni}| \to_P 0$ and $\{\max_{1 \leq i \leq \tau_n} |X_{ni}|^m\}_{n \geq 1}$ is uniformly integrable. Hence (3.18) holds by part (i) and, as a consequence, result (3.20) follows from (3.17) of Lemma 3.1.

(iii) By (3.16) with $Z_{ni} = X_{ni}^2 I(|X_{ni}| \geq \epsilon)$, (3.18) holds with $m = 2$, which yields $\max_{1 \leq i \leq \tau_n} |X_{ni}| \to_P 0$ by part (i).

Let $Y_{ni} = X_{ni}^2 I(|X_{ni}| < \epsilon)$. Since (3.18) and (3.20) both hold with $m = 2$, to prove the equivalence between (3.10) and (3.21), it suffices to show that, if either (3.10) or (3.21) holds, then

$$\Delta_n := \sum_{i=1}^{\tau_n} \left[Y_{ni} - E(Y_{ni} \mid \mathcal{F}_{n,i-1})\right] \to_P 0, \tag{3.22}$$

as $n \to \infty$ first and then $\epsilon \to 0$.

We only prove (3.22) under (3.10). The other is similar. In fact, by defining $\nu_n = \inf\{k \geq 1 : \sum_{i=1}^{k} X_{ni}^2 > 2\}$ and $k_n = (\nu_n - 1) \wedge \tau_n$, it is readily seen that, for any $\delta > 0$,

$$\mathrm{P}(|\Delta_n| \geq \delta) \leq \mathrm{P}\big(\big| \sum_{i=1}^{k_n} \big[Y_{ni} - \mathrm{E}\left(Y_{ni} \mid \mathcal{F}_{n,i-1}\right)\big]\big| \geq \delta\big) + \mathrm{P}(k_n \neq \tau_n)$$

$$\leq \delta^{-2} E \sum_{i=1}^{k_n} \big[Y_{ni} - \mathrm{E}\left(Y_{ni} \mid \mathcal{F}_{n,i-1}\right)\big]^2 + \mathrm{P}\big(\sum_{i=1}^{\tau_n} X_{ni}^2 > 2\big)$$

$$\leq 2\epsilon^2 \delta^{-2} E \sum_{i=1}^{k_n} X_{ni}^2 + \mathrm{P}\big(\sum_{i=1}^{\tau_n} X_{ni}^2 > 2\big)$$

$$\leq 4\epsilon^2 \delta^{-2} + o(1),$$

which implies (3.22).   □

Corollary 3.1. *Suppose $\{X_{ni}, \mathcal{F}_{n,i}\}_{n\geq 1, i\geq 1}$ is a martingale array. Suppose one of the following three sets of conditions holds:*

(a) (3.20) with $m = 2$ and either (3.10) or (3.21);
(b) (3.10) and $E \max_{1\leq i \leq \tau_n} |X_{ni}| \to 0$;
(c) $\max_{1\leq i\leq \tau_n} |X_{ni}| \to_{\mathrm{P}} 0$, $\sum_{i=1}^{\tau_n} E[X_{ni}^2 \mathrm{I}(|X_{ni}| \geq 1) \mid \mathcal{F}_{n,i-1}] \to_{\mathrm{P}} 0$ and either (3.10) or (3.21).

Then conclusions of Theorem 3.7 remain true.

Using Lemma 3.2 and Theorem 3.7, the proof of Corollary 3.1 is simple, which is left to readers.

3.2.2 Functional central limit theorems for martingales

There is a functional version of Theorem 3.7, which is stated as follows.

Theorem 3.8. *Suppose that f is a measurable function such that $\int_0^t f^2(s)ds < \infty$ for all $t > 0$. Suppose that, for all $t > 0$,*

$$\sum_{i=1}^{\tau_n(t)} X_{ni}^2 \to_{\mathrm{P}} \int_0^t f^2(s)ds, \tag{3.23}$$

$$\sum_{i=1}^{\tau_n(t)} \big| \mathrm{E}\left[X_{ni} \mathrm{I}(|X_{ni}| \leq 1) \mid \mathcal{F}_{n,i-1}\right] \big| \to_{\mathrm{P}} 0. \tag{3.24}$$

Then, on $D_{\mathrm{R}}[0, \infty)$,

$$\sum_{i=1}^{\tau_n(t)} X_{ni} \Rightarrow \int_0^t f^2(s)dB_s \tag{3.25}$$

where $B = \{B_s\}_{s\geq 0}$ is a Brownian motion. If in addition $\mathcal{F}_{ni}$ is nested, i.e., $\mathcal{F}_{ni} \subseteq \mathcal{F}_{n+1,i}$ for $n \geq 1, i \geq 1$, then $\sum_{i=1}^{\tau_n(t)} X_{ni} \Rightarrow \int_0^t f^2(s)dB_s$ (mixing), on $D_{\mathrm{R}}[0, \infty)$.

Proof. First show that (3.23) implies that, for all $t > 0$,

$$\max_{1\leq i\leq \tau_n(t)} |X_{ni}| \to_{\mathrm{P}} 0. \tag{3.26}$$

In fact, for any $T > 0$ and integer $q \geq 1$, we have

$$\max_{1\leq i\leq \tau_n(T)} X_{ni}^2 \leq I(q) + I_n(q),$$

where $I(q) = \max_{1\leq j\leq Tq} \int_{(j-1)/q}^{j/q} f^2(s)ds$ and

$$I_n(q) = \max_{1\leq j\leq Tq} \left| \sum_{i=\tau_n((j-1)/q)+1}^{\tau_n(j/q)} X_{ni}^2 - \int_{(j-1)/q}^{j/q} f^2(s)ds \right|.$$

(3.26) follows from $I(q) \to 0$, as $q \to \infty$, since $\int_0^T f^2(s)ds < \infty$, and the fact that, for each integer $q \geq 1$, $I_n(q) \to_{\mathrm{P}} 0$ due to (3.23), as $n \to \infty$.

Now let $Y_{ni} = X_{ni}\mathrm{I}(|X_{ni}| \leq 1)$ and $Z_{ni} = Y_{ni} - \mathrm{E}\left(Y_{ni} \mid \mathcal{F}_{n,i-1}\right)$. Using (3.26) and Lemma 3.2(i), (3.23) implies

$$\sum_{i=1}^{\tau_n(t)} X_{ni}^2 \mathrm{I}(|X_{ni}| \leq 1) \to_{\mathrm{P}} \int_0^t f^2(s)ds.$$

This, together with (3.24), yields $\sum_{i=1}^{\tau_n(t)} Z_{ni}^2 \to_{\mathrm{P}} \int_0^t f^2(s)ds$, for all $t > 0$. Hence, by the classical functional martingale limit theorems [see, e.g., Jecod and Shiryaev (2003)], we have

$$\sum_{i=1}^{\tau_n(t)} Z_{ni} \Rightarrow \int_0^t f^2(s)dB_s. \tag{3.27}$$

on $D_{\mathrm{R}}[0, \infty)$, which, in turn, implies (3.25) by some routine calculations. The proof for mixing convergence is similar. We omit the details. $\square$

Corollary 3.2. *Suppose $\{X_{ni}, \mathcal{F}_{n,i}\}_{n\geq 1, i\geq 1}$ is a martingale array and f is a measurable function such that $\int_0^t f^2(s)ds < \infty$ for all $t > 0$. Suppose one of the following three sets of conditions holds:*

(a) for all $t > 0$ and all $\epsilon > 0$,

$$\sum_{i=1}^{\tau_n(t)} \mathrm{E}\left[X_{ni}^2 \mid \mathcal{F}_{n,i-1}\right] \to_{\mathrm{P}} \int_0^t f^2(s)ds, \tag{3.28}$$

$$\sum_{i=1}^{\tau_n(t)} \mathrm{E}\left[X_{ni}^2 \mathrm{I}(|X_{ni}| \geq \epsilon) \mid \mathcal{F}_{n,i-1}\right] \to_{\mathrm{P}} 0; \tag{3.29}$$

(b) for all $t > 0$, $E \max_{1 \le i \le \tau_n(t)} |X_{ni}| \to 0$ and *(3.23)*;

(c) for all $t > 0$, $\max_{1 \le i \le \tau_n(t)} |X_{ni}| \to_P 0$,

$$\sum_{i=1}^{\tau_n(t)} E\left[X_{ni}^2 I(|X_{ni}| \ge 1) \mid \mathcal{F}_{n,i-1}\right] \to_P 0$$

and either (3.28) or (3.23).

Then conclusions of Theorem 3.8 remain true.

Using Lemma 3.2 and Theorem 3.8, the proof of Corollary 3.2 is simple, which is left to readers.

3.2.3   *Multivariate martingale limit theorem*

Corollary 3.2 can be easily extended to multivariate martingales by using the so-called Cramér-Wold device.

Let $\{Z_{ni}^{(k)}\}_{i \ge 1}, k = 1, \cdots, d$, be d martingale difference arrays with respect to the same σ-field $\{\mathcal{F}_{ni}\}_{i \ge 1}$ and set

$$Z_{ni} = (Z_{ni}^{(1)}, \cdots, Z_{ni}^{(d)}).$$

Theorem 3.9. *Suppose one of the following two sets of conditions holds:*

(a) for all $t > 0$, $\epsilon > 0$ and $k = 1, 2, ..., d$,

$$\sum_{i=1}^{\tau_n(t)} E\left[(Z_{ni}^{(k)})^2 I(|Z_{ni}^{(k)}| \ge \epsilon) \mid \mathcal{F}_{n,i-1}\right] \to_P 0,$$

$$\sum_{i=1}^{\tau_n(t)} E\left[Z_{ni} Z_{ni}' \mid \mathcal{F}_{n,i-1}\right] \to_P \Omega_t,$$

where Z_{ni}' denotes the transpose of Z_{ni} and Ω_t is a continuous, non-negative definite $d \times d$ matrix function;

(b) for all $t > 0$, $E \max_{1 \le i \le \tau_n(t)} |Z_{ni}^{(k)}| \to 0$, $k = 1, 2, ..., d$, and

$$\sum_{i=1}^{\tau_n(t)} Z_{ni} Z_{ni}' \to_P \Omega_t.$$

Then, on $D_{\mathbb{R}^d}[0, \infty)$,

$$\sum_{i=1}^{\tau_n(t)} Z_{ni} \Rightarrow Z_t, \qquad (3.30)$$

where $\{Z_t\}_{t \geq 0}$ *is a d-dimensional Gaussian process with mean zero, covariance matrix* Ω_t *and stationary, independent increments. If in addition* $\mathcal{F}_{ni}$ *is nested, then the convergence in (3.30) is mixing.*

For each $1 \leq k \leq d$, let $\tau_{nk}(t)$ be a stopping time with respect to $\{\mathcal{F}_{ni}\}_{i \geq 1}$ for each $n \geq 1$ and $t > 0$, and the sample paths of $\tau_{nk}(.)$ are integer-valued, non-decreasing and right-continuous with $\tau_{nk}(0) = 0$.

Since $\tilde{\tau}_n(t) = \max_{1 \leq k \leq d} \tau_{nk}(t)$ is a stopping time having the same properties as $\tau_{nk}(t), 1 \leq k \leq d$, and for each $1 \leq k \leq d$,

$$S_{nk}(t) := \sum_{i=1}^{\tau_{nk}(t)} Z_{ni}^{(k)} = \sum_{i=1}^{\tilde{\tau}_n(t)} Z_{ni}^{(k)} \mathrm{I}[\tau_{nk}(t) \geq i],$$

an application of Theorem 3.9 yields the following corollaries.

Corollary 3.3. *Suppose that* $f_k, 1 \leq k \leq d$, *are measurable functions such that* $\int_0^t f_k^2(s)ds < \infty$ *for all* $t > 0$. *Suppose one of the following two sets of conditions holds:*

(a) for all $t > 0$, $\epsilon > 0$ *and* $1 \leq k \neq l \leq d$,

$$\sum_{i=1}^{\tau_{nk}(t)} \mathrm{E}\,[(Z_{ni}^{(k)})^2 \mid \mathcal{F}_{n,i-1}] \to_{\mathrm{P}} \int_0^t f_k^2(s)ds,$$

$$\sum_{i=1}^{\tau_{nk}(t)} \mathrm{E}\,[(Z_{ni}^{(k)})^2 \mathrm{I}(|Z_{ni}^{(k)}| \geq \epsilon) \mid \mathcal{F}_{n,i-1}] \to_{\mathrm{P}} 0,$$

and $\sum_{i=1}^{\tau_{nk}(t) \wedge \tau_{nl}(t)} \mathrm{E}\,[Z_{ni}^{(k)} Z_{ni}^{(l)} \mid \mathcal{F}_{n,i-1}] \to_{\mathrm{P}} 0$;

(b) for all $t > 0$ *and* $1 \leq k \neq l \leq d$,

$$\mathrm{E} \sup_{1 \leq i \leq \tau_{nk}(t)} |Z_{ni}^{(k)}| \to_{\mathrm{P}} 0, \qquad \sum_{i=1}^{\tau_{nk}(t)} (Z_{ni}^{(k)})^2 \to_{\mathrm{P}} \int_0^t f_k^2(s)ds,$$

and $\sum_{i=1}^{\tau_{nk}(t) \wedge \tau_{nl}(t)} Z_{ni}^{(k)} Z_{ni}^{(l)} \to_{\mathrm{P}} 0$.

Then, on $D_{\mathrm{R}^d}[0, \infty)$, *we have*

$$(S_{n1}(t), \cdots, S_{nd}(t)) \Rightarrow \left(\int_0^t f_1^2(s)dB_{1s}, \cdots, \int_0^t f_d^2(s)dB_{ds} \right), \quad (3.31)$$

where $(B_{1s}, \cdots, B_{ds})_{s \geq 0}$ *is a standard d-dimensional Brownian motion. If in addition* $\mathcal{F}_{ni}$ *is nested, then the convergence in (3.31) is mixing.*

Corollary 3.4. *Suppose f is a measurable function such that $\int_0^t f^2(s)ds < \infty$ for all $t > 0$. Suppose one of the following two sets of conditions holds:*

(a) $\{Z_{ni}\}_{n\geq 1, i\geq 1}$ satisfies condition (a) of Theorem 3.9, $\{X_{ni}\}_{n\geq 1, i\geq 1}$ satisfies condition (a) of Corollary 3.2 but $\tau_n(t)$ is replaced by $\tau_{n1}(t)$ and for $k = 1, 2, ..., d$ and all $t > 0$,

$$\sum_{i=1}^{\tau_n(t)\wedge\tau_{n1}(t)} \mathrm{E}\left[X_{ni} Z_{ni}^{(k)} \mid \mathcal{F}_{n,i-1}\right] \to_P 0;$$

(b) $\{Z_{ni}\}_{n\geq 1, i\geq 1}$ satisfies condition (b) of Theorem 3.9, $\{X_{ni}\}_{n\geq 1, i\geq 1}$ satisfies condition (b) of Corollary 3.2 but $\tau_n(t)$ is replaced by $\tau_{n1}(t)$, and for $k = 1, 2, ..., d$ and all $t > 0$,

$$\sum_{i=1}^{\tau_n(t)\wedge\tau_{n1}(t)} X_{ni} Z_{ni}^{(k)} \to_P 0.$$

Then, on $D_{\mathbf{R}^{d+1}}[0, \infty)$, we have

$$\left(\sum_{i=1}^{\tau_{n1}(t)} X_{ni}, \sum_{i=1}^{\tau_n(t)} Z_{ni}\right) \Rightarrow \left(\int_0^t f^2(s)dB_s, Z_t\right), \tag{3.32}$$

where $\{Z_t\}_{t\geq 0}$ is a d-dimensional Gaussian process given in Theorem 3.9 and $\{B_t\}_{t\geq 0}$ is a Brownian motion. Furthermore, $\{Z_t\}_{t\geq 0}$ and $(B_s)_{s\geq 0}$ are independent.

3.2.4 Convergence to a stable Lévy process

The conditions in Corollary 3.2 are not necessary for the asymptotics of $\sum_{i=1}^{\tau_n(t)} X_{ni}$. Let, for $\delta > 0$,

$$S_{nt}(\delta) = \sum_{i=1}^{\tau_n(t)} X_{ni}\mathrm{I}(|X_{ni}| \leq \delta),$$

$$A_{nt}(\delta) = \sum_{i=1}^{\tau_n(t)} \mathrm{E}\left[X_{ni}\mathrm{I}(|X_{ni}| \leq \delta) \mid \mathcal{F}_{n,i-1}\right].$$

Let $L(x)$ be a real valued function defined on $R - \{0\}$, which is non-increasing on $(0, \infty)$, non-decreasing on $(-\infty, 0)$ and $\lim_{x\to\infty} L(x) = \lim_{x\to-\infty} L(x) = 0$. Let, for $x \neq 0$,

$$L_{nt}(x) = \begin{cases} \sum_{i=1}^{\tau_n(t)} P[X_{ni} > x \mid \mathcal{F}_{n,i-1}], & \text{if } x > 0, \\ \sum_{i=1}^{\tau_n(t)} P[X_{ni} \leq x \mid \mathcal{F}_{n,i-1}], & \text{if } x < 0. \end{cases}$$

Theorem 3.10. *Suppose that, for all $t > 0$ and all $\epsilon > 0$,* *(i)* $\max_{1 \le i \le \tau_n(t)} P\big[|X_{ni}| > \epsilon \mid \mathcal{F}_{n,i-1}\big] \to_P 0$;

(ii) $\lim_{\delta \downarrow 0} \limsup_{n \to \infty} P\Big(\sup_{0 \le s \le t} |S_{ns}(\delta) - A_{ns}(\delta)| \ge \epsilon\Big) = 0$;

(iii) $\sum_{i=1}^{\tau_n(t)} \big| \operatorname{E}[X_{ni}\mathrm{I}(|X_{ni}| \le \gamma) \mid \mathcal{F}_{n,i-1}]\big| \to_P 0$, *for some* $\gamma > 0$, *where* $\pm\gamma$ *are continuous points of* $L(x)$;

(iv) $L_{nt}(x) \to_P t L(x)$, *for all continuous points* $x \ne 0$ *of* $L(x)$.

Then $\sum_{i=1}^{\tau_n(t)} X_{ni} \Rightarrow Z_t$ *on* $D_{\mathrm{R}}[0, \infty)$, *where* $Z = \{Z_s\}_{s \ge 0}$ *is a process with stationary independent increments, satisfying* $\operatorname{E}e^{i\theta Z_s} = e^{s\psi(\theta)}$, *and where*

$$\psi(\theta) = \int_{-\infty}^{\infty} \big(e^{i\theta y} - 1 - i\theta y \mathrm{I}\,(|y| < \gamma)\big)\, dL(y).$$

Remark 3.3. Theorem 3.10 provides the convergence of $\sum_{i=1}^{\tau_n(t)} X_{ni}$ to a Lévy process, without assuming (3.28) and (3.29), which was proved in Durrett and Resnick (1978). See, also, Jacod and Shiryaev (2003).

If $L(x) = \begin{cases} px^{-\alpha}, & x > 0, \\ q|x|^{-\alpha}, & x < 0, \end{cases}$ where $p, q \ge 0, p + q > 0$ and $0 < \alpha < 2$, then $Z = \{Z_s\}_{s \ge 0}$ is a stable Lévy process with index α, i.e., $|\operatorname{E}e^{i\theta Z_s}| = \exp\{-bs|\theta|^{\alpha}\}$ for some constant $b > 0$.

Remark 3.4. Let $\widetilde{S}_{nt}(\delta) = \sum_{i=1}^{\tau_n(t)} \widetilde{X}_{ni}$, where

$$\widetilde{X}_{ni} = X_{ni}\mathrm{I}(|X_{ni}| \le \delta) - \operatorname{E}[X_{ni}\mathrm{I}(|X_{ni}| \le \delta) \mid \mathcal{F}_{n,i-1}].$$

Note that, for each $n \ge 1$, $\{\widetilde{S}_{nt}(\delta), \mathcal{F}_{n,\tau_n(t)}\}_{t \ge 0}$ forms a square integrable martingale. For any stopping time T with respect to $\mathcal{F}_{n,\tau_n(t)}$, we have

$$\operatorname{E}\big[\widetilde{S}_{nT}(\delta)\big]^2 = E\Big(\sum_{i=1}^{\tau_n(T)} \widetilde{X}_{ni}^2\Big) = EV_{nT}^2(\delta), \tag{3.33}$$

where $V_{nt}^2(\delta) = \sum_{i=1}^{\tau_n(t)} \operatorname{E}\big(\widetilde{X}_{ni}^2 \mid \mathcal{F}_{n,i-1}\big)$. Due to (3.33), the Lenglart-Rebolledo inequality [see Theorem 1.9.3 of Liptser and Shiryaev (1989)] yields that, for any $b > 0$,

$$P\Big(\sup_{0 \le s \le t} |S_{ns}(\delta) - A_{ns}(\delta)| \ge \epsilon\Big) = P\Big(\sup_{0 \le s \le t} |\widetilde{S}_{ns}(\delta)|^2 \ge \epsilon^2\Big)$$

$$\le P(V_{nt}^2(\delta) \ge b) + \frac{1}{\epsilon^2}\operatorname{E}\big(V_{nt}^2(\delta) \wedge b\big). \tag{3.34}$$

Hence, the condition (ii) will follow from

(ii)' for all $t > 0$ and all $\epsilon > 0$,

$$\lim_{\delta \downarrow 0} \limsup_{n \to \infty} P(V_{nt}^2(\delta) \ge \epsilon) = 0. \tag{3.35}$$

The condition $(ii)'$ provides an easily verified sufficient condition for (ii). Furthermore, if (3.35) is strengthened to

$$\lim_{\delta\downarrow 0} \limsup_{n\to\infty} P(|V_{nt}^2(\delta) - c^2 t| \geq \epsilon) = 0, \qquad (3.36)$$

for all $t > 0$ and all $\epsilon > 0$, where $c \in \mathrm{R}$ is a constant, then Theorem 3.10 still holds if we replace Z_t by $Z'_t = cB_t + Z_t$, where $B = \{B_s\}_{s\geq 0}$ is a Brownian motion independent of Z.

Remark 3.5. Theorem 3.10 can be similarly extended to multivariate martingales. We refer to Chapter VIII of Jacod and Shiryayev (2003) for more details.

3.3   Convergence to a mixture of normal distributions: martingale arrays

A random variable X is said to be *a mixture of normal distributions* if there exists an a.s. finite random variable η^2 such that

$$\mathrm{E}\,e^{itX} = \mathrm{E}\,e^{-\eta^2 t^2/2}, \quad \text{for } t \in \mathrm{R}.$$

We may write $X = \eta\,N$, where N is a normal variate independent of η.

Let $\{X_{ni}\}_{i\geq 1, n\geq 1}$ be an array of random variables. Let $0 < k_n \to \infty$ be a sequence of constants. This section investigates the sufficient conditions such that $\sum_{i=1}^{k_n} X_{ni} \to_D X$.

3.3.1   *A framework*

Let $B = \{B_s\}_{s\geq 0}$ be a Brownian motion. For each $n \geq 1$, let $V_{ni}^2, i \geq 1$, be an increasing sequence of random variables. Set

$$\eta_n(t) = \inf\{k : V_{nk}^2 \geq t\} \quad \text{and} \quad \eta_n^{-1}(k) = \inf\{t \geq 0 : \eta_n(t) \geq k\}.$$

Theorem 3.11. *Suppose that*

$$V_{nk_n}^2 - V_{n,k_n-1}^2 + |X_{nk_n}| \to_P 0. \qquad (3.37)$$

Suppose there exists an a.s. finite random variable $\eta^2 > 0$, which is independent of B, such that

$$\left(V_{nk_n}^2, \sum_{i=1}^{\eta_n(t)} X_{ni}\right) \Rightarrow (\eta^2, B_t), \qquad (3.38)$$

on $D_{\mathbb{R}^2}[0, \infty)$. We have

$$\left(V^2_{nk_n}, \sum_{i=1}^{k_n} X_{ni} \right) \to_D (\eta^2, \eta N), \tag{3.39}$$

where N is a standard normal variate independent of η^2. Furthermore, if the convergence in (3.38) is stable, so is (3.39).

Proof. Using (3.37), we have

$$|\eta_n^{-1}(k_n) - V^2_{nk_n}| \leq V^2_{nk_n} - V^2_{n,k_n-1} \to_P 0 \tag{3.40}$$

and

$$\left| \sum_{i=1}^{k_n} X_{ni} - \sum_{i=1}^{\eta_n[\eta_n^{-1}(k_n)]} X_{ni} \right| \leq |X_{nk_n}| \to_P 0. \tag{3.41}$$

It follows from (3.38) and (3.40) that

$$\left(\eta_n^{-1}(k_n), \sum_{i=1}^{\eta_n(t)} X_{ni} \right) \Rightarrow (\eta^2, B_t),$$

on $D_{\mathbb{R}^2}[0, \infty)$. This, together with the continuous mapping theorem, yields

$$\left(\eta_n^{-1}(k_n), \sum_{i=1}^{\eta_n[\eta_n^{-1}(k_n)]} X_{ni} \right) \Rightarrow (\eta^2, B_{\eta^2}). \tag{3.42}$$

See, e.g., (17.7)–(17.9) of Billingsley (1968) with minor improvement. Recall that η is independent of B. We have $(\eta^2, B_{\eta^2}) =_D (\eta^2, \eta N)$, where N is a standard normal variate independent of η^2. Now the result (3.39) follows from (3.40)-(3.42).

By recalling Theorem 3.3, the proof for stable convergence is similar. We omit the details. □

3.3.2 Limit theorem for martingale: under conditional variance condition (CVC)

For each $n \geq 1$, let $\{X_{ni}, \mathcal{F}_{ni}\}_{i \geq 1}$ be a sequence of martingale differences, where the σ-fields are increasing: $\mathcal{F}_{nk} \subseteq \mathcal{F}_{n,k+1}$ and $\mathcal{F}_{n0} = \{\phi, \Omega\}$. Let $0 < k_n \to \infty$ be a sequence of constants and denote by

$$V^2_{nk_n} = \sum_{i=1}^{k_n} \mathrm{E}\left(X^2_{ni} \mid \mathcal{F}_{n,i-1} \right) \tag{3.43}$$

the conditional variance for martingale $\sum_{i=1}^{k_n} X_{ni}$.

Let $f_n, n \geq 1$, and f be measurable mappings from $D_{\mathbf{R}^d}[0, \infty)$ to R. Denote by E the set of x such that $f_n(x_n) \not\to f(x)$ for some $x_n \to x$. Let $\{Z_{ni}^{(k)}\}_{n \geq 1, i \geq 1}, k = 1, \cdots, d$, be d martingale difference arrays with respect to the same σ-fields $\{\mathcal{F}_{ni}\}_{n \geq 1, i \geq 1}$ and write

$$Z_{ni} = (Z_{ni}^{(1)}, \cdots, Z_{ni}^{(d)}).$$

We say $V_{nk_n}^2$ satisfies conditional variance condition 1 (**CVC1**) if there exist $f_n, n \geq 1, f$ and $\{Z_{ni}^{(k)}\}_{n \geq 1, i \geq 1}, k = 1, \cdots, d$, such that

(i) $\{Z_{ni}\}_{n \geq 1, i \geq 1}$ satisfies the conditions of Theorem 3.9 with some stopping time $\tau_n(t)$, i.e., there is a d-dimensional Gaussian process $G = \{G_t\}_{t \geq 0}$ with mean zero and independent increments such that

$$Z_n(t) := \sum_{i=1}^{\tau_n(t)} Z_{ni} \Rightarrow G_t, \quad \text{on } D_{\mathbf{R}^d}[0, \infty);$$

(ii) $P(E) = 0$ and $V_{nk_n}^2 - f_n(Z_n) = o_P(1)$, where $Z_n = \{Z_n(t)\}_{t \geq 0}$.

We say $V_{nk_n}^2$ satisfies conditional variance condition 2 (**CVC2**) if there exist $f_n, n \geq 1, f$ and $\{Z_{ni}^{(k)}\}_{n \geq 1, i \geq 1}, k = 1, \cdots, d$, such that

(i) there is a d-dimensional Lévy process $L = \{L_t\}_{t \geq 0}$ with triplet $(0, 0, \nu)$ such that

$$Z_n(t) := \sum_{i=1}^{\tau_n(t)} Z_{ni} \Rightarrow L_t, \quad \text{on } D_{\mathbf{R}^d}[0, \infty);$$

(ii) $P(E) = 0$ and $V_{nk_n}^2 - f_n(Z_n) = o_P(1)$, where $Z_n = \{Z_n(t)\}_{t \geq 0}$.

Theorem 3.12. *Suppose, for all $\epsilon > 0$,*

$$\sum_{i=1}^{k_n} \mathrm{E}\left[X_{ni}^2 \mathrm{I}(|X_{ni}| \geq \epsilon) \mid \mathcal{F}_{n,i-1}\right] \to_P 0. \qquad (3.44)$$

Suppose one of the following three sets of conditions holds:

(i) *The σ-fields are nested: $\mathcal{F}_{ni} \subseteq \mathcal{F}_{n+1,i}$ for $i \geq 1, n \geq 1$, and for an a.s. finite random variable Z^2,*

$$V_{nk_n}^2 = \sum_{i=1}^{k_n} \mathrm{E}\left[X_{ni}^2 \mid \mathcal{F}_{n,i-1}\right] \to_P Z^2. \qquad (3.45)$$

(ii) $V_{nk_n}^2$ *satisfies* **CVC1** *and for* $l = 1, 2, ..., d,$

$$\max_{1 \le k \le k_n} \| \sum_{i=1}^{k} \mathrm{E}\, (X_{ni} Z_{ni}^{(l)} \mid \mathcal{F}_{n,i-1}) \| = o_{\mathrm{P}}(1). \qquad (3.46)$$

(iii) $V_{nk_n}^2$ *satisfies* **CVC2**.

In either case, by letting $Z^2 = f(G)$ *or* $f(L)$ *in the case (ii) or (iii) respectively, we have*

$$(\sum_{i=1}^{k_n} X_{ni}, V_{nk_n}^2) \to_D (X, Z^2), \qquad (3.47)$$

where X *is a random variable having the characteristic function* $\mathrm{E}\, e^{-Z^2 t^2/2}$, *i.e.,* $X = Z N$, *where* N *is a standard normal variate independent of* Z.

Proof. We establish Theorem 3.12 by applying Corollary 3.4 and Theorem 3.11. To do this, we need to introduce an associate martingale array $\{X_{ni}^*, \mathcal{F}_{ni}^*\}_{i \ge 1, n \ge 1}$ that is used as an alternative of the original $\{X_{ni}, \mathcal{F}_{ni}\}_{i \ge 1, n \ge 1}$ in the proof.

Let $\epsilon_k, k \ge 1$, be a sequence of i.i.d. $N(0,1)$ random variables, which are independent of $\{X_{ni}, \mathcal{F}_{ni}\}_{i \ge 1, n \ge 1}$. For each $n \ge 1$, let

$$X_{ni}^* = \begin{cases} X_{ni}, & \text{if } 1 \le i \le k_n, \\ \epsilon_i / \sqrt{n}, & \text{if } i \ge k_n + 1, \end{cases}$$

and

$$\mathcal{F}_{ni}^* = \begin{cases} \mathcal{F}_{ni}, & \text{if } 1 \le i \le k_n, \\ \sigma\{\mathcal{F}_{ni}, \epsilon_{k_n+1}, \cdots, \epsilon_i\}, & \text{if } i \ge k_n + 1. \end{cases}$$

Obviously, $\{X_{ni}^*, \mathcal{F}_{ni}^*\}_{n \ge 1, i \ge 1}$ forms a martingale difference array, $\mathcal{F}_{ni}^* \subseteq \mathcal{F}_{n,i+1}^*$ and if the σ-fields $\mathcal{F}_{ni}$ are nested, then so are $\mathcal{F}_{ni}^*$.

For this new martingale difference array $\{X_{ni}^*, \mathcal{F}_{ni}^*\}_{i \ge 1, n \ge 1}$, set $\eta_n(t) = \inf\{k : V_{nk}^{*2} \ge t\}$, where

$$V_{nk}^{*2} = \sum_{i=1}^{k} \mathrm{E}\, (X_{ni}^{*2} \mid \mathcal{F}_{n,i-1}^*)$$

$$= \begin{cases} V_{nk}^2, & \text{if } 1 \le k \le k_n, \\ V_{nk}^2 + \frac{1}{n} \sum_{i=k_n+1}^{k} \epsilon_i^2, & \text{if } k \ge k_n + 1. \end{cases} \qquad (3.48)$$

For each $n \ge 1$ and $t > 0$, $\eta_n(t)$ is a stopping time, satisfying

$$1 \le \eta_n(t) \le k_n + 2nt, \quad a.s., \qquad (3.49)$$

since, by recalling $\epsilon_k \sim$ i.i.d. $N(0,1)$, $\frac{1}{n}\sum_{k=k_n+1}^{k_n+[2nt]} \epsilon_k^2 \geq t, a.s.$ Note that

$$\max_{1 \leq i \leq k_n} \mathrm{E}\left[X_{ni}^2 \mid \mathcal{F}_{n,i-1}\right] + \max_{1 \leq i \leq k_n} |X_{ni}| \to_\mathrm{P} 0, \tag{3.50}$$

due to (3.44) and Lemma 3.2 (iii). It follows from (3.44), (3.49)–(3.50) that

$$0 \leq \sum_{i=1}^{\eta_n(t)} \mathrm{E}\left(X_{ni}^{*2} \mid \mathcal{F}_{n,i-1}^*\right) - t \leq \mathrm{E}\left(X_{n,\eta_n(t)}^{*2} \mid \mathcal{F}_{n,\eta_n(t)-1}^*\right)$$

$$\leq \max_{1 \leq i \leq k_n} \mathrm{E}\left(X_{ni}^2 \mid \mathcal{F}_{n,i-1}\right) + \frac{1}{n} \max_{k_n \leq i \leq k_n+2nt} \epsilon_i^2, \quad a.s.,$$

$$= o_\mathrm{P}(1), \tag{3.51}$$

for all $t > 0$, and

$$\sum_{i=1}^{\eta_n(t)} \mathrm{E}\left(X_{ni}^{*2}\mathrm{I}(|X_{ni}^*| \geq \epsilon) \mid \mathcal{F}_{n,i-1}^*\right)$$

$$\leq \sum_{i=1}^{k_n} \mathrm{E}\left[X_{ni}^2\mathrm{I}(|X_{ni}| \geq \epsilon) \mid \mathcal{F}_{n,i-1}\right] + 2t\,\mathrm{E}\epsilon_1^2\mathrm{I}(|\epsilon_1| \geq \sqrt{n}\epsilon), \quad a.s.$$

$$= o_\mathrm{P}(1), \tag{3.52}$$

for all $t > 0$ and all $\epsilon > 0$.

We are now ready to prove (3.47). By Theorem 3.11 and (3.48), it suffices to show that, on $D_{\mathrm{R}^2}[0,\infty)$,

$$\left(V_{nk_n}^2, \sum_{i=1}^{\eta_n(t)} X_{ni}^*\right) \Rightarrow (Z^2, B_t), \tag{3.53}$$

where $B = \{B_s\}_{s \geq 0}$ is a Brownian motion independent of Z^2, under each of the condition sets (i), (ii) and (iii).

First assume (i). By (3.51)–(3.52) and noting that the σ-fields $\{\mathcal{F}_{ni}^*, i \geq 1, n \geq 1\}$ are nested, it follows from Corollary 3.2 (a) with $f(s) = 1$ that

$$\sum_{i=1}^{\eta_n(t)} X_{ni}^* \Rightarrow B_t \quad (mixing),$$

on $D_\mathrm{R}[0,\infty)$, where $B = \{B_s\}_{s \geq 0}$ is a Brownian motion. This, together with (3.45), yields (3.53) by Theorem 3.3 (ii).

Next assume (ii). Recalling Theorem 3.5, to prove (3.53), it suffices to show

$$\left(\sum_{i=1}^{\tau_n(t)} Z_{ni}, \sum_{i=1}^{\eta_n(t)} X_{ni}^*\right) \Rightarrow (G_t, B_t), \tag{3.54}$$

on $D_{\mathbb{R}^{d+1}}[0, \infty)$, where $B = \{B_s\}_{s \geq 0}$ is a Brownian motion independent of $G = \{G_s\}_{s \geq 0}$. Note that properties of Z_{ni} do not change if $\mathcal{F}_{ni}$ is replaced by $\mathcal{F}_{ni}^*$. Furthermore, condition (3.46), together with the definitions of X_{ni}^* and $\mathcal{F}_{ni}^*$, yields

$$\left| \sum_{i=1}^{\tau_n(t) \wedge \eta_n(t)} \mathrm{E}\left(X_{ni}^* Z_{ni}^{(l)} \mid \mathcal{F}_{n,i-1}^*\right) \right|$$

$$\leq \max_{1 \leq k \leq k_n} \| \sum_{i=1}^{k} \mathrm{E}\left(X_{ni} Z_{ni}^{(l)} \mid \mathcal{F}_{n,i-1}\right) \| = o_\mathrm{P}(1), \qquad (3.55)$$

for $l = 1, 2, ..., d$ and all $t > 0$. By virtue of (3.51)–(3.52) and (3.55), result (3.54) follows from Corollary 3.4.

Finally assume (iii). As in (ii), it suffices to show

$$\left(\sum_{i=1}^{\tau_n(t)} Z_{ni}, \sum_{i=1}^{\eta_n(t)} X_{ni}^* \right) \Rightarrow (L_t, B_t), \qquad (3.56)$$

on $D_{\mathbb{R}^{d+1}}[0, \infty)$, where $B = \{B_s\}_{s \geq 0}$ is a Brownian motion independent of $L = \{L_s\}_{s \geq 0}$. This is obvious under **CVC2** as Lévy process with triplet $(0, 0, \nu)$ is automatically independent of a Brownian motion whenever they have the same filtration. See, e.g., Theorem 3.6. □

3.3.3 Examples and remarks

Let $\epsilon_k, k \geq 1$, be i.i.d. random variables with $\mathrm{E}\epsilon_1 = 0$ and $\mathrm{E}\epsilon_1^2 = 1$. Denote by $\mathcal{F}_{ni}$ the σ-field generated by $\epsilon_1, \epsilon_2, ..., \epsilon_i, 1 \leq i \leq n$, and set $S_0 = 0, S_i = \sum_{j=1}^{i} \epsilon_j, i \geq 1$, and for $1 \leq m(n) \leq n$,

$$X_{ni} = \begin{cases} \epsilon_i/\sqrt{n}, & 1 \leq i \leq n, \\ \mathrm{I}(S_{m(n)} > 0)\epsilon_i/\sqrt{n}, & 1 + n \leq i \leq 2n, \end{cases}$$

Then $\{X_{ni}, \mathcal{F}_{ni}\}$ is a martingale difference array. It is easy to see that the σ-fields are nested: $\mathcal{F}_{ni} \subseteq \mathcal{F}_{n+1,i}$ for $i \geq 1, n \geq 1$, (3.44) holds and the conditional variance V_n^2 for the martingale $\sum_{i=1}^{2n} X_{ni}$ satisfies

$$V_n^2 = 1 + \mathrm{I}(S_{m(n)} > 0)$$

$$\to_D \begin{cases} 1 + \mathrm{I}(B_1 > 0), & \text{if } m(n) = n, \\ 1 + \mathrm{I}(N > 0), & \text{if } m(n) \to \infty \text{ and } m(n)/n \to 0, \end{cases}$$

where $B = \{B_s\}_{s \geq 0}$ is a Brownian motion and N is a normal variate that is chosen to be independent of B for later use. The convergence in distribution here cannot be strengthened to convergence in probability.

Example 3.3. Suppose that $m(n) = n$. Then

$$\sum_{i=1}^{2n} X_{ni} = S_n/\sqrt{n} + \mathrm{I}(S_n/\sqrt{n} > 0)(S_{2n} - S_n)/\sqrt{n}$$

$$\to_D X = B_1 + \mathrm{I}(B_1 > 0)(B_2 - B_1).$$

The distribution of X is not a mixture of normal distributions. Furthermore, V_n^2 satisfies **CVC1** with

$$Z_{ni} = \begin{cases} \epsilon_i/\sqrt{n}, & 1 \leq i \leq n, \\ 0, & i \geq n+1, \end{cases}$$

$\tau_n(t) = n$ and $f_n(X) = f(X) = 1 + \mathrm{I}(X > 0)$ for X in $D_{\mathrm{R}}[0, \infty]$, but

$$\max_{1 \leq k \leq 2n} \| \sum_{i=1}^{k} \mathrm{E}\left(X_{ni} Z_{ni} \mid \mathcal{F}_{n,i-1}\right) \| = 1 \neq o_{\mathrm{P}}(1).$$

Example 3.4. Suppose that $m(n)/n \to 0$. Then

$$\sum_{i=1}^{2n} X_{ni} = \frac{S_{m(n)}}{\sqrt{n}} + \frac{S_n - S_{m(n)}}{\sqrt{n}} + \mathrm{I}\left(\frac{S_{m(n)}}{\sqrt{m(n)}} > 0\right) \frac{S_{2n} - S_n}{\sqrt{n}}$$

$$\to_D X = B_1 + \mathrm{I}(N > 0)(B_2 - B_1).$$

Note that $X =_D \eta^2 B_1$, where $\eta^2 = 1 + \mathrm{I}(N > 0)$. The distribution of X is a mixture of normal distributions. Furthermore, V_n^2 satisfies **CVC1** with

$$Z_{ni} = \begin{cases} \epsilon_i/\sqrt{n}, & 1 \leq i \leq m(n), \\ 0, & i \geq m(n)+1 \end{cases}$$

$\tau_n(t) = m(n)$ and $f_n(X) = f(X) = 1 + \mathrm{I}(X > 0)$ for X in $D_{\mathrm{R}}[0, \infty]$, and

$$\max_{1 \leq k \leq 2n} \| \sum_{i=1}^{k} \mathrm{E}\left(X_{ni} Z_{ni} \mid \mathcal{F}_{n,i-1}\right) \| = m(n)/n \to 0.$$

Example 3.3 explains the convergence in probability in (3.45) cannot be replaced by the convergence in distribution and $o_{\mathrm{P}}(1)$ in condition (3.46) cannot be relaxed to $O_{\mathrm{P}}(1)$. Example 3.4 indicates the convergence in probability in (3.45) is not a necessary condition to establish the convergence to a mixture of normal distributions for martingale array, which also provides a non-trivial corollary of Theorem 3.12 (ii).

The condition that $V_{nk_n}^2$ satisfies **CVC1** or **CVC2** implies $V_{nk_n}^2 \to_D$ $f(G)$ or $f(L)$, respectively, which is necessary to establish (3.47). Since verification of (3.45) is usually difficult or impossible, Theorem 3.12 (ii) and (iii) have big advantages in applications, particularly in nonlinear cointegrating regression, which will be investigated in Chapter 4. For a certain class of martingales, extensions of Theorem 3.12 (ii) and (iii) will be considered in next section.

Next example indicates that the nested structure of σ-fields in Theorem 3.12 (i) cannot be removed.

Example 3.5. Let $B = \{B_s\}_{s \geq 0}$ be a Brownian motion and set

$$M_{ni} = \begin{cases} B_{i/n}, & \text{if } 1 \leq i \leq n, \\ B_1 \mathrm{I}(B_1 \leq 0) + B_{i/n} \mathrm{I}(B_1 > 0), & \text{if } n+1 \leq i \leq 2n. \end{cases}$$

Let $\mathcal{F}_{ni}$ be the σ-field generated by B_t for $0 \leq t \leq i/n$, $\{M_{ni}, \mathcal{F}_{ni}, 1 \leq i \leq 2n, n \geq 1\}$, be a martingale array with the difference

$$X_{ni} = \begin{cases} Z_{ni}, & \text{if } 1 \leq i \leq n, \\ Z_{ni} \mathrm{I}(B_1 > 0), & \text{if } n+1 \leq i \leq 2n, \end{cases}$$

where $Z_{ni} = B_{i/n} - B_{(i-1)/n}$, and the condition variance

$$V_n^2 = \sum_{i=1}^{n} Z_{ni}^2 + \sum_{i=n+1}^{2n} Z_{ni}^2 \mathrm{I}(B_1 > 0).$$

Since Z_{ni} are i.i.d. $N(0, 1/n)$ for each $n \geq 1$, it is easy to see that (3.44) holds and $V_n^2 \to_P 1 + \mathrm{I}(B_1 > 0)$, but $\mathcal{F}_{ni} \not\subseteq \mathcal{F}_{n+1,i}$ for $i \geq 1, n \geq 1$. Furthermore,

$$M_{n,2n} = B_1 \mathrm{I}(B_1 \leq 0) + B_2 \mathrm{I}(B_1 > 0),$$

which is not a mixture of normal distributions.

3.4   Martingale limit theorem revisited

As in Section 2.3.3, suppose $\{X_{ni}\}_{n \geq 1, i \geq 1}$ and $\{Z_{ni}^{(k)}\}_{n \geq 1, i \geq 1}, k = 1, \cdots, d$, are martingale difference arrays with respect to the same σ-fields $\{\mathcal{F}_{ni}\}_{n \geq 1, i \geq 1}$ where the σ-fields are increasing: $\mathcal{F}_{nk} \subseteq \mathcal{F}_{n,k+1}$ and $\mathcal{F}_{n0} = \{\phi, \Omega\}$. Let $0 < k_n \to \infty$ be a sequence of constants and denote by

$$V_{nk_n}^2 = \sum_{i=1}^{k_n} \mathrm{E}\left(X_{ni}^2 \mid \mathcal{F}_{n,i-1}\right)$$

the conditional variance for martingale $S_n = \sum_{i=1}^{k_n} X_{ni}$. Furthermore, write

$$Z_{ni} = (Z_{ni}^{(1)}, \cdots, Z_{ni}^{(d)}).$$

When the conditional variance $V_{nk_n}^2$ has certain structure (i.e., satisfies **CVC1**), Theorem 3.12 (ii) provides a martingale limit theorem without imposing the convergence in probability for $V_{nk_n}^2$. This result has a big advantage in many applications since the condition likes (3.45) is difficult or impossible to verify. The condition **CVC1** (ii), however, is still restrictive. It is usually not convenient for checking $P(E) = 0$. Some relaxation of this condition is provided in this section.

3.4.1 *Limit theorem for martingale: under convergence in distribution for conditional variance*

Let $f_n, n \geq 1$, be measurable real functions of their components and specify the $\mathrm{E}\left(X_{ni}^2 \mid \mathcal{F}_{n,i-1}\right), i \geq 2$, to have the form:

$$\mathrm{E}\left(X_{ni}^2 \mid \mathcal{F}_{n,i-1}\right) = f_n\left(Z_{n1}, ..., Z_{n,i-1}; X_{n1}, ..., X_{n,i-1}; \; \xi_{n1}, \xi_{n2}, ...\right),$$

where ξ_{nk} is an arbitrary sequence of random variables satisfying certain measurable conditions. Throughout this section, assume

$$\sum_{i=1}^{k_n} \mathrm{E}\left[X_{ni}^2 \mathrm{I}(|X_{ni}| \geq \epsilon) \mid \mathcal{F}_{n,i-1}\right] \to_\mathrm{P} 0,$$

for all $\epsilon > 0$, $\tau_{nk}(t)$ is a stopping time with respect to $\{\mathcal{F}_{ni}\}_{i \geq 1}$ for each $n \geq 1$ and $t > 0$, and

$$\sum_{i=1}^{\tau_n(t)} \mathrm{E}\left[(Z_{ni}^{(k)})^2 \mathrm{I}(|Z_{ni}^{(k)}| \geq \epsilon) \mid \mathcal{F}_{n,i-1}\right] \to_\mathrm{P} 0, \tag{3.57}$$

$$\sum_{i=1}^{\tau_n(t)} \mathrm{E}\left[Z_{ni} Z_{ni}' \mid \mathcal{F}_{n,i-1}\right] \to_\mathrm{P} \Omega_t, \tag{3.58}$$

for all $t > 0$, $\epsilon > 0$ and $k = 1, 2, ..., d$, where Ω_t is a continuous, non-negative definite $d \times d$ matrix function. Conditions (3.57) and (3.58) are the condition (a) of Theorem 3.9, indicating

$$\sum_{i=1}^{\tau_n(t)} Z_{ni} \Rightarrow Z_t,$$

on $D_{\mathrm{R}^d}[0, \infty)$, where $Z = \{Z_t\}_{t \geq 0}$ is a d-dimensional Gaussian process with mean zero, covariance matrix Ω_t and stationary, independent increments.

Theorem 3.13. *Suppose (i) for $k = 1, 2, ..., d$, and for all $t > 0$,*

$$\sum_{i=1}^{\tau_n(t) \wedge k_n} \mathrm{E}\left(X_{ni} Z_{ni}^{(k)} \mid \mathcal{F}_{n,i-1}\right) = o_\mathrm{P}(1); \tag{3.59}$$

(ii) there exists an a.s. finite functional $f^2(Z)$ of Z such that

$$\left(\sum_{i=1}^{\tau_n(t)} Z_{ni}, V_{nk_n}^2 \right) \Rightarrow \left(Z_t, f^2(Z) \right), \tag{3.60}$$

on $D_{\mathbb{R}^2}[0, \infty)$. Then,

$$\left(\sum_{i=1}^{k_n} X_{ni}, V_{nk_n}^2 \right) \to_D \left(f(Z) \, N, f^2(Z) \right), \tag{3.61}$$

where N is a standard normal variate independent of $f^2(Z)$.

If in addition that $\xi_{nj}, j \geq 1$, are $\mathcal{F}_{n1}$-measurable for each $n \geq 1$ and there exist a continuous process $\Xi = \{\Xi_t\}_{t \geq 0}$ that is independent of Z and an a.s. finite functional $f_1^2(Z, \Xi)$ such that

$$\left(\sum_{i=1}^{\tau_n(t)} Z_{ni}, \sum_{j=1}^{[nt]} \xi_{nj}, V_{nk_n}^2 \right) \Rightarrow \left(Z_t, \, \Xi_t, \, f_1^2(Z, \Xi) \right), \tag{3.62}$$

on $D_{\mathbb{R}^3}[0, \infty)$. Then,

$$\left(\sum_{i=1}^{k_n} X_{ni}, V_{nk_n}^2 \right) \to_D \left(f_1(Z, \Xi) \, N, f_1^2(Z, \Xi) \right), \tag{3.63}$$

where N is a standard normal variate independent of $f_1^2(Z, \Xi)$.

The proof of Theorem 3.13 is given in Section 3.4.3.

Let $\{\eta_i\}_{i \geq 1}, \{Z_i^{(k)}\}_{i \geq 1}, k = 1, 2, ..., d$, be $d+1$ martingale difference with respect to the same σ-fields $\{\mathcal{F}_i\}_{i \geq 0}$, satisfying, for all $1 \leq k \neq l \leq d$,

$$\mathrm{E}\left(\eta_i^2 \mid \mathcal{F}_{i-1}\right) \to_{a.s.} 1,$$

$$\mathrm{E}\left[(Z_i^{(k)})^2 \mid \mathcal{F}_{i-1}\right] \to_{a.s.} 1, \quad E[Z_i^{(k)} Z_i^{(l)} \mid \mathcal{F}_{i-1}] \to_{a.s.} 0,$$

and, as $K \to \infty$,

$$\sup_{i \geq 1} \mathrm{E}\left\{ \left[\eta_i^2 \mathrm{I}(|\eta_i| \geq K) + \sum_{k=1}^{d} (Z_i^{(k)})^2 \mathrm{I}(|Z_i^{(k)}| \geq K) \right] \mid \mathcal{F}_{i-1} \right\} = o_{\mathrm{P}}(1).$$

Let $f_n(...)$ be a real functional of its components and write $Z_i = (Z_i^{(1)}, ..., Z_i^{(d)})$. Consider a class of martingale S_n defined by

$$S_n = \sum_{k=1}^{n} x_{nk} \eta_k,$$

where $x_{nk} = f_n\left(Z_1, ..., Z_{k-1}; \eta_1, ..., \eta_{k-1}; \zeta_k, \zeta_{k-1}, ...\right)$ and $\{\zeta_i, i \leq k\}$ is a sequence of arbitrary random variables such that x_{nk} is adapted to $\mathcal{F}_{k-1}$.

Let $B_n(t) = \frac{1}{\sqrt{n}} \sum_{j=1}^{[nt]} Z_j$. It follows from Theorem 3.9 that

$$B_n(t) \Rightarrow B_t, \quad \text{on } D_{\mathrm{R}^d}[0, \infty)$$

where $B = \{B_t\}_{t \geq 0}$ is a d-dimensional Browinan motion. The following theorem is a direct consequence of Theorem 3.13.

Theorem 3.14. *Suppose (a)* $\max_{1 \leq k \leq n} |x_{nk}| = o_{\mathrm{P}}(1)$ *and*

$$\frac{1}{\sqrt{n}} \sum_{k=1}^{n} x_{nk} \, \mathrm{E}\left(\eta_k Z_k \mid \mathcal{F}_{k-1}\right) = o_{\mathrm{P}}(1); \tag{3.64}$$

(b) there exists an a.s. finite functional $F^2(B)$ *of B such that*

$$\left(B_n(t), \sum_{k=1}^{n} x_{nk}^2\right) \Rightarrow \left(B_t, F^2(B)\right), \tag{3.65}$$

on $D_{\mathrm{R}^{d+1}}[0, \infty)$. *Then, as* $n \to \infty$,

$$\left(S_n, \sum_{k=1}^{n} x_{nk}^2\right) \to_D \left\{F(B)\,N, \; F^2(B)\right\}, \tag{3.66}$$

where N is a standard normal variate independent of $F(B)$.

In order to establish (3.66), the convergence to a mixture of normal distributions, condition (3.64) cannot be removed as indicated in Example 3.3. Conditions (3.64) and (3.65) are achievable for many useful random sequences x_{nk} such as those appeared in Theorems 2.19 and 2.22. Two illustrated examples are given in next section.

3.4.2   *Two examples*

Throughout the section, suppose $\{\epsilon_j\}_{j \in \mathbb{Z}}$ is a sequence of i.i.d. random variables with $\mathrm{E}\epsilon_0 = 0$, $\mathrm{E}\epsilon_0^2 = 1$ and $\lim_{|t| \to \infty} |t|^\eta |\mathrm{E}\, e^{it\epsilon_0}| < \infty$ for some $\eta > 0$.

In our first example, let $\tau \geq 0$ be a constant, $\gamma = 1 - \tau/n$, $y_{n0} = 0$,

$$y_{nk} = \gamma\, y_{n,k-1} + \xi_k,$$

where ξ_k is a linear process defined as in Example 2.12, i.e., $\xi_j = \sum_{k=0}^{\infty} \phi_k \, \epsilon_{j-k}$, where the coefficients $\phi_k, k \geq 0$, satisfy one of the following conditions:

LM. $\phi_k \sim k^{-\mu} \rho(k)$, where $1/2 < \mu < 1$ and $\rho(k)$ is a function slowly varying at ∞.

SM. $\sum_{k=0}^{\infty} |\phi_k| < \infty$ and $\phi \equiv \sum_{k=0}^{\infty} \phi_k \neq 0$.

Write $d_n^2 = \mathrm{E}\,(\sum_{j=1}^{n} \xi_j)^2$ and define $Z = \{Z_t\}_{t \geq 0}$ as in (2.82). As indicated in Section 2.3.3, Z_t is a farctional Ornstein-Uhlenbeck process, having a continuous local time $L_Z(t, x)$.

Corollary 3.5. *Suppose*

(a) $\{\eta_i, \mathcal{F}_i\}_{i \geq 1}$, where $\mathcal{F}_i = \sigma(\eta_i, ..., \eta_1; \epsilon_{i+1}, \epsilon_i, ...)$, is a martingale difference such that $\mathrm{E}\,(\eta_i^2 \mid \mathcal{F}_{i-1}) \to_{a.s.} 1$, as $i \to \infty$, and

$$\sup_{i \geq 1} \mathrm{E}\,\left[\eta_i^2 \mathrm{I}(|\eta_i| \geq K) \mid \mathcal{F}_{i-1}\right] = o_{\mathrm{P}}(1),$$

as $K \to \infty$;

(b) $g(x)$ is a bounded real function satisfying $\int_{-\infty}^{\infty} |g(x)| dx < \infty$.

Then, for any $h = O(1)$ and $nh/d_n \to \infty$, we have

$$\left\{ \left(\frac{d_n}{nh}\right)^{1/2} \sum_{k=1}^{n} g\big[(y_{nk} + x)/h\big]\,\eta_k,\ \frac{d_n}{nh} \sum_{k=1}^{n} g^2\big[(y_{nk} + x)/h\big] \right\}$$

$$\to_D (V\,N,\ V^2), \tag{3.67}$$

where x is fixed, N is a normal variate independent of $V^2 = \int_{-\infty}^{\infty} g^2(x) dx\, L_Z(1, 0)$.

Proof. It suffices to verify the conditions of Theorem 3.14 with $Z_k = (\epsilon_k, \epsilon_{-k})$ and $x_{nk} = \left(\frac{d_n}{nh}\right)^{1/2} g\big[(y_{nk} + x)/h\big]$. This is a simple application of Theorem 2.21 and hence the details are omitted. $\qquad\square$

Remark 3.6. If $g(x)$ has compact support (say $g(x) = 0$ if $|x| \geq M$), then

$$g(z + x/h)g(z + y/h) = 0 \tag{3.68}$$

for all $z \in \mathrm{R}$ and $|x - y| \geq Mh$ whenever h is sufficiently small. This, together with (3.67), implies that, for any $x, y \in [-A, A]$ and $|x - y| \geq Mh$,

$$\frac{d_n}{nh} \sum_{k=1}^{n} \left\{ \alpha g\big[(y_{nk} + x)/h\big] + \beta g\big[(y_{nk} + y)/h\big] \right\}^2 \to_D \tau\, L_Z(1, 0),$$

where $A > 0$ is a constant and $\tau = (\alpha^2 + \beta^2) \int_{-\infty}^{\infty} g^2(x) dx$. Hence, by the Cramér-Wold device, we have

$$\left\{ S_n(x),\ S_n(y) \right\} \to_D \left\{ N(0, V^2), N_1(0, V^2) \right\}, \tag{3.69}$$

for any $x, y \in [-A, A]$, where $S_n(x) = \left(\frac{d_n}{nh}\right)^{1/2} \sum_{k=1}^{n} g\big[(y_{nk} + x)/h\big]\,\eta_k$, N and N_1 are independent standard normal variate, and both are independent of V^2.

Functional limit theorem for $S_n(x)$ on $C[-A, A]$, however, does not exist since, if we take $|x - y| = \sqrt{h} \to 0$, it follows from (3.69) that

$$P(|S_n(x) - S_n(y)| \geq \epsilon) \to 1,$$

as $n \to \infty$ first and then $\epsilon \to 0$, which indicates $S_n(x)$ is not tight.

For the functional of $S_n(x)$, we have the following result, which is useful in nonlinear cointegrating regression.

Theorem 3.15. *Let $|\pi(x)|$ be a bounded integrable real function and, in addition to conditions of Corollary 3.5, $g(x)$ has compact support satisfying $|g(x) - g(y)| \leq C|x - y|$ for all $x, y \in \mathbb{R}$. Then, for any h satisfying $nh/d_n \to \infty$ and $nh^2/d_n \to 0$, we have*

$$\frac{d_n}{nh} \int_{-\infty}^{\infty} \left\{ \sum_{k=1}^{n} g[(y_{nk} + x)/h]\eta_k \right\}^2 \pi(x)dx \to_D \tau_0 \, L_Z(1,0), \tag{3.70}$$

where $\tau_0 = \int_{-\infty}^{\infty} g^2(x)dx \int_{-\infty}^{\infty} \pi(x)dx$.

Proof.   First assume $|\eta_k| \leq A$. Let $z_k = \int_{-\infty}^{\infty} g^2[(y_{nk} + x)/h] \, \pi(x)dx$ and

$$z_{kj} = \int_{-\infty}^{\infty} g[(y_{nk} + x)/h]g[(y_{nj} + x)/h] \, \pi(x)dx.$$

We may write

$$\int_{-\infty}^{\infty} \left\{ \sum_{k=1}^{n} g[(y_{nk} + x)/h]\eta_k \right\}^2 \pi(x)dx$$

$$= \sum_{k=1}^{n} z_k \mathrm{E} \left(\eta_k^2 \mid \mathcal{F}_{k-1} \right) + \sum_{k=1}^{n} z_k \left[\eta_k^2 - \mathrm{E} \left(\eta_k^2 \mid \mathcal{F}_{k-1} \right) \right] + 2 \sum_{k=2}^{n} \eta_k \sum_{j=1}^{k-1} \eta_j \, z_{kj}$$

$$= R_{1n} + R_{2n} + R_{3n}, \quad \text{say.} \tag{3.71}$$

As in the proof of Corollary 2.5, we have

$$\frac{d_n}{nh} \sum_{k=1}^{n} z_k \to_D \tau_0 \, L_Z(1,0),$$

which implies $\frac{d_n}{nh} R_{1n} \to_D \tau_0 \, L_Z(1,0)$ due to $\mathrm{E} \left(\eta_k^2 \mid \mathcal{F}_{k-1} \right) \to 1, a.s.$ and $z_k, k \geq 1$, are non-negative. Note that, by (3.68),

$$|\mathrm{E} \left(z_{kj} \, z_{kl} \right)| \leq \mathrm{E} \left\{ \int_{-\infty}^{\infty} \int_{-\infty}^{\infty} \left| g[(y_{nk} + x)/h]g[(y_{nk} + y)/h] \right| \right.$$

$$\left. \left| g[(y_{nj} + x)/h]g[(y_{nl} + y)/h] \right| \, |\pi(x)| \, |\pi(y)| dxdy \right\}$$

$$\leq \int_{-\infty}^{\infty} \int_{|y-x| \leq Mh} \mathrm{E} \left\{ \left| g[(y_{nk} + x)/h]g[(y_{nj} + x)/h] \right. \right.$$

$$\left. \left. g[(y_{nl} + y)/h] \right| \right\} |\pi(x)| \, |\pi(y)| \, dxdy.$$

As z_{kj} is $\mathcal{F}_{k-1}$-measurable, it is readily seen from $\mathrm{E}(\eta_k \mid \mathcal{F}_{k-1}) = 0$, $|\eta_k| \leq A$ and Lemma 2.5 that

$$
\mathrm{E}R_{3n}^2 \leq 4A^2 \int_{-\infty}^{\infty} \int_{|y-x|\leq Mh} \mathrm{E} \sum_{1\leq k\neq j,l\leq n} \Big\{ \big| g[(y_{nk}+x)/h]
$$

$$
g[(y_{nj}+x)/h]g[(y_{nl}+y)/h] \big| \Big\} |\pi(x)| \, |\pi(y)| \, dxdy
$$

$$
\leq C \sup_x \mathrm{E} \Big(\sum_{k=1}^{n} \big| g[(y_{nk}+x)/h] \big| \Big)^3 \int_{-\infty}^{\infty} \int_{|y-x|\leq Mh} |\pi(x)||\pi(y)| \, dxdy
$$

$$
\leq C \, (n/d_n)^3 h^4,
$$

which implies $R_{3n} = o_{\mathrm{P}}(nh/d_n)$ due to $nh^2/d_n \to 0$. Similarly, we have $R_{2n} = o_{\mathrm{P}}(nh/d_n)$. Taking these estimates into (3.71), result (3.70) is proved under $|\eta_k| \leq A$.

Removal of $|\eta_k| \leq A$ is trivial. Indeed, by letting

$$
\widetilde{\eta}_k = \eta_k I(|\eta_k| \leq A) - \mathrm{E}\,[\eta_k I(|\eta_k| \leq A) \mid \mathcal{F}_{k-1}],
$$

(3.70) holds if we replace η_k by $\widetilde{\eta}_k$, and hence claim follows from:

$$
\int_{-\infty}^{\infty} \mathrm{E}\Big\{ \sum_{k=1}^{n} g[(y_{nk}+x)/h](\eta_k - \widetilde{\eta}_k)^2 \Big\}^2 \pi(x)dx
$$

$$
\leq C \sup_k \mathrm{E}\eta_k^2 \mathrm{I}(|\eta_k| \geq A) \sup_x \mathrm{E} \sum_{k=1}^{n} g^2[(y_{nk}+x)/h]
$$

$$
= o_{\mathrm{P}}(1),
$$

as $n \to \infty$ first and then $A \to \infty$. $\square$

We next consider another example. Let $\kappa \geq 0$ be a constant, $x_0 = 0$, $\gamma = 1 - \kappa/n$ and

$$
x_t = \gamma \, x_{t-1} + \xi_t, \quad t \geq 1,
$$

where $\xi_t = \sum_{k=0}^{\infty} \phi_k \epsilon_{t-k}$ with $\phi \equiv \sum_{k=0}^{\infty} \phi_k \neq 0$ and $\sum_{k=0}^{\infty} k^{1+\delta}|\phi_k| < \infty$ for some $\delta > 0$.

Corollary 3.6. *Suppose*

(i) $\{\eta_i, \mathcal{F}_i\}_{i\geq 1}$*, where* $\mathcal{F}_i = \sigma(\eta_i, ..., \eta_1; \epsilon_{i+1}, \epsilon_i, ...)$*, is a martingale difference such that* $\mathcal{F}_i$ *is independent of* $\epsilon_k, k \geq i+2$*,* $\mathrm{E}(\eta_i^2|\mathcal{F}_{i-1}) \to_{a.s.} 1$*,* $\sup_{i\geq 1} \mathrm{E}(|\eta_i|^4 \mid \mathcal{F}_{i-1}) < \infty$ *and* $\mathrm{E}(\eta_i\epsilon_i|\mathcal{F}_{i-1}) \to_{a.s.} C_0$ *for some constant* C_0.

(ii) $g(x)$ *is a bounded real function satisfying* $\int_{-\infty}^{\infty} |g(x)|dx < \infty$.

Then, for any h satisfying $nh^2 \to \infty$ and $h \log^2 n \to 0$, we have

$$\left(\frac{1}{d_n} \sum_{s,t=1,s\neq t}^{n} \eta_t \eta_s \, g\big[(x_t - x_s)/h\big], \; \frac{1}{d_n^2} \sum_{\substack{t,s=1 \\ t\neq s}}^{n} g^2\big[(x_t - x_s)/h\big] \right)$$

$$\to_D (V\,N, \; V^2), \tag{3.72}$$

where N is a standard normal variate which is independent of V^2,

$$d_n^2 = (\sqrt{2}\,\phi)^{-1} n^{3/2} h \int_{-\infty}^{\infty} g^2(x)dx,$$

and, with $G = \{G_t\}_{t\geq 0}$ where $G_t = \int_0^t e^{\kappa(t-s)} dB_s$,

$$V^2 = \lim_{\varepsilon \to 0} \frac{1}{2\varepsilon} \int_0^1 \int_0^1 \mathbf{1}\big(|G_x - G_y| < \varepsilon \big) dx dy$$

i.e., the self intersection local time generated by the process G (see Section 2.4 for a definition).

Proof.   Without loss of generality, assume $|\eta_k| \leq A$. This restriction can be removed by a similar argument as that in Wang and Phillips (2012, pages 754-756). Write $x_{nt} = 2y_{nt}/d_n$, where $y_{nt} = \sum_{i=1}^{t-1} \eta_i g\big[(x_t - x_i)/h\big]$. Then,

$$U_n \equiv \frac{1}{d_n} \sum_{s,t=1,s\neq t}^{n} \eta_t \eta_s \, g\big[(x_t - x_s)/h\big] = \sum_{k=2}^{n} x_{nk}\eta_k$$

is a martingale having a similar structure as that in S_n. To prove (3.72), it suffices to verify the conditions of Theorem 3.14 with $x_{nk} = 2y_{nk}/d_n$ and $Z_k = \epsilon_k$, which are based on the following results:
    if $h \log^2 n \to 0$, then

$$E y_{nk}^2 \leq C\,(1 + h\sqrt{k}), \quad \sum_{k=1}^{n} y_{nk} = O_{\mathrm{P}}(n^{5/4}h^{3/4}), \tag{3.73}$$

$$\frac{4}{d_n^2} \sum_{k=1}^{n} y_{nk}^2 - \frac{1}{d_n^2} \sum_{\substack{t,s=1 \\ t\neq s}}^{n} g^2\big[(x_t - x_s)/h\big] = o_{\mathrm{P}}(1); \tag{3.74}$$

if in addition $|\eta_k| \leq A$, then

$$E y_{nk}^4 \leq C\,(1 + h^3\,k^{3/2}). \tag{3.75}$$

The proofs of (3.73-3.75) are tedious, but elementary, which is given in Section 6 of Wang and Phillips (2012). Explicitly (3.74), (3.73) and (3.75) follow from (6.16), Propositions 6.2 and 6.5 of the paper, respectively.

Due to (3.75), it is routine to see that

$$\max_{1\le t\le n} |x_{nt}| \le \frac{2}{d_n} \Big(\sum_{k=1}^n y_{nk}^4 \Big)^{1/4}$$

$$= O_P \big[(n^{5/2}h^3)^{1/4}/(n^{3/2}h)^{1/2} \big] = o_P(1).$$

By (3.73) and recalling $\mathrm{E}\,(\epsilon_k \eta_k \mid \mathcal{F}_{k-1}) \to_{a.s.} C_0$, we have

$$\frac{1}{\sqrt{n}} \sum_{k=1}^n x_{nk} \, \mathrm{E}\,(\epsilon_k \eta_k \mid \mathcal{F}_{k-1})$$

$$= \frac{1}{\sqrt{n}} \sum_{k=1}^n x_{nk} \big[\mathrm{E}\,(\epsilon_k \eta_k \mid \mathcal{F}_{k-1}) - C_0 \big] + \frac{C_0}{d_n \sqrt{n}} \sum_{k=1}^n y_{nk}$$

$$= o_P(1),$$

where we have used (3.73) and the fact $\frac{1}{\sqrt{n}} \sum_{k=1}^n |x_{n,k}| = O_P(1)$, due to

$$\frac{1}{\sqrt{n}} \sum_{k=1}^n \mathrm{E}\,|x_{nk}| \le \frac{2}{\sqrt{n}d_n} \sum_{k=1}^n (Ey_{nk}^2)^{1/2}$$

$$\le \frac{C}{\sqrt{n}d_n} \sum_{k=1}^n (1 + h\sqrt{k})^{1/2} \le C.$$

Furthermore it follows from (3.74) and Theorem 2.25 that

$$\Big(\frac{1}{\sqrt{n}} \sum_{j=1}^{[nt]} \epsilon_j, \ \sum_{k=1}^n x_{nk}^2 \Big) = \Big(\frac{1}{\sqrt{n}} \sum_{j=1}^{[nt]} \epsilon_j, \ \frac{1}{d_n^2} \sum_{\substack{t,s=1 \\ t\ne s}}^n g^2\big[(x_t - x_s)/h\big] \Big) + o_P(1)$$

$$\Rightarrow \big\{ B_t, \ V^2 \big\},$$

on $D_{\mathrm{R}^2}[0,\infty)$. We have now verified all conditions of Theorem 3.14, and hence (3.72) follows. $\qquad\square$

3.4.3 Proof of Theorem 3.13

We assume $d = 1$ and only prove (3.61). The extensions to $d > 1$ and (3.63) are straightforward. See, e.g., Wang (2014), for more details.

Suppose first $\sigma^2 := f^2(Z)$ is a.s. bounded so that, for some $\lambda > 1$,

$$P(\sigma^2 \le \lambda) = 1. \tag{3.76}$$

Let $x_{nk}^2 = \mathrm{E}\,(X_{nk}^2 \mid \mathcal{F}_{n,k-1})$, $\sigma_k^2 = \sum_{j=1}^k x_{nj}^2$,

$$X_{nk}^* = X_{nk}\mathrm{I}(\sigma_k^2 \le 2\lambda), \quad S_n^* = \sum_{k=1}^{k_n} X_{nk}^* \ \text{ and } \ G_n^2 = \sum_{k=1}^{k_n} x_{nk}^2 \mathrm{I}(\sigma_k^2 \le 2\lambda).$$

Note that $S_n^* = S_n$ and $V_{nk_n}^2 = G_n^2$ on the set $V_{nk_n}^2 \leq 2\lambda$ and $P(V_{nk_n}^2 > 2\lambda) \to 0$ due to (3.60) and (3.76). It is readily seen that, for all $\alpha, \beta \in \mathbb{R}$,

$$E\left| e^{i\alpha S_n^* + i\beta G_n^2} - e^{i\alpha S_n + i\beta V_{nk_n}^2} \right| \leq 2P(V_{nk_n}^2 > 2\lambda) \to 0,$$

and hence (3.61) will follow if we prove

$$\mathrm{E}\, e^{i\alpha S_n^* + i\beta G_n^2} \to \mathrm{E}\, e^{(-\frac{\alpha^2}{2} + i\beta)\sigma^2}. \tag{3.77}$$

To prove (3.77), we start with some preliminaries. For $\beta_k \in \mathbb{R}, k = 1, 2, ..., N$, and $0 = t_0 < t_1 < ... < t_N < \infty$, let

(i) $m_n = \tau_n(t_N) \vee k_n$;
(ii) $\beta_k^* = \beta_j$ for $\tau_n(t_{j-1}) < k \leq \tau_n(t_j)$;
(iii) $Z_{nk} = 0$ for $k > \tau_n(T_N)$ and $X_{nk} = 0$ for $k > k_n$;
(iv) $Z_{nk}^* = Z_{nk}\mathrm{I}(s_k^2 \leq 2\Omega_{t_N})$, where $s_k^2 = \sum_{i=1}^k \mathrm{E}\left(Z_{ni}^2 \mid \mathcal{F}_{n,i-1} \right)$.

Define $V_n = \sum_{k=1}^N \beta_k[Z_n^*(t_k) - Z_n^*(t_{k-1})]$ where $Z_n^*(t) = \sum_{i=1}^{\tau_n(t)} Z_{ni}^*$, and

$$\Gamma_n = \sum_{k=1}^{m_n} \mathrm{E}\left\{ \left(e^{i\beta_k^* Z_{nk}^* + i\alpha X_{nk}^*} - 1 \right) \mid \mathcal{F}_{n,k-1} \right\}.$$

Since $P(s_{\tau_n(t_N)}^2 > 2\Omega_{t_N}) + P(V_{nk_n}^2 > 2\lambda) \to 0$ by (3.58), (3.60) and (3.76), and on the sets $s_{\tau_n(t_N)}^2 \leq 2\Omega_{t_N}$ and $V_{nk_n}^2 \leq 2\lambda$,

$$(V_n,\ G_n^2) = \left(\sum_{k=1}^N \beta_k\, [Z_n(t_k) - Z_n(t_{k-1})],\ V_{nk_n}^2 \right),$$

where $Z_n(t) = \sum_{i=1}^{\tau_n(t)} Z_{ni}$, it is readily seen from (3.60) that

$$(V_n,\ G_n^2) \to_D (V, \sigma^2), \tag{3.78}$$

where $V = \sum_{k=1}^N \beta_k(Z_{t_k} - Z_{t_{k-1}})$. Furthermore, we have the following two propositions under (3.76). The proofs of these propositions will be given in the end of this section.

Proposition 3.1. *For any $\alpha, \beta_k \in \mathbb{R}, k = 1, 2, ..., N$, and $0 = t_0 < t_1 < ... < t_N < \infty$, we have that $e^{|\Gamma_n|}$ is uniformly integrable and*

$$\{V_n,\ G_n^2,\ \Gamma_n\} \to_D \{V,\ \sigma^2,\ \Gamma\}, \tag{3.79}$$

where $\Gamma = -\frac{1}{2}\sum_{k=1}^N \beta_k^2(\Omega_{t_k} - \Omega_{t_{k-1}}) - \frac{1}{2}\alpha^2\sigma^2$.

Proposition 3.2. *For any $\alpha, \beta_k \in \mathbb{R}, k = 1, 2, ..., N$, and $0 = t_0 < t_1 < ... < t_N < \infty$, we have*

$$\mathrm{E}\, e^{i\alpha S_n^* + iV_n - \Gamma_n} \to 1. \tag{3.80}$$

We are now ready to prove (3.77). First note that $\{S_n^*\}_{n\geq 1}$ is tight, due to $\mathrm{E}S_n^{*2} = EG_n^2$ and

$$G_n^2 \leq V_{nk_n}^2 \mathrm{I}(V_{nk_n}^2 \leq 2\lambda) + 2\lambda \mathrm{I}(V_{nk_n}^2 > 2\lambda) \leq 4\lambda. \qquad (3.81)$$

This, together with (3.79), yields that

$$\{S_n^*,\ V_n,\ G_n^2,\ \Gamma_n\}_{n\geq 1}$$

is tight. Hence, for each $\{n'\} \subseteq \{n\}$, there exists a subsequence $\{n''\} \subseteq \{n'\}$ such that

$$\{S_{n''}^*,\ V_{n''},\ G_{n''}^2,\ \Gamma_{n''}\} \to_D \{S,\ V,\ \sigma^2,\ \Gamma\}, \qquad (3.82)$$

where S is a limit random variable of $S_{n''}^*$. Since $\mathrm{E}\,e^{i\alpha S_{n''}^* + iV_{n''} - \Gamma_{n''}}$ is uniformly integrable due to Proposition 3.1, we have

$$\mathrm{E}\,e^{i\alpha S_{n''}^* + iV_{n''} - \Gamma_{n''}} \to \mathrm{E}\,e^{i\alpha S + iV - \Gamma}.$$

This, together with (3.80), yields $\mathrm{E}\,e^{i\alpha S + iV - \Gamma} = 1$. It follows that

$$\mathrm{E}\,e^{i\alpha S + iV + \frac{1}{2}\alpha^2\sigma^2} = e^{-\frac{1}{2}\sum_{k=1}^N \beta_k^2(\Omega_{t_k} - \Omega_{t_{k-1}})} = \mathrm{E}\,e^{iV}. \qquad (3.83)$$

Due to the arbitrariness of $\beta_k \in \mathrm{R}, k = 1,...,N$ and $N \geq 1$, result (3.83) implies

$$E\left(\left[e^{i\alpha S + \frac{1}{2}\alpha^2\sigma^2} - 1\right] \mid \mathcal{F}\right) = 0,$$

where $\mathcal{F} = \sigma\{Z_s, 0 \leq s < \infty\}$. Hence, for all $\alpha, \beta \in \mathrm{R}$,

$$\mathrm{E}\left(e^{i\alpha S + i\beta\sigma^2} \mid \mathcal{F}\right) = e^{(-\frac{1}{2}\alpha^2 + i\beta)\sigma^2},$$

due to $\sigma^2 \in \mathcal{F}$. Now, by using (3.82) again, for each $\{n'\} \subseteq \{n\}$, there exists a subsequence $\{n''\} \subseteq \{n'\}$ such that

$$\lim_{n''\to\infty} \mathrm{E}\,e^{i\alpha S_{n''}^* + i\beta G_{n''}^2} = \mathrm{E}\,e^{i\alpha S + i\beta\sigma^2} = \mathrm{E}\,e^{(-\frac{\alpha^2}{2} + i\beta)\sigma^2}.$$

This establishes (3.77) as the limit does not depend on the choice of the subsequence, and also completes the proof in the special case when (3.76) holds.

It remains to remove the boundedness condition (3.76). To this end, for given $\delta > 0$, choose a continuous point λ of the distribution function of σ^2 such that $\mathrm{P}(\sigma^2 \geq \lambda) \leq \delta$. Let $\sigma_\lambda^2 = \sigma^2\mathrm{I}(\sigma^2 \leq \lambda) + \lambda\mathrm{I}(\sigma^2 > \lambda)$,

$$X'_{nk} = X_{nk}\mathrm{I}(\sigma_k^2 \leq \lambda),\ \ S'_n = \sum_{k=1}^{k_n} X'_{nk},\ \ G_n^{'2} = \sum_{k=1}^{k_n} x_{nk}^2\mathrm{I}(\sigma_k^2 \leq \lambda)$$

and $\widetilde{V}_{nk_n}^2 = V_{nk_n}^2 \mathrm{I}(V_{nk_n}^2 \leq \lambda) + \lambda\,\mathrm{I}(V_{nk_n}^2 > \lambda)$.

It is easy to see from (3.44) and (3.59) that, for any $\epsilon > 0$,

$$\sum_{i=1}^{k_n} \mathrm{E}\,[X_{ni}^{'2}\mathrm{I}(|X_{ni}'| \geq \epsilon) \mid \mathcal{F}_{n,i-1}] \to_\mathrm{P} 0,$$

and for all $t > 0$,

$$\Delta_n := \sum_{i=1}^{\tau_n(t)\wedge k_n} \mathrm{E}\,(X_{ni}'Z_{ni} \mid \mathcal{F}_{n,i-1})$$

$$= \sum_{i=1}^{\tau_n(t)\wedge k_n} \mathrm{I}(\sigma_i^2 \leq \lambda)\mathrm{E}\,(X_{ni}Z_{ni} \mid \mathcal{F}_{n,i-1}) = o_\mathrm{P}(1). \qquad (3.84)$$

Indeed, to see (3.84), by noting that $\Delta_n = \sum_{i=1}^{\tau_n(t)\wedge k_n} \mathrm{E}\,(X_{ni}Z_{ni} \mid \mathcal{F}_{n,i-1})$ on the set $V_{nk_n}^2 \leq \lambda_n$, where $\lambda \leq \lambda_n \to \infty$, the claim follows from (3.59) and $\mathrm{P}(V_{nk_n}^2 > \lambda_n) \to 0$.

On the other hand, by using the identity:

$$\sum_{k=1}^{k_n} x_{nk}^2 I(\sigma_k^2 \leq \lambda) = \sum_{k=1}^{k_n-1} x_{nk}^2 I(\sigma_k^2 \leq \lambda < \sigma_{k+1}^2) + V_{nk_n}^2 I(V_{nk_n}^2 \leq \lambda),$$

we have

$$\widetilde{V}_{nk_n}^2 = G_n^{'2} + \mathrm{I}(V_{nk_n}^2 > \lambda)\Big\{\lambda - \sum_{k=1}^{k_n} x_{nk}^2 I(\sigma_k^2 \leq \lambda)\Big\}$$

$$\leq G_n^{'2} + \max_{1\leq k\leq k_n} x_{nj}^2,$$

which, together with $G_n^{'2} \leq \widetilde{V}_{nk_n}^2$, implies $\widetilde{V}_{nk_n}^2 - G_n^{'2} = o_\mathrm{P}(1)$, since, due to (3.44),

$$\max_{1\leq k\leq k_n} x_{nk}^2 = \max_{1\leq k\leq k_n} \mathrm{E}\,(X_{nk}^2 \mid \mathcal{F}_{n,k-1}) = o_\mathrm{P}(1).$$

Hence, by (3.60) and the continuous mapping theorem,

$$\Big(\sum_{i=1}^{\tau_n(t)} Z_{ni},\ G_n^{'2}\Big) = \Big(\sum_{i=1}^{\tau_n(t)} Z_{ni},\ \widetilde{V}_{nk_n}^2\Big) + o_\mathrm{P}(1) \Rightarrow \big(Z_t, \sigma_\lambda^2\big).$$

Now, by recalling σ_λ^2 is a.s. bounded, the first part of the proof shows that

$$\mathrm{E}\,e^{i\alpha S_n' + i\beta G_n^{'2}} \to \mathrm{E}\,e^{(-\frac{\alpha^2}{2}+i\beta)\sigma_\lambda^2}.$$

This, together with the facts that $S'_n = S_n$ and $G'^2_n = V^2_{nk_n}$ on the set $V^2_{nk_n} \leq \lambda$ and $P(V^2_{nk_n} > \lambda) \to P(\sigma^2 > \lambda)$ due to (3.60), yields that, for all $\delta > 0$,

$$\limsup_{n \to \infty} \left| E\, e^{i\alpha\, S_n + i\beta V^2_{nk_n}} - E\, e^{(-\frac{\alpha^2}{2} + i\beta)\sigma^2} \right|$$

$$\leq \limsup_{n \to \infty} \left| E\, e^{i\alpha\, S'_n + i\beta G'^2_n} - E\, e^{(-\frac{\alpha^2}{2} + i\beta)\sigma^2_\lambda} \right|$$

$$+ \limsup_{n \to \infty} P(V^2_{nk_n} > \lambda) + P(\sigma^2 > \lambda)$$

$$\leq 4\delta.$$

It follows that $E\, e^{i\alpha\, S_n + i\beta V^2_{nk_n}} \to E\, e^{(-\frac{\alpha^2}{2} + i\beta)\sigma^2}$, which implies (3.61). This completes the proof of Theorem 3.13. $\square$

We now prove Propositions 3.1 and 3.2, starting with the following lemma. Recall $m_n = \tau_n(t_N) \vee k_n$ and write

$$Y_{nm} = \beta^*_m Z^*_{nm} + \alpha\, X^*_{nm}, \quad \eta_{nm} = E\left[\left(e^{i\,Y_{nm}} - 1 \right) \mid \mathcal{F}_{n,m-1} \right],$$

$$\Delta_{1n} = \max_{1 \leq m \leq m_n} E\left(Y^2_{nm} \mid \mathcal{F}_{n,m-1} \right), \quad \Delta_{2n} = \sum_{m=1}^{m_n} E\left(Y^2_{nm} \mid \mathcal{F}_{n,m-1} \right).$$

Lemma 3.3. *We have, by letting* $C_0 = |\alpha| + \max_{1 \leq k \leq N} |\beta_k|$,

$$|\eta_{nm}| \leq \frac{1}{2} E\left(Y^2_{nm} \mid \mathcal{F}_{n,m-1} \right), \quad \Delta_{1n} \leq \Delta_{2n}, \tag{3.85}$$

$$\Delta_{1n} \leq 2C_0 \Big[\max_{1 \leq k \leq \tau_n(t_N)} E\left(Z^2_{nk} \mid \mathcal{F}_{n,k-1} \right)$$

$$+ \max_{1 \leq k \leq k_n} E\left(X^2_{nk} \mid \mathcal{F}_{n,k-1} \right) \Big] = o_P(1), \tag{3.86}$$

$$\Delta_{2n} \leq 8C_0(\Omega_{t_N} + \lambda), \tag{3.87}$$

$$\Delta_{2n} = \sum_{m=1}^{N} \beta^2_m \left[\Omega_{t_m} - \Omega_{t_{m-1}} \right] + \alpha^2 G^2_n + o_P(1). \tag{3.88}$$

Proof. (3.85) is simple. (3.86) follows from (3.44) and (3.57). By noting that

$$\Delta_{2n} \leq 2C_0 \Big[\sum_{m=1}^{\tau_n(t_N)} I(s^2_m \leq 2\Omega_{t_N}) E\left(Z^2_{nm} \mid \mathcal{F}_{n,m-1} \right) + G^2_n \Big],$$

(3.87) follows from (3.81) and the fact that, as in the proof of (3.81),

$$\sum_{m=1}^{\tau_n(t_N)} I(s^2_m \leq 2\Omega_{t_N}) E\left(Z^2_{nm} \mid \mathcal{F}_{n,m-1} \right) \leq 4\Omega_{t_N}.$$

We next prove (3.88). In fact, similar to the proofs of (3.78) and (3.84), it follows from (3.58) and (3.59) that

$$
\sum_{m=1}^{\tau_n(t_N)} \beta_m^{*2} \mathrm{E}\left(Z_{nm}^{*2} \mid \mathcal{F}_{n,m-1}\right)
$$

$$
= \sum_{k=1}^{N} \beta_k^2 \sum_{m=\tau_n(t_{k-1})+1}^{\tau_n(t_k)} \mathrm{I}(s_m^2 \le 2\Omega_{t_N}) \mathrm{E}\left(Z_{nm}^2 \mid \mathcal{F}_{n,m-1}\right)
$$

$$
= \sum_{k=1}^{N} \beta_k^2 \sum_{m=\tau_n(t_{k-1})+1}^{\tau_n(t_k)} \mathrm{E}\left(Z_{nm}^2 \mid \mathcal{F}_{n,m-1}\right) + o_P(1)
$$

$$
= \sum_{k=1}^{N} \beta_k^2 \left[\Omega_{t_m} - \Omega_{t_{m-1}}\right] + o_P(1);
$$

$$
\Big| \sum_{m=1}^{m_n} \alpha\,\beta_m^* \mathrm{E}\left(X_{nm}^* Z_{nm}^* \mid \mathcal{F}_{n,m-1}\right) \Big|
$$

$$
= \Big| \sum_{m=1}^{m_n} \alpha\,\beta_m^* \mathrm{I}(\sigma_m^2 \le 2\lambda)\,\mathrm{I}(s_m^2 \le 2\Omega_{t_N}) \mathrm{E}\left(X_{nm} Z_{nm} \mid \mathcal{F}_{n,m-1}\right) \Big|
$$

$$
\le \sum_{k=1}^{N} |\alpha\,\beta_k| \Big| \sum_{m=\tau_n(t_{k-1})+1}^{\tau_n(t_k)} \mathrm{E}\left(X_{nm} Z_{nm} \mid \mathcal{F}_{n,m-1}\right) \Big| + o_P(1)
$$

$$
= o_P(1).
$$

We hence have

$$
\Delta_{2n} = \sum_{m=1}^{\tau_n(t_N)} \beta_m^{*2} \mathrm{E}\left(Z_{nm}^{*2} \mid \mathcal{F}_{n,m-1}\right)
$$

$$
+2 \sum_{m=1}^{m_n} \alpha\,\beta_m^* \mathrm{E}\left(X_{nm}^* Z_{nm}^* \mid \mathcal{F}_{n,m-1}\right) + \alpha^2 \sum_{m=1}^{k_n} \mathrm{E}\left(X_{nm}^{*2} \mid \mathcal{F}_{n,m-1}\right)
$$

$$
= \sum_{m=1}^{N} \beta_m^2 \left[\Omega_{t_m} - \Omega_{t_{m-1}}\right] + \alpha^2 G_n^2 + o_P(1),
$$

which yields (3.88).   □

Proof of Proposition 3.1. Recall (3.44) and (3.57). There exists a sequence of constants $0 < \epsilon_n \to 0$ such that

$$I_{1n} := \sum_{i=1}^{k_n} E\left[X_{ni}^2 I(|X_{ni}| \geq \epsilon_n) \mid \mathcal{F}_{n,i-1}\right] \to_P 0, \qquad (3.89)$$

$$I_{2n} := \sum_{i=1}^{\tau_n(t_N)} E\left[Z_{ni}^2 I(|Z_{ni}| \geq \epsilon_n) \mid \mathcal{F}_{n,i-1}\right] \to_P 0. \qquad (3.90)$$

Write $\widehat{\Omega}_{nm} = \Omega - \Omega_{nm}$, where

$$\Omega_{nm} = \{|Z_{nm}| \leq \epsilon_n \text{ and } |X_{nk}| \leq \epsilon_n\}.$$

Using the Taylor's expansion of e^{ix}: $|e^{ix} - 1 - ix| \leq x^2/2$ on $\widehat{\Omega}_{nm}$ and $|e^{ix} - 1 - ix + x^2/2| \leq |x|^3/6$ on Ω_{nm}, we have

$$E\left\{(e^{iY_{nm}} - 1 - iY_{nm}) \mid \mathcal{F}_{n,m-1}\right\}$$

$$= -\frac{1}{2}E\left[Y_{nm}^2 I(\Omega_{nm}) \mid \mathcal{F}_{n,m-1}\right] + E\left\{(e^{iY_{nm}} - 1 - iY_{nm})I(\widehat{\Omega}_{nm}) \mid \mathcal{F}_{n,m-1}\right\}$$

$$+ E\left\{(e^{iY_{nm}} - 1 - iY_{nm} + \frac{1}{2}Y_{nm}^2)I(\Omega_{nm}) \mid \mathcal{F}_{n,m-1}\right\}$$

$$= -\frac{1}{2}E\left[Y_{nm}^2 \mid \mathcal{F}_{n,m-1}\right] + R_{nm},$$

where $|R_{nm}| \leq E\left[Y_{nm}^2 I(\widehat{\Omega}_{nm}) \mid \mathcal{F}_{n,m-1}\right] + E\left[|Y_{nm}|^3 I(\Omega_{nm}) \mid \mathcal{F}_{n,m-1}\right]$. Note that $|\alpha| + \max_{1 \leq k \leq N} |\beta_k| \leq C$ and

$$E\left[Y_{nm}^2 I(\widehat{\Omega}_{nm}) \mid \mathcal{F}_{n,m-1}\right] \leq CE\left[Z_{nm}^2 I(|Z_{nm}| \geq \epsilon_n) \mid \mathcal{F}_{n,m-1}\right]$$

$$+ CE\left[X_{nm}^2 I(|X_{nm}| \geq \epsilon_n) \mid \mathcal{F}_{n,m-1}\right].$$

It follows from $|Y_{nm}| \leq C\epsilon_n$ on Ω_{nm}, (3.89)-(3.90) and (3.87) that

$$\sum_{m=1}^{m_n} |R_{nm}| \leq C\epsilon_n \sum_{m=1}^{m_n} E\left(Y_{nm}^2 \mid \mathcal{F}_{n,m-1}\right) + CI_{1n} + CI_{2n}$$

$$= o_P(1).$$

Now, by recalling (3.88) and $E(Y_{nm} \mid \mathcal{F}_{n,m-1}) = 0$, we obtain

$$\Gamma_n = \sum_{m=1}^{m_n} E\left\{(e^{iY_{nm}} - 1 - iY_{nm}) \mid \mathcal{F}_{n,m-1}\right\}$$

$$= -\frac{1}{2}\sum_{m=1}^{m_n} E\left[Y_{nm}^2 \mid \mathcal{F}_{n,m-1}\right] + o_P(1)$$

$$= -\frac{1}{2}\sum_{m=1}^{N} \beta_m^2 \left[\Omega_{t_m} - \Omega_{t_{m-1}}\right] - \frac{\alpha^2}{2}G_n^2 + o_P(1). \qquad (3.91)$$

Due to (3.78) and (3.91), for any $\alpha_j \in R$, $j = 1, 2, 3$, we have

$$\alpha_1 V_n + \alpha_2 G_n^2 + \alpha_3 \Gamma_n$$

$$= \alpha_1 V_n + \left(\alpha_2 - \alpha_3 \alpha^2/2\right) G_n^2 - \frac{\alpha_3}{2} \sum_{m=1}^{N} \beta_m^2 (\Omega_{t_m} - \Omega_{t_{m-1}}) + o_P(1)$$

$$\to_D \alpha_1 V_1 + (\alpha_2 - \alpha_3 \alpha^2/2) \sigma^2 - \frac{\alpha_3}{2} \sum_{m=1}^{N} \beta_m^2 (\Omega_{t_m} - \Omega_{t_{m-1}})$$

$$= \alpha_1 V_1 + \alpha_2 \sigma^2 + \alpha_3 \Gamma.$$

This proves (3.79).

Note that $\left| E\left\{(e^{iY_{nm}} - 1) \mid \mathcal{F}_{n,m-1}\right\}\right| \leq \frac{1}{2} E\left(Y_{nm}^2 \mid \mathcal{F}_{n,m-1}\right)$. The uniform integrability of $e^{|\Gamma_n|}$ is obvious by (3.87) of Lemma 3.3. The proof of Proposition 3.1 is now completed. □

Proof of Proposition 3.2. Recalling that

$$V_n = \sum_{k=1}^{N} \beta_k [Z_n^*(t_k) - Z_n^*(t_{k-1})] = \sum_{k=1}^{\tau_n(t_N)} \beta_k^* Z_{nk}^*,$$

we have

$$\left| E\, e^{i\,\alpha S_n^* + iV_n - \Gamma_n} - 1\right| = \left| E\, e^{i\,\sum_{k=1}^{m_n} Y_{nk} - \sum_{k=1}^{m_n} \eta_{nk}} - 1\right|$$

$$\leq \left| E\left[e^{i\,\sum_{k=1}^{m_n-1} Y_{nk} - \sum_{k=1}^{m_n} \eta_{nk}} \left(e^{i\,Y_{n,m_n}} - e^{\eta_{n,m_n}}\right)\right]\right|$$

$$+ \left| E\, e^{i\,\sum_{k=1}^{m_n-1} Y_{nk} - \sum_{k=1}^{m_n-1} \eta_{nk}} - 1\right|$$

$$\leq \dots \leq$$

$$\leq \sum_{m=2}^{m_n} \left| E\left[e^{i\,\sum_{k=1}^{m-1} Y_{nk} - \sum_{k=1}^{m} \eta_{nk}} \left(e^{i\,Y_{nm}} - e^{\eta_{nm}}\right)\right]\right|$$

$$+ \left| E\, e^{i\,Y_{n1} - \eta_{n1}} - 1\right|$$

$$:= I_{3n} + I_{4n}. \tag{3.92}$$

It is readily seen that

$$I_{4n} \leq E\left| e^{-\eta_{n1}} - 1\right| + \left| E\, e^{iY_{n1}} - 1\right|$$

$$\leq E\left(\Delta_{1n} e^{\Delta_{1n}}\right) + E\Delta_{1n} \to 0,$$

as $\Delta_{1n} = o_P(1)$ and Δ_{1n} is bounded by a constant due to Lemma 3.3.

To estimate I_{3n}, write

$$u_{n,m} = e^{i\,\sum_{k=1}^{m-1} Y_{nk} - \sum_{k=1}^{m} \eta_{nk}},$$

$$v_{n,m} = e^{i\,Y_{nm}} - e^{\eta_{nm}}.$$

Using Lemma 3.3 again, we have

$$|u_{n,m}| \le e^{\sum_{k=1}^{m} |\eta_{nk}|} \le e^{\sum_{k=1}^{m} \mathrm{E}(Y_{nk}^2 | \mathcal{F}_{n,k-1})} \le e^{\Delta_{2n}},$$

and, due to $|e^x - 1 - x| \le |x|^2 e^{|x|}$,

$$|\mathrm{E}(v_{n,m} | \mathcal{F}_{n,m-1})| = |\eta_{nm} + 1 - e^{\eta_{nm}}| \le |\eta_{nm}|^2 e^{|\eta_{nm}|}$$
$$\le \Delta_{1n} e^{\Delta_{2n}} \mathrm{E}(Y_{nm}^2 | \mathcal{F}_{n,m-1}).$$

It follows that

$$\Delta_{3n} := \sum_{m=2}^{m_n} \left[|u_{n,m}| \, |\mathrm{E}(v_{n,m} | \mathcal{F}_{n,m-1})| \right]$$

$$\le \Delta_{1n} e^{2\Delta_{2n}} \sum_{k=2}^{m_n} \mathrm{E}(Y_{nk}^2 | \mathcal{F}_{n,k-1}) = \Delta_{1n} \Delta_{2n} e^{2\Delta_{2n}}.$$

Now, by recalling $\mathcal{F}_{n,m-1} \subseteq \mathcal{F}_{n,m}$ for any $n \ge m \ge 1$ and $n \ge 1$, $\Delta_{1n} = o_P(1)$, Δ_{1n} and Δ_{2n} are bounded by a constant due to Lemma 3.3, we obtain $I_{3n} \le E\Delta_{3n} \to 0$. Taking these facts into (3.92), we prove (3.80). $\square$

3.5 Convergence to a mixture of normal distributions: beyond martingale arrays

Let $\eta_i \equiv (\epsilon_i, \nu_i)'$, $i \in \mathbb{Z}$, be a sequence of i.i.d. random vectors with $\mathrm{E}\,\eta_0 = 0$, $\mathrm{E}(\eta_0 \eta_0') = \Sigma$ and $\mathrm{E}\|\eta_0\|^4 < \infty$. Suppose $\mathrm{E}\,\epsilon_0^2 = 1$ and $\lim_{|t| \to \infty} |t|^{\eta} |\mathrm{E}\, e^{it\epsilon_0}| < \infty$ for some $\eta > 0$. We make use of the following assumptions in this section.

Assumption 3.1. $y_{nk} = \gamma\, y_{n,k-1} + \xi_k$, where $y_{n0} = 0$, $\gamma = 1 - \tau/n$ with $\tau \ge 0$ being a constant and $\{\xi_j, j \ge 1\}$ is a linear process defined as in Section 2.3.3, i.e., $\xi_j = \sum_{k=0}^{\infty} \phi_k\, \epsilon_{j-k}$, where the coefficients ϕ_k, $k \ge 0$, satisfy one of the following conditions:

LM. $\phi_k \sim k^{-\mu} \rho(k)$, where $1/2 < \mu < 1$ and $\rho(k)$ is a function slowly varying at ∞.

SM. $\sum_{k=0}^{\infty} |\phi_k| < \infty$ and $\phi \equiv \sum_{k=0}^{\infty} \phi_k \neq 0$.

Assumption 3.2. $u_k = \sum_{j=0}^{\infty} \psi_j\, \eta_{k-j}$, where the coefficient vector $\psi_k = (\psi_{k1}, \psi_{k2})$ satisfies $\sum_{k=0}^{\infty} k^{1/4}(|\psi_{1k}| + |\psi_{2k}|) < \infty$ and $\sum_{k=0}^{\infty} \psi_k \neq 0$.

Assumption 3.3. (i) $\widetilde{g}(x) = \sup_{0 \le s \le 1} |g(s,x)|$ is bounded and integrable on R. (ii) $G(s) = \int_{-\infty}^{\infty} g^2(s,x)dx$ is Riemann integrable on $[0,1]$. (iii) $\int_{-\infty}^{\infty} \sup_{0 \le s \le 1} |\hat{g}(s,t)| dt < \infty$, where $\hat{g}(s,t) = \int_{-\infty}^{\infty} e^{itx} g(s,x)dx$.

Write $Z = \{Z_t\}_{t \geq 0}$, where Z_t is defined as in (2.82), denote by $L_Z(u, s)$ the local time process of Z and set $d_n^2 = var(\sum_{j=1}^{n} \xi_j)$. Recalling Theorem 2.21, for any fixed x and any $h \to 0$ satisfying $nh/d_n \to \infty$, we have

$$\left\{ \frac{1}{\sqrt{n}} \sum_{j=1}^{[nt]} \epsilon_j, \ \frac{1}{\sqrt{n}} \sum_{j=1}^{[nt]} \epsilon_{-j}, \ \frac{d_n}{nh} \sum_{j=1}^{n} g^2[j/n, (y_{nj} + x)/h] \right\}$$

$$\Rightarrow \left\{ B_{1t}, B_{2t}, \int_0^1 G(s)dL_Z(ds, 0) \right\} \tag{3.93}$$

on $D_{\mathrm{R}^3}[0, \infty)$, where $B = (B_{1t}, B_{2t})_{t \geq 0}$ is a standard 2-dimensional Brownian motion and $L_Z(u, s)$ is a functional of B.

Theorem 3.16. *Suppose Assumptions 3.1-3.3 hold. Then, for fixed $x \in \mathrm{R}$ and any $h \to 0$ (h$\log n \to 0$ under* **SM***) satisfying $nh/d_n \to \infty$, we have*

$$\left\{ \left(\frac{d_n}{nh}\right)^{1/2} \sum_{j=1}^{n} g[j/n, (y_{nj} + x)/h] \, u_j, \ \frac{d_n}{nh} \sum_{j=1}^{n} g^2[j/n, (y_{nj} + x)/h] \right\}$$

$$\to_D \left\{ (\mathrm{E}\, u_0^2)^{1/2} \, N \, V, \ V^2 \right\}, \tag{3.94}$$

where N is a normal variate independent of $V^2 = \int_0^1 G(s)dL_Z(ds, 0)$.

Remark 3.7. Assumption 3.2 allows the equation error u_t to be cross correlated with the y_{ns} for all $s \leq t$, thereby inducing endogeneity. By a simple calculation

$$\mathrm{E}\, u_0^2 = \sum_{k=0}^{\infty} \psi_k \Sigma \psi_k', \quad \text{where } \sum = \begin{pmatrix} 1 & \mathrm{E}\epsilon_0 \nu_0 \\ \mathrm{E}\epsilon_0 \nu_0 & \mathrm{E}\nu_0^2 \end{pmatrix}.$$

We may have $cov(u_t, y_{nt}) \neq 0$ under Assumptions 3.1 and 3.2, which differs from previous section where we often assumed that y_{nt} is adapted to $\mathcal{F}_{t-1}$ and that $(u_t, \mathcal{F}_t)$ forms a martingale difference sequence. In that case, $cov(u_t, y_{nt}) = \mathrm{E}\,[y_{nt}\mathbb{E}(u_t|\mathcal{F}_{t-1})] = 0$.

Remark 3.8. Recall (2.86) and note that, in the proof of Corollary 2.3, we in fact established

$$\frac{d_n}{nh} \sum_{j=1}^{n} g^2[j/n, (y_{nj} + x)/h](u_j^2 - \mathrm{E}\, u_0^2) = o_P(1).$$

This, together with (3.94), implies

$$\frac{1}{V_n} \sum_{j=1}^{n} g[j/n, (y_{nj} + x)/h] \, u_j \to_D N(0, 1), \tag{3.95}$$

where $V_n^2 = \frac{d_n}{nh} \sum_{j=1}^{n} g^2[j/n, (y_{nj} + x)/h]u_k^2$.

Remark 3.9. Results (3.94) and (3.95) still hold if we replace u_k by $\sigma_k = \sigma(\eta_k, \eta_{k-1}, ..., \eta_{k-m_0})$ with $E\sigma_1 = 0$ and $E\sigma_1^4 < \infty$, where m_0 is fixed and $\sigma(...)$ is an arbitrary measurable function. For a result in this regard, see Wang and Phillips (2009b).

Remark 3.10. When $h = 1$, similar arguments show

$$\frac{d_n}{n} \sum_{k=1}^n \left| g(k/n, y_{nk}) \right| (1 + |u_k|) = O_P(1), \qquad (3.96)$$

$$\left(\frac{d_n}{n}\right)^{1/2} \sum_{k=1}^n g(k/n, y_{nk}) u_k$$

$$= \begin{cases} O_P(1), & \text{under } \mathbf{LM}, \\ O_P(\log^{1/2} n), & \text{under } \mathbf{SM}. \end{cases} \qquad (3.97)$$

See Wang and Phillips (2014). The establishment of limit distribution is very challenge. We refer to an unpublished manuscript Jeganathan (2008).

Remark 3.11. As noticed in Remark 3.6, functional limit theorem for $S_n(x) = \left(\frac{d_n}{nh}\right)^{1/2} \sum_{j=1}^n g\left[j/n, (y_{nj} + x)/h\right] u_j$ on $C[-A, A]$ does not exist since $S_n(x)$ is not tight. As in Theorem 3.15, however, we have the following result, which is useful in nonlinear cointegrating regression.

Theorem 3.17. *Let $|\pi(x)|$ be a bounded integrable real function and $g(x)$ has compact support satisfying $|g(x) - g(y)| \le C|x - y|$ for all $x, y \in R$. Suppose Assumptions 3.1-3.2 hold. Then, for any h satisfying $nh/d_n \to \infty$ and $nh^2 \log n/d_n \to 0$, we have*

$$\frac{d_n}{nh} \int_{-\infty}^{\infty} \left\{ \sum_{k=1}^n g[(y_{nk} + x)/h] u_k \right\}^2 \pi(x) dx \to_D \tau_0 L_Z(1, 0), \qquad (3.98)$$

where $\tau_0 = E u_0^2 \int_{-\infty}^{\infty} g^2(x) dx \int_{-\infty}^{\infty} \pi(x) dx$.

Proof. We only provide an outline of the proof. More details can be found in Wang and Phillips (2015). Let

$$\lambda_k = \eta_k I(\|\eta_k\| \le A) - E\eta_k I(\|\eta_k\| \le A)$$

and $\Lambda_k = \sum_{j=0}^{m_0} \psi_j \lambda_{k-j}$, where m_0 and A are fixed constants. By using Lemma 2.2, tedious but routine calculations show that (3.98) will follow if we prove

$$\frac{d_n}{nh} \int_{-\infty}^{\infty} \left\{ \sum_{k=1}^n g[(y_{nk} + x)/h] \Lambda_k \right\}^2 \pi(x) dx \to_D \widetilde{\tau}_0 L_Z(1, 0), \qquad (3.99)$$

where $\widetilde{\tau}_0 = \sum_{j=0}^{m_0} \psi_j^2 \, \mathrm{E}\lambda_1^2$. To establish (3.99), the key step is to show

$$R_n \equiv \sum_{1 \leq k < j \leq n} \Lambda_k \Lambda_j \, z_{kj} = o_P(nh/d_n),$$

where $z_{kj} = \int_{-\infty}^{\infty} g[(y_{nk} + x)/h]g[(y_{nj} + x)/h] \, \pi(x)dx$.

We may write

$$\mathrm{E}R_n^2 \leq c_1 \, I_1(n) + c_2 \, I_2(n) + c_3 \, I_3(n), \tag{3.100}$$

for some constants $c_1, c_2, c_3 > 0$, where

$$I_1(n) = \sum_{1 \leq k < j \leq n} \mathrm{E}\left(z_{kj}^2\right), \quad I_2(n) = \sum_{1 \leq k \neq j \neq l \leq n} \mathrm{E}\left\{z_{kj} \, z_{kl}\right\},$$

$$I_3(n) = \sum_{1 \leq k < j < l < m \leq n} \left|\mathrm{E}\left\{\Lambda_k \, \Lambda_j \, \Lambda_l \, \Lambda_m \, z_{kj} \, z_{lm}\right\}\right|.$$

As in the proof of Theorem 3.15,

$$I_1(n) + I_2(n) \leq C[(n/d_n)^3 h^4 + (n/d_n)^2 h^3].$$

Noting that, by Lemma 2.1,

$$\left|\mathrm{E}\left(\Lambda_m g[(y_{ml} + y)/h] \mid \mathcal{F}_l\right)\right| \leq \begin{cases} Ch/d_{m-l}, & \text{if } m - l \leq m_0 + 1, \\ Ch/d_{m-l}^2, & \text{if } m - l > m_0 + 1, \end{cases}$$

we have

$$A_{ln} := \sup_x \sum_{m=l+1}^{n} \left|\mathrm{E}\left(\Lambda_m \, g[(y_{nm} + y)/h] \mid \mathcal{F}_l\right)\right|$$

$$\leq C \, h \sum_{m=l+1}^{l+m_0} d_{m-l} + C \, h \sum_{m=l+m_0}^{m} d_{m-l}^{-2} \leq C h \log n.$$

This result, together with $|\Lambda_k| \leq C$ and Lemma 2.5, implies that

$$I_3(n) \leq C \int_{-\infty}^{\infty} \int_{-\infty}^{\infty} \sum_{1 \leq k < j < l < n} \mathrm{E}\left\{\left|g[(y_{nk} + x)/h]g[(y_{nj} + x)/h]\right.\right.$$

$$\left.\left. g[(y_{nl} + y)/h] \, A_{ln}\right|\right\} |\pi(x)| \, |\pi(y)| \, dxdy,$$

$$\leq C h \log n \sup_x \mathrm{E}\left(\sum_{k=1}^{n} g[(y_{nk} + x)/h]\right)^3$$

$$\leq C \, (n/d_n)^3 h^4 \, \log n.$$

Taking these estimates into (3.100), result $R_n = o_P(nh/d_n)$ is proved due to $nh^2 \log n/d_n \to 0$. $\qquad\qquad\qquad\qquad\qquad\qquad\qquad\qquad\qquad\qquad \Box$

3.5.1 Proof of Theorem 3.16

Let $v_{k,m_0} = \sum_{j=0}^{m_0} \psi_j \eta_{k-j}$. The result (2.97) in Proposition 2.2 establishes that, as $n \to \infty$ first and then $m_0 \to \infty$,

$$\sum_{k=1}^{n} [u_k - v_{k,m_0}] g[\frac{k}{n}, (y_{nk} + x)/h] = o_P[(nh/d_n)^{1/2}].$$

This implies that (3.94) holds if we prove the following: for any $m_0 > 0$,

$$\left\{ (\frac{d_n}{nh})^{1/2} \sum_{k=1}^{n} v_{k,m_0} g[\frac{k}{n}, (y_{nk} + x)/h], \ \frac{d_n}{nh} \sum_{k=1}^{n} g^2[\frac{k}{n}, (y_{nk} + x)/h] \right\}$$
$$\to_D \{\tau_0 N V, \ V^2\} \tag{3.101}$$

where $\tau_0^2 = E u_0^2(m_0)$.

In order to prove (3.101), we use the following lemma.

Lemma 3.4. *Suppose that $\{\mathcal{F}_t\}_{t \geq 0}$ is an increasing sequence of σ-fields, $q(t)$ is a process that is $\mathcal{F}_t$-measurable for each t and continuous with probability 1, $Eq^2(t) < \infty$ and $q(0) = 0$. Let $\psi(t), t \geq 0$, be a process that is nondecreasing and continuous with probability 1 and satisfies $\psi(0) = 0$ and $E\psi^2(t) < \infty$. Let ξ be a random variable which is $\mathcal{F}_t$-measurable for each $t \geq 0$. If, for any $\gamma_j \geq 0, j = 1, 2, ..., r$, and any $0 \leq s < t \leq t_0 < t_1 < ... < t_r < \infty$,*

$$E\left(e^{-\sum_{j=1}^{r} \gamma_j [\psi(t_j) - \psi(t_{j-1})]} [q(t) - q(s)] \mid \mathcal{F}_s \right) = 0, \quad a.s.,$$
$$E\left(e^{-\sum_{j=1}^{r} \gamma_j [\psi(t_j) - \psi(t_{j-1})]} \{ [q(t) - q(s)]^2 - [\psi(t) - \psi(s)] \} \mid \mathcal{F}_s \right) = 0, \quad a.s.$$

then the finite-dimensional distributions of the process $(q(t), \xi)_{t \geq 0}$ coincide with those of the process $(B_{\psi(t)}, \xi)_{t \geq 0}$, where $B = \{B_s\}_{s \geq 0}$ is a Brownian motion with $EB_s^2 = s$ independent of $\psi(t)$.

Lemma 3.4 is an extension of the well-known Lévy's characterization theorem for Brownian motion processes. A proof can be found in Theorem 3.1 of Borodin and Ibragimov (1995, page 14). See, also, Wang and Phillips (2009b).

We next prove (3.101). Technical details of some subsidiary results that are used in this proof are given in the next section. Set, for $0 \le t \le 1$,

$$S_n(t) = \Big(\frac{d_n}{nh}\Big)^{1/2} \sum_{k=1}^{[nt]} v_{k,m_0}\, g[k/n, (y_{nk} + x)/h],$$

$$\psi_n(t) = \frac{d_n}{nh} \sum_{k=1}^{[nt]} v_{k,m_0}^2\, g^2[k/n, (y_{nk} + x)/h],$$

$$Z_{n1} = \frac{d_n}{nh} \sum_{k=1}^{n} g^2[k/n, (y_{nk} + x)/h],$$

and for all $\alpha_i, \beta_j \in \mathrm{R}$, $0 \le s_0 < s_1 < ... < s_m < \infty$ and $0 \le t_0 < t_1 < ... < t_l < \infty$,

$$Z_{n2} = \sum_{i=1}^{l} \alpha_i \big[\zeta_{n1}(t_i) - \zeta_{n1}(t_{i-1})\big] + \sum_{i=1}^{m} \beta_i \big[\zeta_{n2}(s_i) - \zeta_{n2}(s_{i-1})\big],$$

where $\zeta_{n1}(t) = \frac{1}{\sqrt{n}} \sum_{j=1}^{[nt]} \epsilon_j$ and $\zeta_{n2}(t) = \frac{1}{\sqrt{n}} \sum_{j=1}^{[nt]} \epsilon_{-j}$.

Corollary 2.3 yields $\psi_n(t) \Rightarrow \tau_0^2 \psi(t)$ on $D[0,1]$, where $\psi(t) := \int_0^t G(s)dL_Z(ds, 0)$. By Proposition 3.4, $\{S_n(t)\}_{n\ge 1}$ is tight on $D[0,1]$. These facts, together with (3.93), imply that

$$\{S_n(t),\ \psi_n(t),\ Z_{n1},\ Z_{n2}\}_{n\ge 1}$$

is tight on $D_{\mathrm{R}^4}[0,1]$. Hence, for each $\{n'\} \subseteq \{n\}$, there exists a subsequence $\{n''\} \subseteq \{n'\}$ such that

$$\{S_{n''}(t),\ \psi_{n''}(t),\ Z_{n''1},\ Z_{n''2}\} \Rightarrow \{\eta(t),\ \tau_0^2\psi(t),\ \psi(1),\ Z_2\}. \quad (3.102)$$

on $D_{\mathrm{R}^4}[0,1]$, where

$$Z_2 = \sum_{i=1}^{l} \alpha_i \big(B_{1t_i} - B_{1,t_{i-1}}\big) + \sum_{i=1}^{m} \beta_i \big(B_{2s_i} - B_{2,s_{i-1}}\big)$$

and $\eta(t)$ is a process continuous with probability one, since, by recalling $\mathrm{E}\,||\eta_0||^4 < \infty$, it follows from (2.94) that

$$\Big(\frac{d_n}{nh}\Big)^{1/2} \max_{1\le k\le n} |v_{k,m_0}\, g[k/n, (y_{nk} + x)/h]|$$

$$= o(1) \sum_{j=0}^{m_0} \Big(\frac{d_n}{nh} \sum_{k=1}^{n} ||\eta_{k-j}||^4\, |g[k/n, (y_{nk} + x)/h]|\Big)^{1/4}$$

$$= o_{\mathrm{P}}(1). \quad (3.103)$$

Let $Z_{n3} = \sum_{i=1}^{v} \gamma_i \big[S_n(t_i) - S_n(t_{i-1})\big]$ and $Z_3 = \sum_{i=1}^{v} \gamma_i \big[\eta(t_i) - \eta(t_{i-1})\big]$, where $\gamma_j \in \mathrm{R}$ and $0 \le t_0 < t_1 < ... < t_v \le s$. Since, for each $0 \le t \le 1$,

$S_n(t), S_n^2(t)$ and $\psi_n(t)$ are uniformly integrable (see Proposition 3.3), it follows from (3.102) and Proposition 3.5 that, for any $s < t$,

$$\mathrm{E}\, e^{i(Z_3 + Z_2)} \big[\eta(t) - \eta(s)\big]$$
$$= \lim_{n'' \to \infty} \mathrm{E}\, e^{i(Z_{n''3} + Z_{n''2})} [S_{n''}(t) - S_{n''}(s)] = 0, \qquad (3.104)$$
$$\mathrm{E}\, e^{i(Z_3 + Z_2)} \big\{ [\eta(t) - \eta(s)]^2 - [\psi(t) - \psi(s)] \big\}$$
$$= \lim_{n'' \to \infty} \mathrm{E}\, e^{i(Z_{n''3} + Z_{n''2})} \big\{ [S_{n''}(t) - S_{n''}(s)]^2 - [\psi_{n''}(t) - \psi_{n''}(s)] \big\}$$
$$= 0. \qquad (3.105)$$

See, e.g., Theorem 5.4 of Billingsley (1968). Let $\mathcal{F}_s = \sigma\{B_{1t}, 0 \le t \le 1; B_{2t}, 0 \le t < \infty, \eta(t), 0 \le t \le s\}$. Results (3.104) and (3.105) imply that, for any $0 \le s < t \le 1$,

$$\mathrm{E}\Big(\big[\eta(t) - \eta(s)\big] \mid \mathcal{F}_s \Big) = 0, \quad a.s.,$$
$$\mathrm{E}\Big(\big\{ [\eta(t) - \eta(s)]^2 - \tau_0^2 [\psi(t) - \psi(s)] \big\} \mid \mathcal{F}_s \Big) = 0, \quad a.s.$$

Note that $\mathcal{F}_s \uparrow$, $\eta(s)$ is $\mathcal{F}_s$-measurable for each $0 \le s \le 1$ and $\psi(t)$ (for any fixed $t \in [0,1]$) is $\mathcal{F}_s$-measurable for each $0 \le s \le 1$. It follows from Lemma 3.4 that the finite-dimensional distributions of $(\eta(t), \psi(1))$ coincide with those of $\{\tau_0 N \, \psi^{1/2}(t), \, \psi(1)\}$, where N is normal variate independent of $\psi(t)$. The result (3.101) therefore follows from (3.102), since $\eta(t)$ does not depend on the choice of the subsequence. $\qquad \square$

3.5.2 *Some subsidiary propositions*

In this section we will prove the following propositions required in the proof of Theorem 3.16. Notation will be the same as in the previous section except when explicitly mentioned.

Proposition 3.3. *For any fixed $0 \le t \le 1$, $S_n(t)$, $S_n^2(t)$ and $\psi_n(t)$, $n \ge 1$, are uniformly integrable.*

Proposition 3.4. $\{S_n(t)\}_{n \ge 1}$ *is tight on $D[0,1]$.*

Proposition 3.5. *For any $0 \le s < t \le 1$, we have*

$$\lim_{n \to \infty} \mathrm{E}\, e^{i(Z_{n3} + Z_{n2})} [S_n(t) - S_n(s)] = 0, \quad (3.106)$$
$$\lim_{n \to \infty} \mathrm{E}\, e^{i(Z_{n3} + Z_{n2})} \big\{ [S_n(t) - S_n(s)]^2 - [\psi_n(t) - \psi_n(s)] \big\} = 0. \quad (3.107)$$

To prove Propositions 3.3-3.5, we start with some preliminaries. Without loss of generality, we assume $\phi_0 \neq 0$. Otherwise, we only need some routine modifications.

As in (2.65), for any $s < k$, we may write

$$y_{nk} = x_{sk} + D_{nks}, \qquad (3.108)$$

where $x_{sk} = \sum_{i=s+1}^{k} \epsilon_i a_{k-i}$ with $a_j = \sum_{u=0}^{j} \gamma^{j-u} \phi_u$ and D_{nks} depends only on $\epsilon_k, \epsilon_{k-1}, \dots$ There exist $A_0 > 0$, $0 < \delta_1 < \delta_2 < \infty$ such that $\delta_1 |a_j| \leq |a_k| \leq \delta_2 |a_j|$,

$$\delta_1 / \sqrt{j} \leq |a_k| / d_j \leq \delta_2 / \sqrt{j} \quad \text{and} \quad d_1 d_j^2 \leq \sum_{i=1}^{j} a_i^2 \leq \delta_2 d_j^2,$$

for all $j \geq A_0$ and $j/2 \leq k \leq j$. Similar arguments as in the proof of Theorem 2.21 yield that there exist $\gamma > 0$ and $A_0 > 0$ such that, for all $A_0 \leq k - s \leq n$,

$$\left| \mathbb{E} \, e^{it \sum_{q=s+1}^{k-m_0} \epsilon_q a_{k-q}/h + iu \sum_{q=s+1}^{k-m_0} \epsilon_q / \sqrt{n}} \right|$$

$$\leq \begin{cases} e^{-\gamma(k-s)}, & \text{if } |t| \geq \delta h / |a_{k-s}|, \\ e^{-\gamma t^2 (d_{k-s}/h)^2}, & \text{if } |t| \leq \delta h / |a_{k-s}|, \end{cases} \qquad (3.109)$$

where $m_0, u \in \mathbb{R}$ are fixed and δ is a constant located in (δ_1, δ_2).

Let $\sigma_j = \sigma(\eta_j, \eta_{j-1}, \dots, \eta_{j-m_0})$ with $E\sigma_1 = 0$ and $E\sigma_1^2 < \infty$. Set $g_k = g[k/n, (y_{nk} + x)/h]$ and $\mathcal{F}_s = \sigma(\eta_s, \eta_{s-1,\dots})$. In addition to Lemmas 2.1 and 2.2, we have the following lemmas.

Lemma 3.5. *There exist $\gamma > 0$ and $A_0 > 0$ such that*

$$\Lambda_{jk}(x) := \left| \mathbb{E} \left\{ g_k \, \sigma_k \, \exp \left(i \sum_{q=j+1}^{n} \mu_q \epsilon_q / \sqrt{n} \right) \mid \mathcal{F}_j \right\} \right|$$

$$\leq C(h/d_{k-j}^2 + e^{-\gamma(k-j)}); \qquad (3.110)$$

for all $k - j \geq A_0$ and fixed $x, m_0, \mu_q \in \mathbb{R}$ with $|\mu_q| \leq C$, and

$$\mathbb{E} \left(|g_k| \, |\sigma_k| \mid \mathcal{F}_j \right) \leq C \, h^{1/2} \left[\mathbb{E} \left(\sigma_k^2 \mid \mathcal{F}_j \right) \right]^{1/2}, \qquad (3.111)$$

for all $k - j \geq 1$ and fixed $x, m_0 \in \mathbb{R}$.

Proof. Let $z^{(1)} = t \sum_{q=j+1}^{k-m_0} \epsilon_q a_{k-q} + \sum_{q=j+1}^{k-m_0} \mu_q \epsilon_q / \sqrt{n}$ and

$$z^{(2)} = t \sum_{q=k-m_0+1}^{k} \epsilon_q a_{k-q} + \sum_{q=k-m_0+1}^{k} \mu_q \epsilon_q / \sqrt{n}.$$

Note that, by Assumption 3.3,

$$g_k = \frac{1}{2\pi} \int_{-\infty}^{\infty} e^{it(y_{nk}+x)/h} \hat{g}(k/n, t) dt.$$

It follows from (3.108)-(3.109) and the independence of ϵ_k that

$$\Lambda_{jk}(x) \leq \frac{1}{2\pi} \int \left| \mathrm{E}\, e^{iz^{(1)}/h} \right| \left| \mathrm{E}\left\{ e^{iz^{(2)}/h}\, \sigma_k \right\} \right| |\hat{g}(k/n, t)|\, dt$$

$$\leq \frac{1}{2\pi} \Big(\int_{|t| \geq \delta\, h/|a_{k-j}|} + \int_{|t| \leq \delta\, h/|a_{k-j}|} \Big) \left| \mathrm{E}\, e^{iz^{(1)}/h} \right|$$

$$\left| \mathrm{E}\left\{ e^{iz^{(2)}/h}\, \sigma_k \right\} \right| |\hat{g}(k/n, t)|\, dt$$

$$\leq C\, e^{-\gamma(k-j)} \int |\hat{r}(t)|\, dt$$

$$+ C \int_{|t| \leq \delta\, h/|a_{k-j}|} e^{-\gamma t^2 (d_{k-j}/h)^2} \big(|t| h^{-1} + C/\sqrt{n} \big) dt$$

$$\leq C(h/d_{k-j}^2 + e^{-\gamma(k-j)}),$$

where we have used the fact: due to $\mathrm{E}\, \sigma_k = 0$,

$$\left| \mathrm{E}\left\{ e^{iz^{(2)}/h}\, \sigma_k \right\} \right| = \left| \mathrm{E}\left\{ (e^{iz^{(2)}/h} - 1)\, \sigma_k \right\} \right| \leq C\, (|t h^{-1}| + C/\sqrt{n}).$$

This proves (3.110).

Result (3.111) follows from Lemma 2.1 (iii). Indeed, by recalling that $\phi_0 \neq 0$ and $|g_k| \leq C$, we have

$$\mathrm{E}\left(|g_k|\, |\sigma_k|\, |\, \mathcal{F}_j \right) \leq C \left[\mathrm{E}\left(|g_k|\, |\, \mathcal{F}_j \right) \right]^{1/2} \left\{ \mathrm{E}\left(\sigma_k^2\, |\, \mathcal{F}_j \right) \right\}^{1/2}$$

$$\leq C h^{1/2} \left\{ \mathrm{E}\left(\sigma_k^2\, |\, \mathcal{F}_j \right) \right\}^{1/2},$$

for any $k \geq j + 1$, as required. $\qquad\square$

Lemma 3.6. *For any* $x, \mu_q \in \mathrm{R}$ *with* $|\mu_q| \leq C$ *and any* $0 \leq s < m \leq n$, *we have*

$$\sum_{k=s+1}^{m} \mathrm{E}\left| \mathrm{E}\left(I_k(x, \mu)\, |\, \mathcal{F}_s \right) \right| = o\big[(nh/d_n)^{1/2} \big], \qquad (3.112)$$

$$\sum_{s+1 \leq j < k \leq m} \mathrm{E}\left| \mathrm{E}\left(I_{jk}(x, \mu)\, |\, \mathcal{F}_s \right) \right| = o\big(nh/d_n \big), \qquad (3.113)$$

when $h \to 0$ *(*$h \log n \to 0$ *under* **SM***) and* $nh/d_n \to \infty$, *where*

$$I_k(x, \mu) = g_k\, \sigma_k \exp\Big\{ i \sum_{q=s+1}^{n} u_q \epsilon_q / \sqrt{n} \Big\},$$

$$I_{jk}(x, \mu) = g_k\, g_j\, \sigma_k\, \sigma_j \exp\Big\{ i \sum_{q=s+1}^{n} \mu_q \epsilon_q / \sqrt{n} \Big\}.$$

Proof. Let A_0 be the maximum appeared in Lemma 3.5 and Lemma 2.1. Recall $\mathcal{F}_s \subseteq \mathcal{F}_t$ for any $s \leq t$. It follows from $h \to 0$, (3.110)-(3.111) and Lemma 2.1 that, for $k - j \geq \log h^{-1}$ and $j \geq s + 1$,

$$
\mathrm{E}\left| \mathrm{E}\left(I_{jk}(x,\mu) \mid \mathcal{F}_s \right) \right| \leq \mathrm{E}\left\{ |g_j|\, |\sigma_j|\, \Lambda_{jk}(x) \right\}
$$

$$
\leq C(h/d_{k-j}^2 + e^{-\gamma(k-j)})
\begin{cases}
h^{1/2}, & \text{if } j \leq A_0, \\
h/d_j, & \text{if } j \geq A_0 + 1;
\end{cases}
$$

and for $k - j \leq \log h^{-1}$ and $j \geq s + 1$,

$$
\mathrm{E}\left| \mathrm{E}\left(I_{jk}(x,\mu) \mid \mathcal{F}_s \right) \right| \leq \mathrm{E}\left\{ |g_j|\, |\sigma_j|\, \mathrm{E}\left(|g_k|\, |\sigma_k| \mid \mathcal{F}_j \right) \right\}
$$

$$
\leq C h^{1/2}\, \mathrm{E}\left\{ |g_j|\, |\sigma_j| \left[\mathrm{E}\,(\sigma_k^2 \mid \mathcal{F}_j) \right]^{1/2} \right\}
$$

$$
\leq C
\begin{cases}
h^{1/2}, & \text{if } j \leq A_0, \\
h^{3/2}/d_j, & \text{if } j \geq A_0 + 1.
\end{cases}
$$

Hence, any $x, \mu_q \in \mathrm{R}$ with $|\mu_q| \leq C$ and $0 \leq s < n$, we have (let $\sum_{j=k}^{l} = 0$ if $k > l$ and $\widetilde{A}_0 = \max\{A_0, s\} + 1$)

$$
\sum_{s+1 \leq j < k \leq n} \mathrm{E}\left| \mathrm{E}\left(I_{jk}(x,\mu) \mid \mathcal{F}_s \right) \right|
$$

$$
\leq \left(\sum_{j=s+1}^{A_0} + \sum_{j=\widetilde{A}_0}^{n-1} \right) \left(\sum_{k=j+1}^{j+\log h^{-1}} + \sum_{k=j+\log h^{-1}+1}^{n} \right) \mathrm{E}\left| \mathrm{E}\left(I_{jk}(x,\mu) \mid \mathcal{F}_s \right) \right|
$$

$$
\leq C h^{1/2} \left[\log h^{-1} + \sum_{k=1}^{n} (h/d_k^2 + e^{-\gamma k}) \right] + C h^{3/2} \log h^{-1} \sum_{k=1}^{n} d_k^{-1}
$$

$$
+ C h \sum_{j=1}^{n-1} \sum_{k=j+\log h^{-1}+1}^{n} d_j^{-1} \left(h/d_{k-j}^2 + e^{-\gamma(k-j)} \right)
$$

$$
\leq C h^{1/4} \left(1 + h \sum_{k=1}^{n} d_k^{-2} + nh/d_n \right) + C(nh/d_n) \sum_{j=\log h^{-1}+1}^{n} (h/d_j^2 + e^{-\gamma j})
$$

$$
\leq C(nh/d_n)
\begin{cases}
h^{\min\{1/4,\gamma\}}, & \text{under } \mathbf{LM}, \\
h^{1/4} + h \log n, & \text{under } \mathbf{SM},
\end{cases}
$$

$$
= o(nh/d_n),
$$

due to $h \to 0$ and $h \log n \to 0$ under **SM**, which yields (3.113). The proof of (3.112) is similar and hence the details are omitted. $\square$

Lemma 3.7. *For any $x \in \mathbb{R}$ and $s \in [0,1]$, there exists a sequence of $\alpha_n(\epsilon, \delta)$ satisfying $\lim_{\delta \to 0} \limsup_{n \to \infty} \alpha_n(\epsilon, \delta) = 0$ for each $\epsilon > 0$ such that*

$$\sup_{|t-s| \le \delta} P\Big(\Big| \sum_{k=[ns]+1}^{[nt]} \sigma_k g_k \Big| \ge \epsilon \,(nh/d_n)^{1/2} \mid \mathcal{F}_{[ns]}\Big)$$

$$\le \alpha_n(\epsilon, \delta), \quad a.s., \tag{3.114}$$

*when $h \to 0$ ($h \log n \to 0$ under **SM**) and $nh/d_n \to \infty$.*

Proof. Let A_0 be the maximum appeared in Lemma 3.5 and Lemma 2.1. Similar to the proof of Lemma 3.6, by Lemma 2.1, we have

$$\mathrm{E}\left(g_k^2 \sigma_k^2 \mid \mathcal{F}_m\right) \le Ch/d_{k-m},$$

$$\left|\mathrm{E}\left(g_k g_j \sigma_k \sigma_j \mid \mathcal{F}_m\right)\right| \le \mathrm{E}\Big\{|g_j \sigma_j| \left|\mathrm{E}\left(g_k \sigma_k \mid \mathcal{F}_j\right)\right| \Big| \mathcal{F}_m\Big\}$$

$$\le C \begin{cases} h/d_{j-m}, & \text{if } k - j \le A_0, \\ (h/d_{j-m}) \,(h/d_{k-j}^2 + e^{-\gamma(k-j)}), & \text{if } k - j \ge A_0 + 1, \end{cases}$$

for all $m + A_0 \le j < k \le n$, which implies that

$$\mathrm{E}\Big[\Big(\sum_{k=[ns]+A_0}^{[nt]} \sigma_k g_k \Big)^2 \mid \mathcal{F}_{[ns]}\Big]$$

$$\le Ch \sum_{j=[ns]+1}^{[nt]} d_{j-[ns]}^{-1} \Big[1 + \sum_{k=j+1}^{n} (h/d_{k-j}^2 + e^{-\gamma(k-j)}) \Big]$$

$$\le C(t-s)^\alpha (nh/d_n) \begin{cases} 1, & \text{under } \mathbf{LM}, \\ 1 + h \log n, & \text{under } \mathbf{SM}, \end{cases}$$

for some $\alpha > 0$. Now routine calculations show that (3.114) follows by choosing $\alpha_n(\epsilon, \delta) = 2C\epsilon^{-2}\delta^\alpha$ and the facts that g_k is bounded and

$$\mathrm{E}\left(|\sigma_k| \mid \mathcal{F}_{[ns]}\right) = o(nh/d_n), \quad a.s.$$

for all $[ns] \le k \le [ns] + A_0$. $\qquad \square$

We are now ready to prove the propositions.

Proof of Proposition 3.3. For any fixed $0 \le t \le 1$, it is readily seen from Lemma 2.1 that $\sup_n \mathrm{E}\psi_n^2(t) < \infty$. Hence $\psi_n(t)$ is uniformly integrable. Note that, by using (3.113) with $\sigma_k = u_k(m_0)$ and $s = 0$,

$$\sup_{0 \le t \le 1} \mathrm{E}|\psi_n(t) - S_n^2(t)| \le \frac{2d_n}{nh} \sum_{1 \le k < l \le n} |\mathrm{E}\left(u_k(m_0) \, u_l(m_0) \, g_k \, g_l\right)| = o(1).$$

For any $A > 0$ and fixed t, we have

$$|\mathrm{E}S_n^2(t)\,\mathrm{I}(S_n^2(t) \geq A) - \mathrm{E}\psi_n(t)\,\mathrm{I}(S_n^2(t) \geq A)|$$
$$\leq \sup_{0 \leq t \leq 1} \mathrm{E}\,|\psi_n(t) - S_n^2(t)| = o(1).$$

This implies that $S_n^2(t)$ is uniformly integrable. The integrability of $S_n(t)$ follows from that of $S_n^2(t)$. This completes the proof. $\qquad\square$

Proof of Proposition 3.4. Recall (3.103) and Lemma 3.7. Claim follows from the same argument as that in the proof of Theorem 2.20. $\qquad\square$

Proof of Proposition 3.5. Recalling that Z_{n3} depends only on $\eta_{[ns]}, \eta_{[ns]-1}, \ldots,$ claims follow from Lemma 3.6. $\qquad\square$

3.6 Convergence to a mixture of normal distributions: zero energy functionals

Suppose $\{\epsilon_i\}_{i \in \mathbb{Z}}$ is a sequence of i.i.d. random variables with $\mathrm{E}\,\epsilon_0 = 0$, $\mathrm{E}\,\epsilon_0^2 = 1$ and $\lim_{|t| \to \infty} |t|^\eta |\mathrm{E}\,e^{it\epsilon_0}| < \infty$ for some $\eta > 0$. Let $g(s,x)$: $[0,1] \times \mathrm{R} \to \mathrm{R}$ be a real function and $\{y_{nk}\}_{k \geq 1, n \geq 1}$ be a triangular array given in Assumption 3.1, i.e.,

$$y_{nk} = \gamma\, y_{n,k-1} + \xi_k,$$

where $y_{n0} = 0$, $\gamma = 1 - \tau/n$ with $\tau \geq 0$ being a constant and $\{\xi_j, j \geq 1\}$ is a linear process defined as in Section 2.3.3. Write, for $x \in \mathrm{R}$,

$$G_n(x) = \left(\frac{d_n}{nh}\right)^{1/2} \sum_{k=1}^{n} g[k/n, (y_{nk} + x)/h],$$

where $d_n^2 = var(\sum_{j=1}^{n} \xi_j)$. This section is concerned with the development of a limit theory for the sample function $G_n(x)$ in the zero energy case where $\int_{-\infty}^{\infty} g(s,x)\,dx = 0, s \in [0,1]$.

Theorem 3.18. *In addition to Assumption 3.3,* $\sup_{0 \leq s \leq 1} |\hat{g}(s,t)| \leq C\min\{|t|, 1\}$, *where C is a positive constant. Then, for any $h \to 0$* *($h^2 \log n \to 0$ under* ***SM****) and $nh/d_n \to \infty$, we have*

$$\left\{G_n(x), \frac{d_n}{nh} \sum_{j=1}^{n} g^2[j/n, (y_{nj} + x)/h]\right\} \to_D \left(N\,V,\ V^2\right), \qquad (3.115)$$

where N is a normal variate independent of V^2, which is defined as in Theorem 3.16.

The conditions on $g(s,x)$ imply $\int_{-\infty}^{\infty} g(s,x)dx = 0$ and $\int_{-\infty}^{\infty} g^2(s,x)dx < \infty$ for $s \in [0,1]$. Indeed it follows by dominated convergence theorem that

$$\int_{-\infty}^{\infty} g(s,x)dx = \int_{-\infty}^{\infty} \lim_{t\to 0} e^{itx} g(s,x)dx = \lim_{t\to 0} \hat{g}(s,t) = 0.$$

On the other hand, $|\hat{g}(s,t)| \leq C \min\{|t|,1\}$ is derived from $\int_{-\infty}^{\infty} g(s,x)dx = 0$ and $\int_{-\infty}^{\infty}(1+|x|)|g(s,x)|dx < \infty$ for $s \in [0,1]$.

To Theorem 3.18, we only need to replace $S_n(t)$ and $\psi_n(t)$ by $S_n'(t)$ and $\psi_n'(t)$, where

$$S_n'(t) = \left(\frac{d_n}{nh}\right)^{1/2} \sum_{k=1}^{[nt]} g[k/n,(y_{nk}+x)/h],$$

$$\psi_n'(t) = \frac{d_n}{nh} \sum_{k=1}^{[nt]} g^2[k/n,(y_{nk}+x)/h],$$

and then follow the same lines as those of Theorem 3.16 with some routine modifications. The only essential difference is to replace Lemma 3.5 by the following:

Lemma 3.8. *There exist $\gamma > 0$ and $A_0 > 0$ such that*

$$\Lambda_{jk}(x) := \left| \mathrm{E}\left\{ g\big[k/n,(y_{nk}+x)/h\big] \exp\left(i \sum_{q=j+1}^{n} \mu_q \epsilon_q/\sqrt{n}\right) \,\Big|\, \mathcal{F}_j \right\} \right|$$

$$\leq C(h/d_{k-j}^2 + e^{-\gamma(k-j)}); \tag{3.116}$$

for all $k - j \geq A_0$ and fixed $x, \mu_q \in \mathrm{R}$ with $|\mu_q| \leq C$, and

$$\Lambda_{jk}(x) \leq C\,h/d_{k-j}, \tag{3.117}$$

for all $k - j \geq 1$ and fixed $x, \mu_q \in \mathrm{R}$ with $|\mu_q| \leq C$.

Proof. Let $z = t\sum_{q=j+1}^{k} \epsilon_q\, a_{k-q} + \sum_{q=j+1}^{k} \mu_q \epsilon_q/\sqrt{n}$. As in the proof of Lemma 3.5, it follows from (3.108)-(3.109) and the independence of ϵ_k that

$$\Lambda_{jk}(x) \leq \frac{1}{2\pi} \int \left| \mathrm{E}\, e^{iz/h} \right| |\hat{g}(k/n,t)|\, dt$$

$$\leq \frac{1}{2\pi} \left(\int_{|t|\geq \delta\, h/|a_{k-j}|} + \int_{|t|\leq \delta\, h/|a_{k-j}|} \right) \left| \mathrm{E}\, e^{iz/h} \right| |\hat{g}(k/n,t)|\, dt$$

$$\leq C\, e^{-\gamma(k-j)} \int \sup_{0\leq s\leq 1} |\hat{g}(s,t)|\, dt$$

$$+ C \int_{|t|\leq \delta\, h/|a_{k-j}|} e^{-\gamma t^2 (d_{k-j}/h)^2} \min\{|t|,1\}dt$$

$$\leq C(h/d_{k-j}^2 + e^{-\gamma(k-j)}),$$

which yields (3.116). Result (3.117) follows from Lemma 2.1 (iii). $\square$

3.7 Uniform convergence for a class of martingales

Let $g(x)$ be a real function on R^m, $m \geq 1$, and $\{u_k, y_{nk}\}_{k\geq1, n\geq1}$ be a triangular array on $\mathrm{R} \times \mathrm{R}^m$. Write, for $x \in \mathrm{R}$,

$$M_n(x) = \sum_{k=1}^{n} u_k \, g[(y_{nk} + x)/h], \quad x \in \mathrm{R}^m$$

where $h = h_n \to 0$. This section is concerned with the uniform asymptotics of $M_n(x)$ on both h and x. Throughout the section, assume that $\{u_t, \mathcal{F}_t\}_{t\geq1}$ is a martingale difference satisfying

$$\sup_{t\geq1} \mathrm{E}(|u_t|^{2p} \mid \mathcal{F}_{t-1}) < \infty, \tag{3.118}$$

for some $p \geq 1$, and y_{nt} is adapted to $\mathcal{F}_{t-1}$ [define $\mathcal{F}_0 = \sigma(\phi, \Omega)$] for each $n \geq 1$.

3.7.1 *A framework and applications*

Write, for some constant sequence $h_n > 0$,

$$\Omega_1 = \Big\{ f : \sup_{y\in\mathrm{R}^m} |f(y)| \leq h_n^{-1/2} \Big\}.$$

The following theorem provides a framework on the uniform bounds for a class of martingales.

Theorem 3.19. *Let Ψ_1 be a subset of Ω_1 such that $\#\Psi_1 \leq n^k$ for some $k > 0$. Suppose there exists a sequence of positive constants $\gamma_n \to \infty$ such that*

$$\sup_{f\in\Psi_1} \sum_{t=1}^{n} f^2(y_{nt}) = O_\mathrm{P}(\gamma_n). \tag{3.119}$$

Then, for any h_n and γ_n satisfying $n\,(h_n\gamma_n)^{-p} \log^{p-1} n = O(1)$, where p is defined as in (3.118), we have

$$\sup_{f\in\Psi_1} \Big| \sum_{t=1}^{n} u_t \, f(y_{nt}) \Big| = O_\mathrm{P}\big[(\gamma_n \log n)^{1/2}\big]. \tag{3.120}$$

Proof. Write $u'_t = u_t \mathrm{I}[|u_t| \leq b_n]$ and $u^*_t = u'_t - \mathrm{E}(u'_t \mid \mathcal{F}_{t-1})$, where $b_n = (h_n\gamma_n/\log n)^{1/2}$. Simple calculations show that, under (3.118) and $n\,(h_n\gamma_n)^{-p} \log^{p-1} n = O(1)$,

$$\sup_{f \in \Psi_1} \sum_{t=1}^{n} |u_t - u_t^*| \, |f(y_{nt})|$$

$$\leq h_n^{-1/2} \sum_{t=1}^{n} \left\{ |u_t| \mathrm{I}(|u_t| > b_n) + E\big[|u_t| \mathrm{I}(|u_t| > b_n) | \mathcal{F}_{t-1}\big] \right\}$$

$$\leq h_n^{-1/2} b_n^{-2p+1} \sum_{t=1}^{n} \big[|u_t|^{2p} + \mathrm{E}\left(|u_t|^{2p} \mid \mathcal{F}_{t-1} \right) \big]$$

$$= O_{\mathrm{P}}\big[(\gamma_n \log n)^{1/2} \big].$$

As a consequence, to prove (3.120), it suffices to show

$$\lambda_n := \sup_{f \in \Psi_1} \big| \sum_{t=1}^{n} u_t^* f(y_{nt}) \big| = O_{\mathrm{P}}\big[(\gamma_n \log n)^{1/2} \big],$$

In fact, by noting $\sup_{t \geq 1} E[(u_t^*)^2 \mid \mathcal{F}_{t-1}] < \infty, a.s.$ by (3.118), it follows from condition (3.119) that, for any $\eta > 0$, there exists an $M_0 > 0$ such that

$$\mathrm{P}\big(\sup_{f \in \Psi_1} \sum_{t=1}^{n} \sigma_t^2(f) \geq M_0 \gamma_n \big) \leq \eta.$$

where $\sigma_t^2(f) = f^2(y_{nt}) E[(u_t^*)^2 \mid \mathcal{F}_{t-1}]$, whenever n is sufficiently large. This, together with $|u_t^* f(y_{nt})| \leq (\gamma_n / \log n)^{1/2}$ and the well-known martingale exponential inequality (see, e.g., de La Pena (1999)), implies that, for any $\eta > 0$, there exists an $M_0 > 4k$ such that

$$P[\lambda_n \geq M_0 (\gamma_n \log n)^{1/2}]$$

$$\leq P\Big[\lambda_n \geq M_0 (\gamma_n \log n)^{1/2}, \sup_{f \in \Psi_1} \sum_{t=1}^{n} \sigma_t^2(f) \leq M_0 \gamma_n \Big] + \eta$$

$$\leq \sum_{f \in \Psi_1} P\Big[\sum_{t=1}^{n} u_t^* f(y_{nt}) \geq M_0 (\gamma_n \log n)^{1/2}, \sum_{t=1}^{n} \sigma_t^2(f) \leq M_0 \gamma_n \Big] + \eta$$

$$\leq n^k \exp \big\{ -\frac{1}{4} M_0 \log n \big\} + \eta \leq 2\eta, \tag{3.121}$$

whenever n is sufficiently large. This yields $\lambda_n = O_{\mathrm{P}}\big[(\gamma_n \log n)^{1/2} \big]$, as required. $\qquad \square$

Theorem 3.19 has many applications, as indicated in the following corollaries.

Corollary 3.7. *Suppose* $f_n(y, x), n \geq 1$, *is a sequence of real functions on* $\mathrm{R}^m \times \mathrm{R}^{m_1}$, *where* $m, m_1 \geq 1$, *satisfying* $\sup_{n,x,y} |f_n(y, x)| < \infty$ *and there exists an* $\alpha > 0$ *such that, whenever* $||a||$ *is sufficiently small,*

$$\sup_{n,x,y} |f_n(y, x + a) - f_n(y, x)| \leq C\, n^\alpha\, ||a||. \tag{3.122}$$

If there is a sequence of constants $0 < \gamma_n \to \infty$ *such that*

$$\sup_{||x|| \leq b_n} \sum_{t=1}^n f_n^2(y_{nt}, x) = O_\mathrm{P}(\gamma_n),$$

where $b_n = O(n^k)$ *for some* $k > 0$, *then*

$$\sup_{||x|| \leq b_n} \Big| \sum_{t=1}^n u_t\, f_n(y_{nt}, x) \Big| = O_\mathrm{P}\big[(\gamma_n \log n)^{1/2}\big], \tag{3.123}$$

for any $n\,\gamma_n^{-p} \log^{p-1} n = O(1)$.

Proof. We split set $A_n = \{x : ||x|| \leq b_n\}$ into m_n balls of the form

$$A_{nj} = \{x : ||x - x_j|| \leq 1/m_n'\}$$

where $m_n' = [n^{1+\alpha}/(\gamma_n \log n)^{1/2}]$, $m_n = (b_n m_n')^m$ and x_j are chosen so that $A_n \subset \bigcup A_{nj}$. It follows that

$$\sup_{||x|| \leq b_n} \Big| \sum_{t=1}^n u_t\, f_n(y_{nt}, x) \Big|$$

$$\leq \max_{0 \leq j \leq m_n} \sup_{x \in A_{nj}} \sum_{t=1}^n |u_t|\, \big|f_n(y_{nt}, x) - f_n(y_{nt}, x_j)\big|$$

$$+ \max_{0 \leq j \leq m_n} \Big| \sum_{t=1}^n u_t\, f_n(y_{nt}, x_j) \Big|$$

$$:= \lambda_{1n} + \lambda_{2n}. \tag{3.124}$$

Recalling (3.122) and $\frac{1}{n}\sum_{k=1}^n |u_k| = O_\mathrm{P}(1)$ due to $\sup_{t \geq 1} \mathrm{E}\,(|u_t|\,|\mathcal{F}_{t-1}) < \infty$, a.s., it is readily seen that

$$\lambda_{1n} \leq \sum_{t=1}^n |u_t| \max_{0 \leq j \leq m_n} \sup_{x \in A_{nj}} \big|f_n(y_{nt}, x) - f_n(y_{nt}, x_j)\big|$$

$$\leq C\,(n^\alpha m_n')^{-1} \sum_{t=1}^n |u_t|$$

$$\leq C\,(\gamma_n \log n)^{1/2}\, \frac{1}{n} \sum_{t=1}^n |u_t| = O_\mathrm{P}[(\gamma_n \log n)^{1/2}].$$

On the other hand, an application of Theorem 3.19 yields $\lambda_{2n} = O_\mathrm{P}\big[(\gamma_n \log n)^{1/2}\big]$ for any $n\,\gamma_n^{-p} \log^{p-1} n = O(1)$. Taking these estimates into (3.124), we prove (3.123). $\qquad \square$

Corollary 3.8. *Let $g(x)$ be a bounded real function on $\mathbb{R}^m$ satisfying the Lipschitz condition, i.e.,*

$$|g(x) - g(y)| \leq C||x - y||, \quad x, \ y \in \mathbb{R}^m.$$

If there exist positive constant sequences $\gamma_n \to \infty$ and $b_n = O(n^k)$ for some $k > 0$ such that

$$\sup_{||x|| \leq b_n} \sum_{t=1}^{n} g^2[(y_{nt} + x)/h] = O_{\mathrm{P}}(\gamma_n). \tag{3.125}$$

then, for any h satisfying $nh \to \infty$ and $n\gamma_n^{-p} \log^{p-1} n = O(1)$,

$$\sup_{||x|| \leq b_n} \left| \sum_{t=1}^{n} u_t \, g[(y_{nt} + x)/h] \right| = O_{\mathrm{P}}\big[(\gamma_n \log n)^{1/2}\big]. \tag{3.126}$$

Furthermore, if $n \sup_{||x|| > b_n/2} |g(x/h)| = O\big[(\gamma_n \log n)^{1/2}\big]$ and there exists a $k_0 > 0$ such that

$$b_n^{-k_0} \sum_{t=1}^{n} \mathrm{E} \, ||y_{nt}||^{k_0} = O\big[(\gamma_n \log n)^{1/2}\big],$$

then, for any h satisfying $nh \to \infty$ and $n\gamma_n^{-p} \log^{p-1} n = O(1)$,

$$\sup_{x \in \mathbb{R}^d} \left| \sum_{t=1}^{n} u_t \, g[(y_{nt} + x)/h] \right| = O_{\mathrm{P}}\big[(\gamma_n \log n)^{1/2}\big]. \tag{3.127}$$

Proof. Note that, due to $nh \to \infty$,

$$\sup_{x} |g(x + a/h) - g(x)| \leq C||a||/h \leq C \, n \, ||a||.$$

Result (3.126) follows from Corollary 3.7 with $f_n(y, x) = g[(x + y)/h]$. In order to prove (3.127), we may write

$$\sum_{t=1}^{n} u_t \, g[(y_{nt} + x)/h] = \sum_{t=1}^{n} u_t \, g[(y_{nt} + x)/h]\mathrm{I}(||y_{nt}|| \leq b_n/2)$$

$$+ \sum_{t=1}^{n} u_t \, g[(y_{nt} + x)/h]\mathrm{I}(||y_{nt}|| > b_n/2)$$

$$= \lambda_{3n}(x) + \lambda_{4n}(x), \quad \text{say}. \tag{3.128}$$

It is readily seen from (3.126) and

$$n \sup_{||x|| > b_n/2} |f(x/h)| = O\big[(\gamma_n \log n)^{1/2}\big]$$

that

$$\sup_{x \in \mathbb{R}^d} |\lambda_{3n}(x)| \leq \sup_{||x|| \leq b_n} |\lambda_{3n}(x)| + \sup_{||x|| > b_n} |\lambda_{3n}(x)|$$

$$\leq O_P\big[(\gamma_n \log n)^{1/2}\big] + \sup_{||x|| > b_n/2} |f(x/h)| \sum_{t=1}^{n} |u_t|$$

$$\leq O_P\big[(\gamma_n \log n)^{1/2}\big],$$

since $\frac{1}{n} \sum_{t=1}^{n} |u_t| = O_P(1)$. As for $\lambda_{4n}(x)$, we have

$$\mathrm{E} \sup_{x \in \mathbb{R}^d} |\lambda_{4n}(x)| \leq C \sum_{t=1}^{n} \mathrm{E}\big[|u_t|\, \mathrm{I}(||y_{nt}|| > b_n/2)\big]$$

$$\leq C \sum_{t=1}^{n} \mathrm{P}(||y_{nt}|| > b_n/2) \leq C\, b_n^{-k_0} \sum_{t=1}^{n} E||y_{nt}||^{k_0}$$

$$= O\big[(\gamma_n \log n)^{1/2}\big],$$

which yields $\sup_{x \in \mathbb{R}^d} |\lambda_{4n}(x)| = O_P\big[(\gamma_n \log n)^{1/2}\big]$. Taking these estimates into (3.128), we obtain (3.127). $\qquad\square$

Corollary 3.9. *Let $\sigma(x)$ be a real function on $\mathbb{R}$ such that $\inf_{x \in \mathbb{R}^d} \sigma(x) > 0$ and for any $||a||$ sufficiently small,*

$$\sup_{x \in \mathbb{R}^d} \frac{|\sigma(x+a) - \sigma(x)|}{\sigma(x)} \leq C\, ||a||, \tag{3.129}$$

where C is a positive constant. Further, in addition to the conditions of Corollary 3.8 on $g(x)$, $g(x)$ has a compact support and $\mathrm{P}(\max_{1 \leq t \leq n} |y_{nt}| \geq b_n/2) = o(1)$. Then, for any h satisfying $nh \to \infty$ and $n\gamma_n^{-p} \log^{p-1} n = O(1)$,

$$\sup_{x \in \mathbb{R}^d} \Big| \sum_{t=1}^{n} \frac{\sigma(y_{nt})}{\sigma(x)}\, g[(y_{nt} - x)/h]\, u_t \Big| = O_P\big[(\gamma_n \log n)^{1/2}\big]. \tag{3.130}$$

Proof. Let $f_n(y, x) = \frac{\sigma(y)}{\sigma(x)} g\big(\frac{y-x}{h}\big)$. First note that, for some η_0 sufficiently small,

$$C_0 := \sup_{x \in \mathbb{R}^d, |y-x| \leq \eta_0} \frac{\sigma(y)}{\sigma(x)} < \infty,$$

due to (3.129). This, together with $g(z) = 0$ for $||z|| \geq A$, implies that

$$|f_n(y, x)| \leq C_0 \Big| g\big(\frac{y-x}{h}\big) \Big|, \tag{3.131}$$

uniformly for n, x and y, whenever h is sufficiently small. Similarly, for $||y - x|| \geq hA + ||a||$, we have

$$g\left(\frac{y - x - a}{h}\right) - g\left(\frac{y - x}{h}\right) = 0,$$

and hence it follows from (3.129) and $|g(x) - g(y)| \leq C||x - y||$ that

$$|f_n(y, x + a) - f_n(y, x)| \leq \frac{\sigma(y)}{\sigma(x + a)} \left| g\left(\frac{y - x - a}{h}\right) - g\left(\frac{y - x}{h}\right) \right|$$

$$+ |f_n(y, x)| \frac{|\sigma(x) - \sigma(x + a)|}{\sigma(x + a)}$$

$$\leq C(h^{-1} + 1)||a|| \leq C n||a||, \qquad (3.132)$$

uniformly for n, x and y, whenever $nh \to \infty$ and $||a||$ is sufficiently small. By virtue of (3.131) and (3.132), it is readily seen from Corollary 3.7 that

$$\sup_{||x|| \leq b_n} \left| \sum_{t=1}^{n} \frac{\sigma(y_{nt})}{\sigma(x)} g[(y_{nt} - x)/h] u_t \right| = O_P\left[(\gamma_n \log n)^{1/2}\right].$$

Note that $g[(y_{nt} - x)/h] = 0$ if $||x|| \geq b_n$, $\max_{1 \leq k \leq n} |y_{nk}| < b_n/2$ and h is sufficiently small. Result (3.130) hence follows from the fact: for any $\eta > 0$, there exists an $M_0 > 0$ such that

$$P\left(\sup_{x \in \mathbb{R}^d} \left| \sum_{t=1}^{n} \frac{\sigma(y_{nt})}{\sigma(x)} g[(y_{nt} - x)/h] u_t \right| \geq M_0(\gamma_n \log n)^{1/2} \right)$$

$$\leq P\left(\sup_{||x|| \leq b_n} \left| \sum_{t=1}^{n} \frac{\sigma(y_{nt})}{\sigma(x)} g[(y_{nt} - x)/h] u_t \right| \geq M_0(\gamma_n \log n)^{1/2} \right)$$

$$+ P\left(\max_{1 \leq k \leq n} |y_{nk}| \geq b_n/2 \right) \leq 2\eta,$$

whenever n is sufficiently large. $\qquad \square$

3.7.2 *Remarks and extension*

Condition $n (h_n \gamma_n)^{-p} \log^{p-1} n = O(1)$ used in Theorem 3.19 provides a trade off among the boundary of $f(x)$, the moment of u_k and the uniform convergence rate for conditional variance of $\sum_{k=1}^{n} u_k f(y_{nk})$. In many applications, we have $h_n \gamma_n = O(n^\alpha)$ for some $0 < \alpha < 1$, which implies that the required moment condition is

$$\sup_{t \geq 1} E(|u_t|^{2p} \mid \mathcal{F}_{t-1}) < \infty, \quad \text{for some } p > 1/\alpha.$$

The convergence rate in (3.120) is dominated by its conditional variance, namely (3.119), and the term $\log^{1/2} n$ in (3.120) is optimal. If we allow for a slightly weaker bound, the condition $n (h_n \gamma_n)^{-p} \log^{p-1} n = O(1)$ can be improved, as indicated in the following theorem.

Theorem 3.20. *Under the condition of Theorem 3.19, we have*

$$\sup_{f \in \Psi_1} \left| \sum_{t=1}^{n} u_t \, f(y_{nt}) \right| = O_P \left(\gamma_n^{1/2} \log n \right), \tag{3.133}$$

for any h_n and γ_n satisfying $n (h_n \gamma_n)^{-p} = O(1)$, where p is defined as in (3.118).

Proof. Define u' and u^* as in Theorem 3.19, but the b_n is replaced by $b_n' = (h_n \gamma_n)^{1/2}$. Similar arguments as that in the proof of Theorem 3.19 yield $\sup_{f \in \Psi_1} \sum_{t=1}^{n} |u_t - u_t^*| \, |f(y_{nt})| = O_P(\gamma_n^{1/2})$, and hence it suffices to show

$$\sup_{f \in \Psi_1} \left| \sum_{t=1}^{n} u_t^* \, f(y_{nt}) \right| = O_P \left(\gamma_n^{1/2} \log n \right).$$

This is a simple application of Lemma 2.4 and the details are omitted. $\square$

Corollaries 3.7-3.9 can be similarly modified as that in Theorem 3.20. It is interesting to notice that we do not impose any conditions on y_{nk} in Corollary 3.8 except (3.125), this enables Corollary 3.8 to cover many classical uniform convergence results. Examples that satisfy (3.125) will be given in next section, including stationary and nonstationary time series.

In many applications, it is useful to allow for the bandwidth h in Corollary 3.8 to be data-dependent (random). This requires uniform convergence of $\sum_{k=1}^{n} u_k \, g[(y_{nk}+x)/h]$ on both h and x. The following corollary provides an extension toward this regard.

As in Corollary 3.8, let $g(x)$ be a bounded real function on $\mathbb{R}^m$ satisfying the Lipschitz condition. Set, for any $\epsilon_0 > 0$, $\alpha > \epsilon_0$ and $\beta > 0$,

$$B_n = \{ (h, x) : n^{-\alpha + \epsilon_0} \leq h \leq h_0, \, \|x\| \leq n^\beta \}.$$

Corollary 3.10. *If $\sup_{t \geq 1} \mathrm{E}(|u_t|^{2p} \mid \mathcal{F}_{t-1}) < \infty$, for some $p > 1/\epsilon_0$, and*

$$\sup_{(h,x) \in B_n} h^{-1} \sum_{t=1}^{n} g^2[(y_{nt} + x)/h] = O_P(n^\alpha), \tag{3.134}$$

then,

$$\sup_{(h,x) \in B_n} \left| h^{-1/2} \sum_{t=1}^{n} u_t \, g[(y_{nt} + x)/h] \right| = O_P \left(n^{\alpha/2} \log^{1/2} n \right). \tag{3.135}$$

Proof. As in the proof of Corollary 3.7, we can split B_n into m_n balls of the form

$$B_{nj} = \{(h,x) : |h - h_j| \le 1/m'_n, \ \|x - x_j\| \le 1/m'_n\}$$

where $m'_n = [n^{1+\alpha+\beta}/(\gamma_n \log n)^{1/2}]$, $m_n = (n^{\alpha+\beta} m'_n)^{m+1}$ and (h_j, x_j) are chosen so that $B_n \subset \bigcup B_{nj}$. Using similar arguments there, it suffices to show

$$\max_{0 \le j \le m_n} \left| h_j^{-1/2} \sum_{t=1}^{n} u_t \, g[(y_{nt} + x_j)/h_j] \right| = O_P\big(n^{\alpha/2} \log^{1/2} n\big). \quad (3.136)$$

Let $f_j(y) = h_j^{-1/2} g[(y + x_j)/h_j], 1 \le j \le m_n$, and $h_n^{-1/2} = Cn^{(\alpha-\delta_0)/2}$. Note that $|f_j(y)| \le h_n^{-1/2}$ and, when $p > 1/\epsilon_0$

$$n \, (h_n n^\alpha)^{-p} \log^{p-1} n \le n^{1-\epsilon_0 p} \log^{p-1} n \to 0.$$

(3.136) follows immediately from an application of Theorem 3.19. $\qquad\square$

Example 3.6. Let $g(x)$ have a compact support satisfying the Lipschitz condition and let $X_{nk} = y_{nk}/d_n$, where $d_n^2 \asymp n^d$ for some $0 < d < 1$, be a strong smooth array (see Definition 2.5) with $\sup_{n,1 \le k \le n} EX_{nk}^2 < \infty$. It follows from (2.38), Corollaries 2.8 and 2.11 that

$$\sup_{(h,x) \in B_{1n}} h^{-1} \sum_{t=1}^{n} g^2[(y_{nt} + x)/h] = O_P(n^{1-d}),$$

where $B_{1n} = \{(h,x) : n^{d-1+\epsilon_0} \le h \le h_0, \ |x| \le n^{d-\delta_0}\}$, for some $0 < \epsilon_0 < 1 - d$ and $0 < \delta_0 < d$. Furthermore, if in addition Assumption 2.2 holds, then

$$\sup_{(h,x) \in B_{2n}} h^{-1} \sum_{t=1}^{n} g^2[(y_{nt} + x)/h] = O_P(n^{1-d}),$$

where $B_{2n} = \{(h,x) : n^{d-1+\epsilon_0} \le h \le h_0, \ x \in R\}$, for some $0 < \epsilon_0 < 1 - d$.

Remark 3.12. Let h_n be a random sequence satisfying, as $n \to \infty$,

$$P\big(h_n \in [n^{d-1+\epsilon_0}, \ h_0]\big) \to 1,$$

where ϵ_0 can be chosen as small as required. Suppose $\sup_{t \ge 1} E(|u_t|^{2p} \mid \mathcal{F}_{t-1}) < \infty$, for some $p > 1/\epsilon_0$.

Corollary 3.10 and Example 3.6 imply that, for any strong smooth array $X_{nk} = y_{nk}/n^d$ where $0 < d < 1$ with $\sup_{n,1 \le k \le n} EX_{nk}^2 < \infty$ and for any real function $g(x)$ having a compact support and satisfying the Lipschitz condition, we have

$$h_n^{-1/2} \sup_{|x| \le n^{d-\delta_0}} \left| \sum_{t=1}^n u_t \, g[(y_{nt} + x)/h_n] \right| = O_P\left(n^{\alpha/2} \log^{1/2} n\right),$$

where δ_0 can be chosen as small as required. Furthermore, if in addition Assumption 2.2 holds, then

$$h_n^{-1/2} \sup_{x \in R} \left| \sum_{t=1}^n u_t \, g[(y_{nt} + x)/h_n] \right| = O_P\left(n^{\alpha/2} \log^{1/2} n\right).$$

3.7.3 Examples: Identification of (3.125)

This section provides several stationary and nonstationary time series examples that satisfy (3.125). Examples 3.7 and 3.8 come from Wu, Huang, and Huang (2010), where more general settings on x_t are established. Example 3.9 discusses a strongly mixing time series. This example comes from Hansen (2008). By making use of other related works such as Peligrad (1991), Ango and Doukhan (2004), Masry (1996), Bosq (1998) and Andrews (1995), similar results can be established for other mixing time series like ρ-mixing and near-epoch-dependent time series. In these examples, we only consider the situation that $d = 1$. The extension to $d > 1$ is straightforward and hence the details are omitted.

Example 3.10 discusses the Harris recurrent Markov chains, which allows for stationary (positive recurrent) or nonstationary (null recurrent) series. Example 3.11 investigates $I(1)$ processes with innovations being linear processes. These results appears in Wang and Chan (2014a, b). More related results can be found in Section 2.5.

Example 3.7. Let $\{x_t\}_{t \ge 0}$ be a linear process defined by

$$x_t = \sum_{k=0}^{\infty} \phi_k \, \epsilon_{t-k},$$

where $\sum_{k=0}^{\infty} |\phi_k| < \infty$, $\phi \equiv \sum_{k=0}^{\infty} \phi_k \ne 0$ and $\{\epsilon_j\}_{j \in \mathbb{Z}}$ is a sequence of i.i.d. random variables with $E\epsilon_0^2 < \infty$ and a density p_ϵ satisfying $\sup_x |p_\epsilon^{(r)}(x)| < \infty$ and

$$\int_R |p_\epsilon^{(r)}(x)|^2 dx < \infty, \quad r = 0, \ 1, 2,$$

where $p_\epsilon^{(r)}(x)$ denotes the rth-order derivative of $p_\epsilon(x)$.

Suppose $g(x)$ has a compact support satisfying $|g(x) - g(y)| \leq C|x - y|$ for $x, y \in R$. It follows from Section 4.1 of Wu, Huang and Huang (2010) that, for any $h \to 0$ and $nh \log^{-1} n \to \infty$,

$$\sup_{x \in R} \left| \frac{1}{n} \sum_{t=1}^{n} \left[g^2[(x_t + x)/h] - Eg^2[(x_t + x)/h] \right] \right|$$

$$= O\left[\sqrt{\frac{\log n}{nh}} + n^{-1/2} l(n) \right] \quad a.s.$$

where $l(n)$ is a slowly varying function. Since x_t is stationary process having a bounded density $p(x)$ under given conditions on ϵ_k, simple calculations show that

$$\sup_{x \in R} \sum_{t=1}^{n} g^2[(x_t + x)/h] = O_P(nh),$$

that is, (3.125) is satisfied with $\gamma_n = nh$ and $y_{nt} = x_t$ for $1 \leq t \leq n$.

Example 3.8. Consider the nonlinear time series of the following form

$$x_k = R(x_{k-1}, \epsilon_k)$$

where $R(x, \epsilon)$ is a bivariate measurable function and ϵ_k are i.i.d. innovations. This is the iterated random function framework that encompasses a lot of popular nonlinear time series models. For example, if $R(x, \epsilon) = a_1 x \mathrm{I}(x < \tau) + a_2 x \mathrm{I}(x \geq \tau) + \epsilon$, it is the threshold autoregressive (TAR) model, (see Tong (1990)). If $R(x, \epsilon) = \epsilon \sqrt{a_1^2 + a_2^2} x$, then it is autoregressive model with conditional heteroscedasticity (ARCH) model. Other nonlinear time series models, including random coefficient model, bilinear autoregressive model and exponential autoregressive model can be fitted in this framework similarly. See Wu and Shao (2004) for details.

In order to identify (3.125), we need some regularity conditions on the initial distribution of x_0 and the function $R(x, \epsilon)$. Define

$$L_\epsilon = \sup_{x \neq x'} \frac{|R(x, \epsilon) - R(x', \epsilon)|}{|x - x'|}.$$

Denote by $p(x|x_0)$ the conditional density of x_1 at x given x_0. Furthermore let $p'(y|x) = \partial p(y|x)/\partial y$,

$$I(x) = \left[\int_R \left| \frac{\partial}{\partial x} p(y|x) \right|^2 dy \right]^{1/2} \text{ and } J(x) = \left[\int_R \left| \frac{\partial}{\partial x} p'(y|x) \right|^2 dy \right]^{1/2}.$$

$I(x)$ and $J(x)$ can be interpreted as prediction sensitivity measures. These quantities measure the change in 1-step predictive distribution of x_1 with respect to change in initial value x_0. Suppose that

(i) there exist α and z_0 such that

$$E\left(|L_{\epsilon_0}|^{\alpha} + |R(z_0, \epsilon_0)|^{\alpha}\right) < \infty, \quad E[\log(L_{\epsilon_0})] < 0 \quad and \quad EL_{\epsilon_0}^2 < 1;$$

(ii) $\sup_x[I(x) + J(x)] < \infty$;

(iii) $g(x)$ has a compact support satisfying $|g(x) - g(y)| \leq C|x - y|$ for $x, y \in \mathrm{R}$.

It follows from Section 4.2 of Wu, Huang and Huang (2010) that, for any $h \to 0$ and $nh \log^{-1} n \to \infty$

$$\sup_{x \in R} \left| \frac{1}{n} \sum_{t=1}^{n} \left[g^2[(x_t + x)/h] - Eg^2[(x_t + x)/h] \right] \right|$$

$$= O\left[\sqrt{\frac{\log n}{nh}} + n^{-1/2} l(n) \right] \quad a.s.$$

where $l(n)$ is a slowly varying function. Note that x_t has a unique and stationary distribution under the given condition (i) and (ii). See Diaconis and Freedman (1999) for instance. Simple calculations show that

$$\sup_{x \in R} \sum_{t=1}^{n} g^2[(x_t + x)/h] = O_{\mathrm{P}}(nh),$$

that is, (3.125) is satisfied with $\gamma_n = nh$ and $y_{nt} = x_t$ for $1 \leq t \leq n$.

Example 3.9. Let $\{x_k\}_{k \geq 0}$ be a strictly stationary time series with density $p(x)$. Suppose that

(i) x_t is strongly mixing with mixing coefficients $\alpha(m)$ that satisfy $\alpha(m) \leq A\,m^{-\beta}$ where $\beta > 2$ and $A < \infty$;

(ii) $\sup_x |x|^q p(x) < \infty$ for some $q \geq 1$ satisfying $\beta > 2 + 1/q$ and there is some $j^* < \infty$ such that for all $j \geq j^*$, $\sup_{x,y} p_j(x, y) < \infty$ where $p_j(x, y)$ is the joint density of $\{x_0, x_j\}$;

(iii) $g(x)$ has a compact support satisfying $|g(x) - g(y)| \leq C|x - y|$ for $x, y \in \mathrm{R}$.

It follows from Theorem 4 (with $Y_i = 1$) of Hansen (2008) that, for any $h \to 0$ and $n^{\theta} h \log^{-1} n \to \infty$ with $\theta = \beta - 2 - 1/q$,

$$\sup_{x \in R} \left| \frac{1}{n} \sum_{t=1}^{n} \left[g^2[(x_t + x)/h] - Eg^2[(x_t + x)/h] \right] \right|$$

$$= O_{\mathrm{P}}\left[\sqrt{\frac{\log n}{nh}} \right].$$

If in addition $E |x_0|^{2q} < \infty$, the result (3.137) can be strengthened to almost surely convergence. Simple calculations show that

$$\sup_{x \in R} \sum_{t=1}^{n} g^2[(x_t + x)/h] = O_P(nh),$$

that is, (3.125) is satisfied with $\gamma_n = nh$ and $y_{nt} = x_t$ for $1 \leq t \leq n$.

Example 3.10. Let $\{x_k\}_{k \geq 0}$ be a Harris recurrent Markov chain with state space $(E, \mathcal{E})$, transition probability $P(x, A)$ and invariant measure π. We denote P_μ for the Markovian probability with the initial distribution μ, E_μ for the corresponding expectation and $P^k(x, A)$ for the k-step transition of $\{x_k\}_{k \geq 0}$. A subset D of E with $0 < \pi(D) < \infty$ is called a D-set of $\{x_k\}_{k \geq 0}$ if for any $A \in \mathcal{E}^+$,

$$\sup_{x \in E} E_x \Big(\sum_{k=1}^{\tau_A} I_D(x_k) \Big) < \infty,$$

where $\mathcal{E}^+ = \{A \in \mathcal{E} : \pi(A) > 0\}$ and $\tau_A = \inf\{n \geq 1 : x_n \in A\}$. As is well-known, D-sets not only exist, but generate the entire sigma algebra $\mathcal{E}$, and for any D-sets C, D and any probability measure ν, μ on $(E, \mathcal{E})$,

$$\lim_{n \to \infty} \sum_{k=1}^{n} \nu P^k(C) / \sum_{k=1}^{n} \mu P^k(D) = \frac{\pi(C)}{\pi(D)}, \tag{3.137}$$

where $\nu P^k(D) = \int_{-\infty}^{\infty} P^k(x, D)\nu(dx)$. See Nummelin (1984) for instance.

Let a D-set D and a probability measure ν on $(E, \mathcal{E})$ be fixed. Define

$$a(t) = \pi^{-1}(D) \sum_{k=1}^{[t]} \nu P^k(D), \quad t \geq 0.$$

By recurrence, $a(t) \to \infty$. By virtue of (3.137), the asymptotic order of $a(t)$ depends only on $\{x_k\}_{k \geq 0}$. As in Chen (2000), a Harris recurrent Markov chain $\{x_k\}_{k \geq 0}$ is called β-regular if

$$\lim_{\lambda \to \infty} a(\lambda t)/a(\lambda) = t^\beta, \quad \forall t > 0, \tag{3.138}$$

where $0 < \beta \leq 1$. Under the condition (3.138), the function $a(t)$ is regularly varying at infinity, i.e., there exists a slowly varying function $l(x)$ such that $a(t) \sim t^\beta l(t)$. Suppose that

(i) $\{x_k\}_{k \geq 0}$ is a β-regular Harris recurrent Markov chain, where the invariant measure π has a bounded density function $p(s)$ on R;

(ii) $g(x)$ has a compact support satisfying $|g(x) - g(y)| \leq C|x - y|$ for $x, y \in R$.

It follows from Wang and Chan (2014) that, for any $h > 0$ satisfying $n^{-\epsilon_0} a(n) h \to \infty$ for some $\epsilon_0 > 0$, we have

$$\sup_{|x| \le n^{m_0}} \sum_{k=1}^{n} g^2[(x_k + x)/h] = O_P[a(n) h].$$

where m_0 can be any finite integer. That is, (3.125) is satisfied with $\gamma_n = a(n)h$, $b_n = n^{m_0}$ and $y_{nt} = x_t$ for $1 \le t \le n$.

Example 3.11. Suppose $\{\epsilon_j\}_{j \in \mathbb{Z}}$ is a sequence of i.i.d. random variables with $E\epsilon_0 = 0$, $E\epsilon_0^2 = 1$ and $\lim_{|t| \to \infty} |t|^\eta |E\, e^{it\epsilon_0}| < \infty$ for some $\eta > 0$. Let

$$y_{nk} = \gamma\, y_{n,k-1} + \xi_k,$$

where $y_{n0} = 0$, $\gamma = 1 - \tau/n$ with $\tau \ge 0$ being a constant and where $\{\xi_j, j \ge 1\}$ is a linear process defined as in Example 2.12, i.e., $\xi_j = \sum_{k=0}^{\infty} \phi_k\, \epsilon_{j-k}$, where the coefficients ϕ_k, $k \ge 0$, satisfy one of the following conditions:

LM. $\phi_k \sim k^{-\mu} \rho(k)$, where $1/2 < \mu < 1$ and $\rho(k)$ is a function slowly varying at ∞.

SM. $\sum_{k=0}^{\infty} |\phi_k| < \infty$ and $\phi \equiv \sum_{k=0}^{\infty} \phi_k \ne 0$.

Write $d_n^2 = var(\sum_{j=1}^{n} \xi_j)$ and suppose $g(x)$ has a compact support, satisfying $|g(x) - g(y)| \le C|x - y|$ for $x, y \in \mathbb{R}$.

Then, for any $h \to 0$, $d_n/nh \to \infty$ and any fixed constant $0 < m_0 < \infty$,

$$\sup_{|x| \le n^{m_0}} \Big| \sum_{k=1}^{n} g[(y_{nk} + x)/h] \Big| = O[(nh/d_n) \log n], \quad a.s. \quad (3.139)$$

We further have, for any h satisfying $h \to 0$ and $n^{1-\delta_0} h/d_n \to \infty$, where $\delta_0 > 0$ can be as small as required,

$$\sup_{|x| \le M_0 d_n / \log^{\gamma_0} n} \Big| \sum_{t=1}^{n} g[(y_{nk} + x)/h] \Big| = O_P(nh/d_n), \quad (3.140)$$

where M_0 is a fixed constant and

$$\gamma_0 = \begin{cases} \frac{4(3-2\mu)}{2\mu-1}, & \text{under } \mathbf{LM} \text{ and } 9/10 < \mu < 1 \ , \\ \frac{(5-2\mu)(3-2\mu)}{(2\mu-1)^2}, & \text{under } \mathbf{LM} \text{ and } 1/2 < \mu \le 9/10, \\ 4, & \text{under } \mathbf{SM}. \end{cases}$$

Proof. Due to Theorem 2.21, y_{nk} satisfies Assumptions 2.2 and 2.3 of Chan and Wang (2014). Results (3.139) and (3.140) follows from Theorems 2.1 and 2.2 of the paper, respectively.   □

3.8 Limit theorems for continuous time martingales

For each $n \geq 1$, let $X^n = \{X^n_t\}_{t \geq 0}$, where $X^n_t = (X_{n1}(t), \cdots, X_{nd}(t))$, be a d-dimensional semimartingale defined on a stochastic basis $\{\Omega_n, \mathcal{F}_n, (F^n_t)_{t \geq 0}, P_n\}$. This section considers the convergence of X^n to a mixture of Gaussian process.

3.8.1 *Convergence to a continuous Gaussian process*

For each semimartingale X, the quadratic variation $[X, X]$ exists and is a cádlág, increasing, adapted process. The co-variation $[X, Y]$ of two semi-martingageles X and Y is defined by

$$[X, Y] = \frac{1}{4}\big([X + Y, X + Y] - [X - Y, X - Y]\big).$$

Let $\{\Omega, \mathcal{F}, P\}$ be a probability space. For each $n \geq 1$, let $\{X^n, F^n\}$ be d-dimensional semimartingale with the decomposition:

$$X^n = M^n + A^n,$$

where $M^n = \{M^n_1(t), \cdots, M^n_d(t)\}_{t \geq 0}$ is a d-dimensional local martingale, $A^n = \{A^n_1(t), \cdots, A^n_d(t)\}_{t \geq 0}$ is a d-dimensional cádlág, adapted process FV process and $F^n = \{\mathcal{F}^n_t\}_{t \geq 0}$ is a filtration satisfying: (i) $\mathcal{F}^n_0$ contains all P-null sets of $\mathcal{F}$ and (ii) $\mathcal{F}^n_t = \cap_{s > t} \mathcal{F}^n_s$.

Let $\tau_n(t), t \geq 0$, be a random change of time, i.e., a stopping time with respect to $\{\mathcal{F}^n_t\}_{t \geq 0}$ for each $n \geq 1$ and $t > 0$, and the sample paths of $\tau_n(.)$ are non-decreasing and right-continuous with $\tau_n(0) = 0$. Let

$$\widetilde{X}^n = \{X^n_{\tau_n(t)}\}_{t \geq 0}.$$

Theorem 3.21. *Suppose that, for each $t > 0$ and all $1 \leq k \neq l \leq d$,*

(i) $\sup_{0 \leq s \leq \tau_n(t)} \left| A^n_k(s) - A_k(s) \right| = o_P(1)$, *where $A = \{A_s\}_{s \geq 0}$ is a d-dimensional cádlág, adapted FV process;*

(ii) $[M^n_k, M^n_l]_{\tau_n(t)} \to_P C^{kl}_t$, *where C^{kl}_t are positive real functions on R;*

(iii) $\{\sup_{s \leq \tau_n(t)} |\Delta_k(t)|\}_{n \geq 1}$ *is uniformly integrable, where $\Delta_k(t) = M^n_k(t) - M^n_k(t-)$.*

Then, on $D_{R^d}[0, \infty)$,

$$\widetilde{X}^n \Rightarrow M + A, \tag{3.141}$$

where $M = \{M_t\}_{t \geq 0}$ is a d-dimensional Gaussian process with mean zero, independent increments and co-variance matrix $\Omega_t = (C^{kl}_t)_{k,l=1,\ldots,d}$.

Furthermore, for any $t_0 > 0$,

$$X^n_{\tau_n(t_0)} \to_D M_{t_0} + A_{t_0}, \quad (stably). \tag{3.142}$$

For a proof of Theorem 3.21, we refer to Chapter III of Jacod and Shiryaev (2003) and Chapter 5 of Liptser and Shiryaev (1999).

For a continuous local martingale $\{X_t, \mathcal{F}_t\}_{t \geq 0}$, write $\tau(t) = \inf\{s : [X, X]_s > t\}$. If $X_0 = 0$ and $[X, X]_\infty = \infty$, then $B = \{X_{\tau(t)}\}_{t \geq 0}$ is a Brownian motion with respect to $\{\mathcal{F}_{\tau(t)}\}_{t \geq 0}$. B is called the DDS (Dambis, Dbins-Schwarz) Brownian motion of X. See, e.g., Revuz and Yor (2005).

Suppose that $M^n = \{M_1^n, ..., M_d^n\}$ is a d-dimensional continuous local martingale, $M_k^n(0) = 0$ and $[M_k^n, M_k^n]_\infty = \infty$ for every n and k. For $k = 1, 2, ..., d$, write $\tau_{nk}(t) = \inf\{s : [M_k^n, M_k^n]_s > t\}$ and $\beta_k^n = \{M_k^n[\tau_n(t)]\}_{t \geq 0}$.

We have the following theorem, which is a consequence of Theorem 3.21. See, also, Chapter XIII of Revuz and Yor (2005).

Theorem 3.22. *If, for each $t > 0$ and all $1 \leq k \neq l \leq d$,*

$$\lim_{n \to \infty} [M_k^n, M_l^n]_{\tau_{nk}(t)} = \lim_{n \to \infty} [M_k^n, M_l^n]_{\tau_{nl}(t)} = 0,$$

in probability, then

$$(\beta_1^n, \cdots, \beta_d^n) \Rightarrow (B_1, \cdots, B_d), \tag{3.143}$$

where $(B_1, \cdots, B_d)$ is a standard d-dimensional Brownian motion.

3.8.2 Convergence to a mixture of Gaussian processes

Two local martingales are called orthogonal if $MN = \{M_t N_t\}_{t \geq 0}$ is a local martingale. It is well-known that, if M and N are continuous local martingales with $E\,(M_t^2 + N_t^2) < \infty$ for $t > 0$, then M and N are orthogonal if and only if $[M, N] = 0$. A process $X = \{X_t\}_{t \geq 0}$ is called *a mixture of Gaussian process* if there exist a Brownian motion $B = \{B_t\}_{t \geq 0}$ and an increasing process non-negative process $V = \{V_t\}_{t \geq 0}$ such that B is independent of V and $X_t = B_{V_t}, t \geq 0$.

Let $\{\Omega, \mathcal{F}, \mathbf{F} = \{\mathcal{F}_t\}_{t \geq 0}, P\}$ be a stochastic basis and $M_n = \{M_n(t)\}_{t \geq 0}$ be a continuous local martingale with $M_n(0) = 0$ for every n. For the convergence of M_n to a mixture of Gaussian processes, a simplified version of Theorem IX.7.3 in Jacod and Shiryaev (2003) is given as follows.

Theorem 3.23. *Suppose that*

(i) there exists an $\mathcal{F}_t$-adapted continuous process $V = \{V_t\}_{t \geq 0}$ such that, for all $t > 0$,

$$[M_n, M_n]_t \to_P V_t; \tag{3.144}$$

(ii) there exists a continuous local martingale $M = \{M_t, \mathcal{F}_t\}_{t \geq 0}$ such that, for all bounded martingale $N = \{N_t, \mathcal{F}_t\}_{t \geq 0}$ which is orthogonal to M,

$$[M_n, M]_t \to_P 0 \quad [M_n, N]_t \to_P 0,$$

for all $t \geq 0$.

Then there exists a Brownian motion $\widetilde{B} = \{\widetilde{B}_t\}_{t \geq 0}$, which is independent of $\mathbf{F}$, such that $M_n \Rightarrow M$ (stably), where $M = \{\widetilde{B}_{V_t}\}_{t \geq 0}$.

Let $B = \{B_t\}_{t \geq 0}$ be a Brownian motion with respect to $\mathbf{F} = \{\mathcal{F}_t\}_{t \geq 0}$ and $\phi_n(t)$ be $\mathcal{F}_t$-measurable for each $n \geq 1$ and $t > 0$. Write

$$M_n(t) = \int_0^t \phi_n(s) dB_s.$$

Corollary 3.11. *Suppose that, for all $t > 0$,*

(i) $\int_0^t \phi_n(s) ds \to_P 0$,
(ii) *there exists a continuous process $\tau = \{\tau_t\}_{t \geq 0}$ such that $\int_0^t \phi_n^2(s) ds \to_P \tau_t$.*

Then there exists a Brownian motion $\widetilde{B} = \{\widetilde{B}_t\}_{t \geq 0}$, which is independent of $\mathbf{F}$, such that $M_n \Rightarrow M$ (stably), where $M = \{\widetilde{B}_{\tau_t}\}_{t \geq 0}$.

Proof. $\{M_n(t), \mathcal{F}_t\}_{t \geq 0}$ forms a continuous local martingale with

$$[M_n, M_n]_t = \int_0^t \phi_n^2(s) ds, \quad [M_n, B]_t = \int_0^t \phi_n(s) ds$$

and $[M_n, B^\perp] = \int_0^t \phi_n(s) d[B, B^\perp] = 0$, where $B^\perp$ denotes the orthogonal martingale of B. An application of Theorem 3.23 provides the result. $\square$

Example 3.12. Let $f(x)$ be a real function satisfying $\int_{-\infty}^{\infty} [|f(x)| + f^2(x)] dx < \infty$. Let $\eta_t = L_B(t, 0)$ be a local time of B. Then,

$$\sqrt{n} \int_0^t f(nB_s) dB_s \Rightarrow \sigma \widetilde{B}_{\eta_t},$$

where $\sigma^2 = \int_{-\infty}^{\infty} f^2(x) dx$ and $\widetilde{B} = \{\widetilde{B}_t\}_{t \geq 0}$ is a Brownian motion independent of B.

Let $(Z_t, \mathcal{F}_t)_{t \geq 0}$ be a continuous local martingale such that $\mathbb{E} Z_t^2 < \infty$ for all $t \geq 0$ and, for each $n \geq 1$, $(\eta_{ni}, \mathcal{F}_{i/n})_{i \geq 1}$ be a martingale difference. The following result from Theorem 7.28 of Jacod and Shiryaev (2003) [see also Podolskij and Vetter (2010)] is particularly useful in high frequency data.

Theorem 3.24. *Suppose there exist continuous $\mathcal{F}_t$ adapted processes C_t and G_t such that, for all $t > 0$ and $\epsilon > 0$,*

(i) $\sum_{i=1}^{[nt]} \mathrm{E} \left(\eta_{ni}^2 \mid \mathcal{F}_{(i-1)/n} \right) \to_\mathrm{P} C_t;$

(ii) $\sum_{i=1}^{[nt]} \mathrm{E} \left[\eta_{ni} \left(Z_{i/n} - Z_{(i-1)/n} \right) \mid \mathcal{F}_{(i-1)/n} \right] \to_\mathrm{P} G_t;$

(iii) $\sum_{i=1}^{[nt]} \mathrm{E} \left(\eta_{ni}^2 \mathrm{I}(|\eta_{ni}| \geq \epsilon) \mid \mathcal{F}_{(i-1)/n} \right) \to_\mathrm{P} 0;$

(iv) $\sum_{i=1}^{[nt]} \mathrm{E} \left[\eta_{ni} \left(N_{i/n} - N_{(i-1)/n} \right) \mid \mathcal{F}_{(i-1)/n} \right] \to_\mathrm{P} 0$, *for all bounded martingale* $\{N_t, \mathcal{F}_t\}_{t \geq 0}$ *which are orthogonal to* Z_t.

Then, on $D_\mathrm{R}[0, \infty)$,

$$\sum_{j=1}^{[nt]} \eta_{nj} \Rightarrow X_t = X'_t + \int_0^t u_s dZ_s$$

with X'_t *is a martingale and* u_t *is a process adapted to* $\mathcal{F}_t$ *satisfying* $< X, X >_t = C_t$ *and* $< X, Z >_t = G_t$. *In particular, if* $Z_t = B_t$ *is a Brownian motion, and*

$$C_t = \int_0^t (\nu_s^2 + w_s^2) ds, \quad G_t = \int_0^t \nu_s ds,$$

then $X_t = \int_0^t \nu_s dB_s + \int_0^t w_s dB'_s$, *where* B'_s *is an Brwonian motion defined on the extension of the original* $\{\Omega, \mathcal{F}, \mathbf{F}, \mathrm{P}\}$ *and independent of* $\mathbf{F}$.

The result is still true if we replace $1/n$ *by any* $\Delta_n \to 0$.

3.9 Bibliographical Notes

Section 3.1. Stable convergence is weaker than convergence in probability and stronger than convergence in distribution, which is crucial in investigating the joint convergence of two random sequences. This section collects some of the most useful results in the area. These results come mainly from Aldous and Eagleson (1978). See also Jacod and Shiryaev (2003). For examples of stable convergence in related to the classical limit theorems, we refer to Eagleson (1976).

Section 3.2. This section provides classical central limit theorems for martingales. Context is largely self-contained, including basic limit theorem for dependent random variables, functional limit theorem for martingales and multivariate martingale limit theorems. Materials were partially taken from Helland (1982), Rootzen (1983) and Durrett and Resnick (1978). For more related results in this area, we refer to Jacod and Shiryaev (2003).

Section 3.3. Using random change of time, this section investigates the convergence to a mixture of normal distributions for a martingale array. We introduce the conditional variance conditions (**CVC1** and **CVC2**), which are less restrictive in comparison with the convergence in probability for conditional variance. The latter condition was used in Hall and Heyde (1980). These results seem to be new to the area.

Section 3.4. This section continues to investigate the martingale limit theorem under less restrictive conditions. It is shown that, for a certain class of martingales, a minor and easily verified additional condition, together with convergence in distribution for the conditional variance, is sufficient to establish the convergence to a mixture of normal distributions. The extension partially removes a barrier in the applications of the classical martingale limit theorem to nonparametric estimation and inference with non-stationarity and enhances the effectiveness of the classical martingale limit theorem as one of the main tools to investigate asymptotics in statistics, econometrics and other fields. Materials in this section are partially taken from a current work by Wang (2014).

Sections 3.5 and 3.6. We contribute to the convergence to a mixture of normal distributions without assuming martingale array structure. Section 3.5 considers sample covariances of functionals of $I(1)$ and $I(0)$ time series and Section 3.6 for zero energy functionals of $I(1)$ processes. Since many applications in econometrics involve a cointegration framework where endogeneity and nonlinearity play a major role, our results provide a foundation in developing nonlinear cointegrating regression. These results are partially taken from a series of papers by Wang and Phillips (2009b, 2011, 2014).

Section 3.7. This section considers uniform convergence for a class of martingales. There are lots of results in the area, but requiring the conditional variance to be a functional of stationary time series. We refer to Andrews (1995), AngoNze and Doukhan (2004), Bosq (1998), Masry (1996), Hansen (2008) and Wu, Huang and Huang(2010). A framework is established in this section, showing that uniform convergence rate for a martingale depends only on that of its conditional variance. As a consequence, our results are useful to stationary, nonstationary time series and to investigate the feasibility of data-dependent bandwidth h. The main result in this section is a modification of Wang and Chan (2014).

Section 3.8. This section collects main results on the convergence to a mixture of Gaussian processes for a continuous martingale. For related results, we refer to Jacod and Shiryaev (2003).

Chapter 4

Convergence to stochastic integrals

Consider a random array $\{x_{nk}, y_{nk}\}_{k \geq 1, n \geq 1}$ constructed from some underlying nonstationary time series and assume that there is a vector process $\{X(t), Y(t), 0 \leq t \leq 1\}$, to which $\{x_{n,[nt]}, y_{n,[nt]}\}$ converges weakly in Skorohod topology on $D_{\mathbf{R}^2}[0, 1]$, where $[a]$ denotes the integer part of a. A common functional of interest S_n of $\{x_{nk}, y_{nk}\}$ is defined by the sample quantity

$$S_n = \int_0^1 f(x_{n,[nt]}) dy_{n,[nt]} = \sum_{k=0}^{n-1} f(x_{nk}) \epsilon_{n,k+1},$$

where $\epsilon_{nk} = y_{nk} - y_{n,k-1}$ and f is a real function on R. This chapter investigates the sufficient conditions in such a way that the distribution of S_n converges to a stochastic integral, namely, $S_n \to_D \int_0^1 f[X(s)] dY(s)$.

Except explicitly mentioned, concept of stochastic processes is defined as in Appendix A. In particular, a (local, semi-, etc.) martingale is always assumed to be cádlág, i.e., its paths are right continuous and admit left-hand limits.

4.1 Definition of stochastic integrals

This section defines the stochastic integral $\int_a^b X_s dY_s$, for certain classes of stochastic processes $X = \{X_s\}_{a \leq s \leq b}$ and $Y = \{Y_s\}_{a \leq s \leq b}$, where $-\infty \leq a < b \leq \infty$.

If Y is a cádlág finite variation (FV) process and X is locally bounded on $[a, b]$, $\int_a^b X_s dY_s$ is naturally understood as the "pathwise" Lebesgue-Stieltjes integral. In particular, when X is continuous and $[c, d]$ is a finite interval, we may define

$$\int_c^d X_s dY_s = \lim_{n \to \infty} \sum_{t_{i-1}, t_i \in \pi_n} X_{t_{i-1}} (Y_{t_i} - Y_{t_{i-1}}), \tag{4.1}$$

where $\{\pi_n\}$ is a sequence of partitions of $[c, d]$ with mesh size going to zero, and the limit is understood as the convergence in probability.

Finite variation is often too restrictive for most of stochastic processes. It is even not the case for a Brownian motion that has an unbounded variation on any compact set. Itô (1944) adopted a different idea to define a stochastic integral. The theory essentially allows for $\int_a^b X_s dY_s$ to be well defined in the situation that X_s is a predictable process and Y_s is a semimartingale.

4.1.1 *Stochastic integrals with respect to a Brownian motion*

Let $\{\Omega, \mathcal{F}, \mathbf{F}, \mathrm{P}\}$ be a stochastic basis and $B = \{B_s\}_{a \le s \le b}$ be a Brownian motion. Denote by Π the space of all adapted processes H satisfying the condition $\int_a^b H_t^2 dt < \infty$ almost surely.

For any $X \in \Pi$, we can define

$$I(X) := \int_a^b X_s dB_s$$

by the L^2-theory (Hilbert space theory), which is established in the following three steps.

Step 1. For a bounded simple predictable (adapted) process $X = \{X_s\}_{a \le s \le b}$, given by

$$X_s = \xi_0 I_0 + \sum_{k=0}^{n-1} \xi_k I_{(t_k, t_{k+1}]}(s), \tag{4.2}$$

where $a = t_0 < t_1 < ... < t_n = b$ and ξ_k's are bounded $\mathcal{F}_{t_k}$-measurable, it is natural to define the stochastic integral $I(X)$ as the sum:

$$I(X) = \sum_{k=0}^{n-1} \xi_k (B_{t_{k+1}} - B_{t_k}).$$

Due to $\mathrm{E}\left[(B_t - B_s)^2 | \mathcal{F}_s\right] = t - s$ for any $s < t$, simple calculations show that, for any bounded simple predictable processes $X, X_1, X_2 \in \Pi$ and $\alpha, \beta \in \mathrm{R}$,

$$\alpha I(X_1) + \beta I(X_2) = I(\alpha X_1 + \beta X_2), \tag{4.3}$$

$$\mathrm{E} \, I^2(X) = \mathrm{E} \int_a^b X_s^2 ds. \tag{4.4}$$

Step 2. Process X satisfies an additional condition $\mathrm{E} \int_a^b X_t^2 dt < \infty$. In this situation, there exists a sequence of bounded simple predictable processes $X^n = \{X_s^n\}_{a \leq s \leq b}$ such that

$$\lim_{n \to \infty} \mathrm{E} \int_a^b |X_s - X_s^n|^2 ds = 0. \tag{4.5}$$

See Liptser and Shiryaev (1989, p97), for instance. Note that $X^m - X^n$ is still a bounded simple predictable process. It follows from (4.3) - (4.5) that, as $m, n \to \infty$,

$$\mathrm{E}\left[I(X^m) - I(X^n)\right]^2 = \mathrm{E} \int_a^b (X_s^m - X_s^n)^2 ds \to 0.$$

Hence $I(X^n)$ is a Cauchy sequence in L^2, where

$$L^2 = \{Z : \int_{-\infty}^{\infty} Z^2 d\mathrm{P} < \infty\},$$

i.e., a space of all random variables $Z \in \{\Omega, \mathcal{F}, \mathrm{P}\}$ satisfying $\mathrm{E}Z^2 < \infty$. Since L^2 is complete with respect to the (pseudo) metric $d(Z_1, Z_2) = [\mathrm{E}(Z_1 - Z_2)^2]^{1/2}$, the limit of $I(X^n)$ exists and is in L^2. We naturally define

$$I(X) = \lim_{n \to \infty} I(X^n), \tag{4.6}$$

where the limit is understood in the sense that

$$\lim_{n \to \infty} \mathrm{E}\big(I(X^n) - I(X)\big)^2 = 0.$$

We have to prove that the $I(X)$ defined by (4.6) is free of the choice for the simple predictable processes $X^n = \{X_s^n\}_{a \leq s \leq b}$. Due to (4.3)-(4.5), this is obvious. Furthermore, it follows easily from (4.5) and (4.6) that (4.3) and (4.4) continue to hold for any $X \in \Pi$ satisfying $\mathrm{E} \int_a^b X_s^2 dt < \infty$.

Step 3. General $X \in \Pi$. Define a sequence $\hat{X}^n = \{\hat{X}_s^n\}_{a \leq s \leq b}$ by

$$\hat{X}_s^n = \begin{cases} X_s, & \text{if } \int_a^s X_t^2 dt \leq n, \\ 0, & \text{otherwise.} \end{cases}$$

For each $n \geq 1$, $\mathrm{E} \int_a^b (\hat{X}_s^n)^2 ds \leq n$, i.e., $I(\hat{X}^n) = \int_a^b \hat{X}_s^n dB_s$ is well defined. If we prove

$$I(\hat{X}^n) - I(\hat{X}^m) = o_{\mathrm{P}}(1), \quad \text{as } m, n \to \infty, \tag{4.7}$$

then $I(\hat{X}^n)$ converges in probability as $n \to \infty$. Consequently, for general $X \in \Pi$, we define

$$I(X) = \lim_{n \to \infty} I(\hat{X}^n), \quad \text{in probability.} \tag{4.8}$$

To prove (4.7), we introduce the following lemma.

Lemma 4.1. *For any $X \in \Pi$ satisfying $\mathrm{E} \int_a^b X_t^2 dt < \infty$, we have*

$$\mathrm{P}(|I(X)| \geq \epsilon) \leq \epsilon/2 + \mathrm{P}\left(\int_a^b X_s^2 ds \geq \epsilon^3/2\right), \qquad (4.9)$$

for any $\epsilon > 0$.

Proof. Let $X^\epsilon = \{X_s^\epsilon\}_{a \leq s \leq b}$ with $X_s^\epsilon = X_s I_{(\int_a^s X_t^2 dt \leq \epsilon^3/2)}$. We have

$$\mathrm{P}(|I(X)| \geq \epsilon) \leq \mathrm{P}(|I(X^\epsilon)| \geq \epsilon) + \mathrm{P}\left(\int_a^b X_s^2 ds \geq \epsilon^3/2\right).$$

The result (4.9) follows from

$$\mathrm{P}(|I(X^\epsilon)| \geq \epsilon) \leq \epsilon^{-2} \mathrm{E}\, I^2(X^\epsilon) = \epsilon^{-2} \mathrm{E} \int_a^b (X_s^\epsilon)^2 ds \leq \epsilon/2. \qquad \square$$

The proof of (4.7) follows from an application of Lemma 4.1. Indeed, by recalling the definition of $\hat{X}_t^n$, we have

$$\int_a^b (\hat{X}_t^n - X_t)^2 dt = o_\mathrm{P}(1),$$

as $n \to \infty$. This implies that

$$\int_a^b (\hat{X}_t^n - \hat{X}_t^m)^2 dt = o_\mathrm{P}(1),$$

as $m, n \to \infty$. Now (4.7) follows (4.9) with $X = \hat{X}^n - \hat{X}^m = \{\hat{X}_s^n - \hat{X}_s^m\}_{a \leq s \leq b}$ and the fact that $I(\hat{X}^n) - I(\hat{X}^m) = I(\hat{X}^n - \hat{X}^m)$ as indicated in Step 2.

4.1.2 *Stochastic integrals with respect to a (local) square integrable martingale*

Let $\{\Omega, \mathcal{F}, \mathbf{F}, \mathrm{P}\}$ be a stochastic basis. Suppose $Y = \{Y_s\}_{a \leq s \leq b}$ is a martingale with $\sup_t \mathrm{E}\, Y_t^2 < \infty$. Since Y^2 is a submartingale, the Doob-Meyer theorem states that there exists a unique decomposition:

$$Y_t^2 = L_t + \langle Y \rangle_t, \qquad (4.10)$$

where L_t is a martingale and $\langle Y \rangle_t$ is a cádlág, predictable increasing process such that $E\langle Y \rangle_a = 0$ and $E\langle Y \rangle_t < \infty$ for all $t \in [a, b]$. The decomposition (4.10) continues to hold for a local square integrable martingale, but L_t is now a local martingale and $\langle Y \rangle_t$ is a local cádlág, predictable increasing process.

Denote by L_P^2 the space of all predictable processes H satisfying the condition $\int_a^b H_t^2 d\langle Y \rangle_t < \infty$ almost surely. For any $X \in L_P^2$, step by step as described in Section 4.1.1, the stochastic integral

$$\int_a^b X_s dY_s$$

is well-defined by the L^2-theory. See, e.g., Chapter IV.4 of Rogers and Williams (2000).

Let τ be a stopping time. For any local martingale $Z = \{Z_s\}_{a \le s \le b}$, we define $Z^\tau = \{Z_{s \wedge \tau}\}_{a \le s \le b}$. It follows from the Doob-Meyer decomposition theorem that $\langle Z^\tau \rangle = \langle Z \rangle^\tau$, i.e.,

$$\{\langle Z^\tau \rangle_s\}_{a \le s \le b} = \{\langle Z \rangle_{s \wedge \tau}\}_{a \le s \le b},$$

provided $\sup_t E Z_{t \wedge \tau}^2 < \infty$. Furthermore, for any H satisfying $E \int_a^{b \wedge \tau} H_t^2 d\langle Y \rangle_t < \infty$,

$$\int_a^{b \wedge \tau} H_s dZ_s = \int_a^b H_s dZ_{s \wedge \tau}. \tag{4.11}$$

See Liptser and Shiryaev (1989, page 97).

Result (4.11) allows us to define $\int_a^b X_s dY_s$ for a local square integrable martingale $Y = \{Y_s\}_{a \le s \le b}$ and $X \in L_P^2$. Indeed, in this situation, there exist stopping times $a = \tau_0 \le \tau_1 \le \tau_2 \le \ldots \le b$ with $\tau_n \to b$ such that for each $n \ge 1$, $\sup_t EY_{t \wedge \tau_n}^2 < \infty$ and $\int_a^{\tau_n} X_t^2 d\langle Y \rangle_t \le n$, implying that $\int_a^{\tau_n} X_s dY_s$ is well defined due to (4.11). Hence we naturally define

$$\int_a^b X_s dY_s = \sum_{j=0}^\infty \int_{\tau_j+}^{\tau_{j+1}} X_s dY_s.$$

4.1.3 Stochastic integrals with respect to a local (semi-) martingale

Let $\{\Omega, \mathcal{F}, \mathbf{F}, \mathrm{P}\}$ be a stochastic basis. For a local martingale $Y = \{Y_s\}_{a \le s \le b}$, there exists a unique decomposition:

$$Y_t = Y_t^c + Y_t^d, \tag{4.12}$$

where $Y^c = \{Y_s^c\}_{a \le s \le b}$ and $Y^d = \{Y_s^d\}_{a \le s \le b}$ are continuous and purely discontinuous local martingales, respectively. Note that Y^c is a local square integrable martingale. The quadratic variation of Y can be written as

$$[Y, Y]_t = \langle Y^c \rangle_t + \sum_{0 \le s \le t} (\Delta Y_s)^2, \qquad a \le t \le b,$$

where $\langle Y^c \rangle = \{\langle Y^c \rangle_s\}_{a \leq s \leq b}$ is the cádlág, predictable increasing process appeared in the the Doob-Meyer decomposition of $(Y^c)^2$ [see (4.10)].

Denote by $\widetilde{L}_P^2(Y)$ the space of all predictable processes H such that $V = \{V_s\}_{a \leq s \leq b}$ is local integrable, where $V_s = (\int_a^s H_t^2 d[Y,Y]_t)^{1/2}$. For any $X \in \widetilde{L}_P^2(Y)$, the stochastic integral

$$\int_a^b X_s dY_s$$

is well-defined. For the details of this definition, we refer to Chapter 2.2 of Liptser and Shiryaev (1989).

Finally, let $Y = \{Y_s\}_{a \leq s \leq b}$ be a semimartingale, having a representation:

$$Y_s = Y_0 + M_s + A_s,$$

where $M = \{M_s\}_{a \leq s \leq b}$ is a local martingale with $M_a = 0$ and $A = \{A_s\}_{a \leq s \leq b}$ is a cádlág FV process. We define the stochastic integral $\int_a^b X_s dY_s$ by

$$\int_a^b X_s dY_s = \int_a^b X_s dM_s + \int_a^b X_s dA_s,$$

where $\int_a^b X_s dA_s$ is understood as a "pathwise" Lebesgue-Stieltjes integral.

4.1.4　*Multivariate stochastic integrals*

Let $Y = \{Y(s)\}_{a \leq s \leq b}$, where $Y(s) = (Y_1(s), ..., Y_d(s))$, be a d-dimensional semi-martingale on a stochastic basis $\{\Omega, \mathcal{F}, \mathbf{F}, P\}$. For a $k \times d$ matrix $X = \{X(s)\}_{a \leq s \leq b}$ of processes, where

$$X(s) = \begin{pmatrix} X_11(s), X_{12}(s), \cdots, X_{1d}(s) \\ X_{21}(s), X_{22}(s), \cdots, X_{2d}(s) \\ \cdots \quad \cdots \\ X_{k1}(s), X_{k2}(s), \cdots, X_{kd}(s) \end{pmatrix},$$

if $X_{ij} = \{X_{ij}(s)\}_{a \leq s \leq b} \in \widetilde{L}_P(Y_j)$ for $i = 1, ..., k$, we define

$$\int_a^b X(s) dY'(s) = Z,$$

where $Z = (Z_1, \cdots, Z_k)$ is a vector-valued process with

$$Z_i = \sum_{j=1}^d \int_a^b X_{ij}(s) dY_j(s).$$

Similarly, if $k = 1$ and $X_{1i} = \{X_{1i}(s)\}_{a \leq s \leq b} \in \tilde{L}_{\mathrm{P}}(Y_j)$ for each $1 \leq i, j \leq d$, we define

$$\int_a^b X'(s) dY(s) = Z,$$

where $Z = (Z_{ij})_{1 \leq i, j \leq d}$ is a matrix of processes with components

$$Z_{ij} = \int_a^b X_{1i}(s) dY_j(s).$$

This definition is natural, but not the most general stochastic integral of a matrix process X with respect to a vector-valued semimartingale Y. The most general extension is related to the quadratic co-variation $[Y_i, Y_j]$, which is given by

$$[Y_i, Y_j] = \frac{1}{4} \left([Y_i + Y_j, Y_i + Y_j] - [Y_i - Y_j, Y_i - Y_j] \right).$$

W refer to Jacod and Shiryaev (2003) for details.

4.1.5 *Properties of stochastic integrals*

Let Y be a semimartingale and X a predictable process such that the stochastic integral $V_t := \int_a^t X_s dY_s$ is well defined for $a \leq t \leq b$. Define

$$X \cdot Y = \{V_t\}_{a \leq t \leq b}.$$

This process has the following properties, as stated in Protter (2005).

(i) If $X = \alpha X_1 + \beta X_2$ and $Y = Y_1 + Y_2$, then

$$X \cdot Y = \alpha (X_1 \cdot Y) + \beta (X_2 \cdot Y), \quad X \cdot Y = X \cdot Y_1 + X \cdot Y_2.$$

(ii) If Y is a square integrable martingale, then so is $X \cdot Y$. Moreover, for $a \leq t \leq b$,

$$\langle X \cdot Y \rangle_t = \int_a^t X_s^2 d\langle Y \rangle_s,$$

$$\mathrm{E} \left(\int_a^t X_s dY_s \right)^2 = \mathrm{E} \int_a^t X_s^2 d\langle Y \rangle_s.$$

(iii) If Y is a semi-(local) martingale, then so is $X \cdot Y$. Moreover, for $a \leq t \leq b$,

$$[X \cdot Y, X \cdot Y]_t = \int_a^t X_s^2 d[Y, Y]_s, \quad \Delta(X \cdot Y)_t = X_t \, \Delta Y_t.$$

In particular, if Y is continuous semi-(local) martingale, then so is $X \cdot Y$.

(iv) If $Y = H \cdot Z$, then $X \cdot Y = (XH) \cdot Z$, i.e., if $Y_t = \int_a^t H_s dZ_s$, then

$$\int_a^t X_s dY_s = \int_a^t X_s H_s dZ_s.$$

(v) **(Dominated Convergence Theorem)** Let H_m be a sequence of predictable processes converging a.s. to a limit H. If $|H_m| \le X$ a.s. for all m, then $H_m \cdot Y$, $H \cdot Y$ are well defined and $H_m \cdot Y$ converges to $H \cdot Y$ uniformly on compact sets with probability one.

4.1.6   *Martingale representation theorem and Itô's formula*

Let $\{\Omega, \mathcal{F}, \mathbf{F}, P\}$ be a stochastic basis and $B = \{B_t\}_{t \ge 0}$ be a Brownian motion. If H is predictable for $\mathbf{F}$ and $\int_0^T H_s^2 ds < \infty$, then $\int_0^t H_s dB_s, 0 \le t \le T$, is a square integrable martingale. The converse of this result is also true, as stated in the following theorem.

Theorem 4.1. (Martingale representation theorem) *Let $\mathcal{F}_t, 0 \le t \le T$, be the filtration generated by B.*

(i) Every square integrable martingale $M = \{M_t\}_{0 \le t \le T}$ with respect to $\mathbf{F} = \{\mathcal{F}_t\}_{0 \le t \le T}$ has a representation

$$M_t = M_0 + \int_0^t H_s dB_s,$$

where $H = \{H_t\}_{0 \le t \le T}$ is predictable and $\mathrm{E} \int_0^T H_s^2 ds < \infty$.

(ii) Every local martingale $M = \{M_t\}_{0 \le t \le T}$ with respect to $\mathbf{F} = \{\mathcal{F}_t\}_{0 \le t \le T}$ has a representation

$$M_t = M_0 + \int_0^t H_s dB_s,$$

where $H = \{H_t\}_{0 \le t \le T}$ is predictable and $\int_0^T H_s^2 ds < \infty$, a.s.

The proof of Theorem 4.1, together with the following Itô's formulas, can be found in Chapter IV of Protter (2005).

Theorem 4.2. (Itô's formula I) *Let X be a semimartingale and let g be a twice differentiable real function. Then $g(X)$ is a semimartingale and*

$$g(X_t) = g(x_0) + \int_{0+}^t g'(X_{s-}) dX_s + \frac{1}{2} \int_{0+}^t g''(X_{s-}) d[X,X]_s^c$$
$$+ \sum_{0 < s \le t} \left\{ g(X_s) - g(X_{s-}) - g'(X_{s-}) \Delta X_s \right\}. \tag{4.13}$$

Theorem 4.3. (Itô's formula II) *Let X be a semimartingale and let g be a real convex function. Then $g(X)$ is a semimartingale and*

$$g(X_t) = g(x_0) + \int_{0+}^{t} g'(X_{s-})dX_s + A_t, \qquad (4.14)$$

where g' is the left derivative of g (i.e., $g'(x) = \lim_{h\uparrow 0} \frac{1}{h}[f(x+h) - f(x)]$), and A is an adapted, right continuous, increasing process. Moreover,

$$\Delta A_t = g(X_t) - g(X_{t-}) - g'(X_{t-})\Delta X_t.$$

If we let μ be the second derivative measure on $(\mathrm{R}, \mathcal{R})$ defined by

$$\mu([a,b)) = g'(b) - g'(a), \quad -\infty < a < b < \infty,$$

then

$$A_t = \frac{1}{2} \int_{\infty}^{\infty} L(t,a)\mu(da) + \sum_{0<s\leq t} \{f(X_s) - f(X_{s-}) - f'(X_{s-})\Delta_s\},$$

where $L = \{L(t,x)\}_{t\geq 0, x\in\mathrm{R}}$ is a local time of X, i.e., $L(t,x)$ satisfies

$$\int_0^t f(X_s)d[X,X]^c = \int_{\infty}^{\infty} f(a)L(t,a)da$$

for all bounded Borel functions.

4.2 Weak convergence of stochastic integrals

Let $\{\Omega^n, \mathcal{F}^n, \mathbf{F}^n, \mathrm{P}^n\}_{n\geq 1}$ be a sequence of stochastic bases on which $Y^n = \{Y_s^n\}_{s\geq 0}$ is a sequence of semimartingales and $X^n = \{X_s^n\}_{s\geq 0}$ is a sequence of cádlág predictable processes. This section investigates the conditions on X^n and Y^n so that

$$\left(X^n,\ Y^n,\ \int_0^{\cdot} X_s^n dY_s^n\right) \Rightarrow \left(X,\ Y,\ \int_0^{\cdot} X_s dY_s\right). \qquad (4.15)$$

Definition 4.1. Let $S^n = \{S_t^n\}_{t\geq 0}, n \geq 1$, be a collection of all simple predictable processes. A sequence of semimartingales $\{Y^n\}_{n\geq 1}$, is said to be uniformly tight (UT) if, for each $t \geq 0$, the set

$$\left\{ \int_0^t H_{s-}^n dY_s^n,\ H^n \in S^n,\ |H^n| \leq 1 \right\}_{n\geq 1}$$

is stochastically bounded.

Condition UT was introduced in Stricker (1985) and later considered by Jakubowski, et al. (1989), indicating the family of all random variables of the form

$$\sum_{i=1}^{m} H_{t_{i-1}}^n (Y_{t_i}^n - Y_{t_{i-1}}^n),$$

is uniformly tight, where $m \in N$, $0 = t_0 < ... < t_m < \infty$, $|H_{t_i}^n| \leq 1$ and H_{t_i} is $\mathcal{F}_{t_i}^n$-measurable.

For a sequence of semimartingales $\{Y^n\}_{n \geq 1}$, where $Y^n = M^n + A^n$, it satisfies UT if, for each $t \geq 0$,

(i) $\sup_n \mathrm{E}^n \{ \sup_{s \leq t} |\Delta M_s^n| \} + \mathrm{E}^n (\int_0^t |dA_s^n|) \} < \infty$; or

(ii) $Y^n \Rightarrow Y$ on $D[0, \infty)$ and

$$\sup_n \{ var(M_t^n) + \mathrm{E}^n (\int_0^t |dA_s^n|) \} < \infty,$$

or equivalently,

$$\sup_n \{ \mathrm{E}^n ([M^n, M^n]_t) + \mathrm{E}^n (\int_0^t |dA_s^n|) \} < \infty.$$

Theorem 4.4. *Suppose $(X^n,\ Y^n) \Rightarrow (X,\ Y)$ on $D_{\mathrm{R}^2}[0, \infty)$ and $\{Y^n\}_{n \geq 1}$ are semimartingales satisfying UT. Then, Y is a semimartingale with respect to the natural filtration generated by (X, Y) and (4.15) holds on $D_{\mathrm{R}^3}[0, \infty)$.*

Theorem 4.4 is a sharp result. A proof can be found in Jakubowski, et al. (1989) or/and Kurtz and Protter (1996). The convergence of $(X^n,\ Y^n)$ can be considered either on $D_{\mathrm{R}}[a, b] \times D_{\mathrm{R}}[a, b]$ or $D_{\mathrm{R}^2}[a, b]$ in Skorokhod topology. The latter convergence is stronger as we require only one sequence $0 \leq \lambda_n(t) \leq 1$ of changes of time such that $(X_{\lambda_n(t)}^n, Y_{\lambda_n(t)}^n)$ converges uniformly to (X_t, Y_t) on $t \in [a, b]$. The following example shows that only $(X^n,\ Y^n) \Rightarrow (X,\ Y)$ on $D_{\mathrm{R}}[a, b] \times D_{\mathrm{R}}[a, b]$ does not suffice to establish (4.15).

Example 4.1. Let $Y_t^n = Y_t = X_t = \mathrm{I}_{[1/2,1]}(t)$, $X_t^n = \mathrm{I}_{[1/2-1/n,1]}(t)$, $n \geq 1$. Then,

$$(X^n,\ Y^n) \Rightarrow (X,\ Y),$$

on $D_{\mathrm{R}}[0, 1] \times D_{\mathrm{R}}[0, 1]$, but $\int_0^1 X_{t-}^n dY_t^n = 1 \neq \int_0^1 X_{t-} dY_t = 0$.

Definition 4.2. A sequence of semimartingales $\{Y^n\}_{n\geq 1}$ is said to be good if, for any cádlág predictable process H^n, $(H^n,\, Y^n) \Rightarrow (H,\, Y)$ on $D_{\mathrm{R}^2}[0,\infty)$ implies that Y is a semi-martingale and

$$\left(H^n,\, Y^n,\, \int_0^{\cdot} H_s^n dY_s^n\right) \Rightarrow \left(H,\, Y,\, \int_0^{\cdot} H_s dY_s\right)$$

on $D_{\mathrm{R}^3}[0,\infty)$.

By Theorem 4.4, a sequence of semimartingales $\{Y^n\}_{n\geq 1}$ is good if it satisfies UT. Furthermore, we have the following results for goodness. These results are given in Kurtz and Protter (1996).

(1) If $\{Y^n\}_{n\geq 1}$ is good and $(H^n,\, Y^n) \Rightarrow (H,\, Y)$ on $D_{\mathrm{R}^2}[0,\infty)$, then $Z^n = \{Z_t^n\}_{t\geq 0}$ is good, where $Z_t^n = \int_0^t H_s^n dY_s^n$.

(2) If $\{Y^n\}_{n\geq 1}$ is good and if $f(x,y) : \mathrm{R} \times \mathrm{R}^+ \to \mathrm{R}$ is twice differentiable on R and differentiable on R^+, then $Z^n = \{Z_t^n\}_{t\geq 0}$ is good, where $Z_t^n = f(Y_t^n, t)$.

(3) If $\{X_n\}_{n\geq 1}$ and $\{Y^n\}_{n\geq 1}$ are good and $(X^n, Y^n)_{n\geq 1}$ is a sequence of semimartingales such that $(X^n,\, Y^n) \Rightarrow (X,\, Y)$ on $D_{\mathrm{R}^2}[0,\infty)$, then

$$(X^n,\, Y^n,\, [X^n, Y^n]) \Rightarrow (X,\, Y,\, [X,Y])$$

and both $[X^n, Y^n]$ and $X^n Y^n$ are good.

4.3 Weak convergence of stochastic integrals: multivariate extension

Let $\Theta_n = \{\Omega^n, \mathcal{F}^n, \mathbf{F}^n, \mathrm{P}^n\}_{n\geq 1}$ be a sequence of stochastic bases. For each $n \geq 1$, suppose that $Y^n = \{Y^n(s)\}_{s\geq 0}$ is a d-dimensional semimartingale, where $Y^n(s) = (Y_1^n(s), ..., Y_d^n(s))$, and $X^n = \{X^n(s)\}_{s\geq 0}$ is a $k \times d$ matrix of cádlág predictable processes, where

$$X^n(s) = \left(X_{ij}^n(s)\right)_{1\leq i\leq k, 1\leq j\leq d}.$$

Further assume stochastic integral $\int_0^{\cdot} X_{ij}^n(s) dY_j^n(s)$ is well defined for each $n \geq 1$, $1 \leq i \leq k$ and $1 \leq j \leq d$. The following result is a corollary of Theorem 4.4 and the Cramér-Wold theorem, providing multivariate extension for weak convergence of stochastic integrals.

Theorem 4.5. *Suppose* $(X^n, Y^n) \Rightarrow (X, Y)$ *on* $D_{\mathbf{R}^{kd} \times \mathbf{R}^d}[0, \infty)$ *and, for each* $1 \leq j \leq d$, $\{Y_j^n\}_{n \geq 1}$ *satisfies UT. Then* Y *is a* d-*dimensional semimartingale with respect to the natural filtration generated by* (X, Y) *and*

$$\left(X^n, Y^n, \int_0^{\cdot} X_n(s) dY_n'(s)\right) \Rightarrow \left(X, Y, \int_0^{\cdot} X(s) dY'(s)\right), \quad (4.16)$$

on $D_{\mathbf{R}^{kd} \times \mathbf{R}^d \times \mathbf{R}^k}[0, \infty)$. *If* $k = 1$, *we also have*

$$\left(X^n, Y^n, \int_0^{\cdot} X_n'(s) dY_n(s)\right) \Rightarrow \left(X, Y, \int_0^{\cdot} X'(s) dY(s)\right), \quad (4.17)$$

on $D_{\mathbf{R}^d \times \mathbf{R}^d \times \mathbf{R}^{d^2}}[0, \infty)$.

4.4 Convergence to stochastic integrals: random arrays

Let $\mathbf{F}_n = \{\mathcal{F}_k^n\}_{k \geq 1}$ be a sequence of filtrations so that, for each n, $\{x_{nk}, y_{nk}\}_{k \geq 1}$ is an $\{\mathcal{F}_k^n\}_{k \geq 1}$-adapted process and $\{y_{nk}\}_{k \geq 1}$ is a d_1-dimensional semimartingale with respect to $\{\mathcal{F}_k^n\}_{k \geq 1}$. Furthermore let each component $y_{nk}^{(i)}$ of y_{nk} has a decomposition:

$$y_{nk}^{(i)} = M_{nk}^{(i)} + A_{nk}^{(i)}, \quad i = 1, \cdots, d_1,$$

where $M_{nk}^{(i)}$ is a martingale and $A_{nk}^{(i)}$ is a predictable process. As a direct consequence of Theorem 4.5, the following theorem provides a result on the weak convergence to stochastic integrals for random arrays with a semimartingale structure.

Theorem 4.6. *Suppose*

(i) $\{x_{n,[nt]}, y_{n,[nt]}\} \Rightarrow \{X(t), Y(t)\}$ *on* $D_{\mathbf{R}^d \times \mathbf{R}^{d_1}}[0, 1]$;
(ii) for each $1 \leq i \leq d_1$,

$$\sup_n \left[\mathrm{E}\,(M_{nn}^{(i)})^2 + \sum_{k=1}^{n-1} \mathrm{E}\,|A_{n,k+1}^{(i)} - A_{nk}^{(i)}|\right] < \infty.$$

Then $Y(t)$ *is a* d_1-*dimensional semimartingale with respect to a filtration to which* $X(t)$ *and* $Y(t)$ *are adapted, and for any continuous functions* $g_1(s)$ *and* $g_2(s)$ *on* $\mathbf{R}^d$,

$$\left\{x_{n,[nt]}, \; y_{n,[nt]}, \; \frac{1}{n}\sum_{k=1}^{[nt]} g_1(x_{nk}), \; \sum_{k=1}^{[nt]} g_2(x_{n,k-1})\,(y_{nk} - y_{n,k-1})\right\}$$

$$\Rightarrow \left\{X(t), \; Y(t), \; \int_0^t g_1[X(s)]ds, \; \int_0^t g_2[X(s)]\,dY(s)\right\}, \quad (4.18)$$

on $D_{\mathbf{R}^{2d+d_1+1}}[0, 1]$, *where we define* $x_{n0} = y_{n0} = 0$.

Proof. Note that $\frac{1}{n}\sum_{k=1}^{[nt]} g_1(x_{nk}) = \int_0^t g_1(x_{n,[ns]})ds + o_P(1)$ and

$$\sum_{k=1}^{[nt]} g_2(x_{n,k-1})\,(y_{nk} - y_{n,k-1}) = \int_0^t g_2(x_{n,[ns]})dy_{n,[ns]} + o_P(1).$$

Since, by the continuous mapping theorem,

$$\left\{x_{n,[nt]},\ y_{n,[nt]},\ g_2(x_{n,[nt]}),\ \frac{1}{n}\sum_{k=1}^{[nt]} g_1(x_{nk})\right\}$$

$$\Rightarrow \left\{X(t),\ Y(t),\ g_2(X(t)),\ \int_0^t g_1[X(s)]ds\right\},$$

result (4.18) follows immediately from Theorem 4.5. $\qquad\square$

Remark 4.1. The continuity assumptions on g_1 and g_2 can be weakened. For instance, Theorem 4.6 holds if both g_1 and g_2 are Riemann integrable functions. Let g_1 and g_2 be measurable mappings from $D_{\mathbb{R}^d}[0,1]$ to R. Denote by D_{g_1} and D_{g_2} the set of discontinuities of g_1 and g_2, respectively. Theorem 4.6 still holds if $P(X \in D_{g_1}) + P(X \in D_{g_2}) = 0$.

Example 4.2. Let $\{\epsilon_t, \mathcal{F}_t\}_{t\geq 0}$ be a sequence of martingale differences such that $\sup_{t\geq 1} E\epsilon_t^2 < \infty$,

$$\frac{1}{n}\sum_{t=2}^n E\left(\epsilon_t^2 \mid \mathcal{F}_{t-1}\right) = 1 + o_P(1), \quad \text{and}$$

$$\frac{1}{n}\sum_{t=2}^n E\left(\epsilon_t^2 I(|\epsilon_t| \geq \alpha\sqrt{n}) \mid \mathcal{F}_{t-1}\right) = o_P(1) \quad \text{for all } \alpha > 0.$$

Let $\beta_n = 1 + \gamma/n$ with $\gamma \in R$, $X_{n0} = 0$ and

$$X_{nt} = \beta_n X_{n,t-1} + \epsilon_t, \quad 1 \leq t \leq n.$$

Then $X_{nk}/\sqrt{n}$ is a semimartingale with decomposition:

$$\frac{1}{\sqrt{n}}X_{nk} = \frac{1}{\sqrt{n}}\sum_{j=1}^k \epsilon_j + \frac{\gamma}{n^{3/2}}\sum_{j=1}^{k-1} X_{nj} := M_{n,k} + A_{n,k}.$$

Simple calculations show that

$$\sup_n \left(EM_{n,n}^2 + \sum_{k=1}^{n-1} E\,|A_{n,k+1} - A_{nk}|\right)$$

$$\leq \sup_{t\geq 1} E\epsilon_t^2 + \sup_{n\geq 2}\frac{1}{n^{3/2}}\sum_{j=1}^n E\,|X_{nj}| < \infty.$$

Furthermore, $\frac{1}{\sqrt{n}} X_{n,[nt]} \Rightarrow \int_0^t e^{\gamma(t-s)} dB_s$ on $D_R[0,1]$, where $B = \{B_s\}_{s \geq 0}$ is a Brownian motion. See, Buchmann and Chan (2007), for instance. As a consequence of Theorem 4.6, for any real functions $f_n(x), n \geq 1$, satisfying

$$\sup_{x \in [0,1]} |f_n(x) - f(x)| = o(1),$$

where $f(x)$ is continuous on $[0,1]$, we have

$$\frac{1}{\sqrt{n}} \sum_{k=1}^n f_n\left(\frac{k}{n}\right) \left(X_{nk} - X_{n,k-1}\right)$$

$$\rightarrow_D \int_0^1 f(t) dB_t + \gamma \int_0^1 f(t) \int_0^t e^{\gamma(t-s)} dB_s dt.$$

Example 4.3. (Kurtz and Protter, 1991). Let $\epsilon_i, i \geq 1$, be a sequence of i.i.d. random variables with $E\epsilon_1 = 0$ and $E\epsilon_1^2 = 1$. Write

$$W_n^{(m)}(t) = \frac{1}{n^{m/2}} \sum_{1 \leq i_1 < \dots < i_m \leq [nt]} \epsilon_{i_1} \cdots \epsilon_{i_m},$$

and $Z_n(t) = (W_n^{(1)}(t), \cdots, W_n^{(m)}(t))$. Note that $W_n^{(m)}(t) = \int_0^t W_n^{(m-1)}(s-) dW_n^{(1)}(s)$. We have

$$Z_n(t) \Rightarrow (B^{(1)}(t), \cdots, B^{(m)}(t))$$

on $D_{R^m}[0,1]$, where $B^{(1)}(t) = B_t$ is a Brownian motion and $B^{(m)}(t) = \int_0^t B^{(m-1)}(s-) dB^{(1)}(s)$.

Example 4.4. (Duffie and Protter, 1992). Let Y_{nk} be a sequence of random variables such that $Z_n(t) = \sum_{k=1}^{[nt]} Y_{nk} \Rightarrow Z(t)$ on $D_R[0,1]$ and $Z^n = \{Z_n(t)\}$ is good. Let

$$S_{nk} = S_{n0} \Pi_{i=1}^k (1 + Y_{ni}).$$

Since $S_{n,k+1} - S_{nk} = S_{nk} Y_{n,k+1}$, we can write

$$S_{n,[nt]} = S_{n0} + \int_0^t S_{n,[ns]-} dZ_n(s).$$

Suppose $S_{n0} \Rightarrow S_0$. It follows from Theorem 4.6 that $S_{n,[nt]} \Rightarrow S(t)$ on $D_R[0,1]$, where the limit process $S(t)$ satisfies

$$S(t) = S_0 + \int_0^t S(s-) dZ(s).$$

When Y_{nk} is the periodic rate of return on a stock with initial price S_0, S_{nk} is the price of the stock after k periods. Hence if $\{\theta_{nk}\}_{k \geq 1}$ represents a trading strategy and $G_{n,[nt]} = \int_0^t \theta_{n,[ns]-} dS_{n,[ns]}$ the resulting "gain" from the strategy $\{\theta_{nk}\}_{k \geq 1}$, we have

$$G_{n,[nt]} \Rightarrow G(t), \quad \text{where } G(t) = \int_0^t \theta_{s-} dS(s),$$

on $D_R[0,1]$.

Example 4.5. (Slominski, 1989) For each n, let $\{H_{nk}, y_{nk}\}_{k \geq 1}$ be an $\{\mathcal{F}_k^n\}_{k \geq 1}$-adapted process and $\{y_{nk}\}$ be an $\{\mathcal{F}_k^n\}$-martingale. Consider a sequence of stochastic difference equations (SDE):

$$x_{nk} = H_{nk} + \sum_{j=0}^{k-1} f(x_{nj})(y_{n,j+1} - y_{nj}), \quad k \geq 1, \quad (4.19)$$

where $f : \mathrm{R} \to \mathrm{R}$ satisfies the Lipschitz condition, i.e., there exists a constant C such that

$$|f(u) - f(v)| \leq C|u - v|, \quad u, v \in \mathrm{R}.$$

It is known that, for each $n \geq 1$, SDE (4.19) has a strong solution. Furthermore, if $\{H_{n,[nt]}, y_{n,[nt]}\} \Rightarrow \{H(t), Y(t)\}$ on $D_{\mathrm{R}^2}[0, 1]$ and $\sup_n Ey_{nn}^2 < \infty$, then

$$\left\{ x_{n,[nt]}, H_{n,[nt]}, y_{n,[nt]} \right\} \Rightarrow \left\{ X(t), H(t), Y(t) \right\} \quad (4.20)$$

on $D_{\mathrm{R}^3}[0, 1]$, where X_t is a strong solution of the SDE:

$$X(t) = H(t) + \int_0^t f(X_{s-})dY(s), \quad t \geq 0.$$

Let $f_n(x)$ be a sequence of real functions satisfying the Lipschitz condition and

if $(x_n, y_n) \Rightarrow (x, y)$ on $D_{\mathrm{R}^2}[0, 1]$, then $(x_n, y_n, f_n(x_n)) \Rightarrow (x, y, f(x))$ on $D_{\mathrm{R}^3}[0, 1]$.

Then result (4.20) still holds if x_{nk} is a sequence of strong solutions of the SDE (4.19) with f_n instead of f.

Remark 4.2. More general results exist. We refer to Kurtz and Protter (1996) and Jacod and Shiryaev (2003) for more details.

4.5 Convergence to stochastic integrals: beyond the semi-martingale

Suppose $\{x_{nk}, y_{nk}\}_{k \geq 1, n \geq 1}$ is a random vector array, where

$$x_{nk} = (x_{nk}^{(1)}, \cdots, x_{nk}^{(d)}), \quad d \geq 1.$$

Let $\epsilon_{nk} = y_{nk} - y_{n,k-1}$ and set $x_{n0} = y_{n0} = 0$. Limit theory for sample quantities

$$\int_0^1 f(x_{n,[ns]})dy_{n,[ns]} = \sum_{k=1}^n f(x_{n,k-1})\epsilon_{nk}$$

in Theorem 4.6 is elegant, but is not sufficiently general to cover many econometric applications, where endogeneity and more general innovation processes are present.

This section extends the analysis to cover linear process with innovations ϵ_{nk}. Explicitly, we will investigate the convergence of sample quantities $\sum_{k=1}^{n} f(x_{n,k-1}) u_{nk}$, where

$$u_{nk} = \sum_{j=0}^{\infty} \varphi_j \, \epsilon_{n,k-j}$$

with $\epsilon_{nk} = x_{nk} - x_{n,k-1}$ if $k \geq 1$, $\varphi = \sum_{j=0}^{\infty} \varphi_j \neq 0$ and $\sum_{j=0}^{\infty} j \, |\varphi_j| < \infty$. We do not need to specify ϵ_{nk} for $k \leq 0$ except certain moment conditions. The array u_{nk} includes all stationary and invertible ARMA time series arrays and may be serially dependent and cross correlated with x_{nk}, which is usually expected in the framework of cointegration.

4.5.1 *LPWW decomposition theorem*

The core component in establishing weak convergence of $\sum_{k=1}^{n} f(x_{n,k-1}) u_{nk}$ is a novel decomposition result, which was originally developed in Liang, Phillips, Wang and Wang (LPWW, 2014).

Theorem 4.7. *Suppose that* $\max_{1 \leq k \leq n} \sum_{i=1}^{d} |x_{nk}^{(i)}| = O_{\mathrm{P}}(1)$ *and*

$$\sup_{j \geq 1, i \in \mathbb{Z}} \frac{1}{j} \sum_{k=1}^{j} \mathrm{E} \, \epsilon_{n,k+i}^2 \to 0, \qquad as \ n \to \infty. \tag{4.21}$$

Then, for any locally bounded function $f(x)$ on R^d, we have

$$\sum_{k=1}^{m} f(x_{n,k-1}) u_{nk} = \varphi \sum_{k=1}^{m} f(x_{n,k-1}) \epsilon_{nk}$$

$$+ \sum_{j=0}^{m-1} \varphi_j \sum_{i=0}^{m-1} [f(x_{n,i+j}) - f(x_{n,i})] \epsilon_{n,i+1} + R(m), \tag{4.22}$$

where $R(m) = o_{\mathrm{P}}(1)$ for each $1 \leq m \leq n$. If in addition

$$\max_{1 \leq k \leq n} |\epsilon_{nk}| = o_{\mathrm{P}}(1), \tag{4.23}$$

then $\max_{1 \leq m \leq n} |R(m)| = o_{\mathrm{P}}(1)$.

Proof. Simple calculations show that

$$
\sum_{k=1}^{m} f(x_{n,k-1})u_{nk} = \sum_{i=1}^{m} f(x_{n,i-1})\Big(\sum_{j=0}^{i-1} + \sum_{j=i}^{\infty}\Big)\varphi_j \epsilon_{n,i-j}
$$

$$
= \sum_{j=0}^{m-1} \varphi_j \sum_{i=1+j}^{m} f(x_{n,i-1})\epsilon_{n,i-j} + \sum_{i=1}^{m} \sum_{j=0}^{\infty} \varphi_{j+i} f(x_{n,i-1})\epsilon_{n,-j}
$$

$$
= \sum_{j=0}^{m-1} \varphi_j \sum_{i=0}^{m-j-1} f(x_{n,i+j})\epsilon_{n,i+1} + \sum_{j=0}^{\infty} \epsilon_{n,-j} \sum_{i=1}^{m} \varphi_{j+i} f(x_{n,i-1})
$$

$$
= \sum_{j=0}^{m-1} \varphi_j \sum_{i=0}^{m-1} f(x_{n,i+j})\epsilon_{n,i+1} - \sum_{j=0}^{m-1} \varphi_j \sum_{i=m-j}^{m-1} f(x_{n,i+j})\epsilon_{n,i+1}
$$

$$
+ \sum_{j=0}^{\infty} \epsilon_{n,-j} \sum_{i=1}^{m} \varphi_{j+i} f(x_{n,i-1})
$$

$$
= \varphi \sum_{k=1}^{m} f(x_{n,k-1})\epsilon_{nk} + \sum_{j=0}^{m-1} \varphi_j \sum_{i=0}^{m-1} [f(x_{n,i+j}) - f(x_{n,i})]\epsilon_{n,i+1}
$$

$$
- R_1(m) - R_2(m) + R_3(m),
$$

where $R_1(m) = \sum_{j=m}^{\infty} \varphi_j \sum_{i=0}^{m-1} f(x_{ni})\epsilon_{n,i+1}$,

$$
R_2(m) = \sum_{j=0}^{m-1} \varphi_j \sum_{i=m-j}^{m-1} f(x_{n,i+j})\epsilon_{n,i+1},
$$

$$
R_3(m) = \sum_{j=0}^{\infty} \epsilon_{n,-j} \sum_{i=1}^{m} \varphi_{j+i} f(x_{n,i-1}).
$$

It suffices to show that, for each $1 \le m \le n$,

$$
|R_j(m)| = o_{\mathrm{P}}(1), \quad j = 1, 2, 3, \tag{4.24}
$$

and if in addition $\max_{1 \le k \le n} |\epsilon_{nk}| = o_{\mathrm{P}}(1)$,

$$
\max_{1 \le m \le n} |R_j(m)| = o_{\mathrm{P}}(1), \quad j = 1, 2, 3. \tag{4.25}
$$

To this end, write $\Omega_K = \{x_{ni} : \max_{1 \le k \le n} \sum_{i=1}^{d} |x_{nk}^{(i)}| \le K\}$. As $f(x)$ is a locally bounded function on R^d, we have $\max_{1 \le k \le n} |f(x_{nk})| \le A_K$, on Ω_K, for some $A_K > 0$. Also note that, under (4.21),

$$
\sup_{-\infty < i < j < \infty} \frac{1}{j-i} \sum_{k=i+1}^{j} \mathrm{E}\,|\epsilon_{nk}| \le \sup_{j \ge 1, i \in \mathbb{Z}} j^{-1/2} \Big(\sum_{k=1}^{j} \mathrm{E}\,\epsilon_{n,k+i}^2\Big)^{1/2} \to 0,
$$

as $n \to \infty$, due to Hölder's inequality. Combining these facts and $\sum_{j=0}^{\infty} j \, |\varphi_j| < \infty$, we have

$$\mathrm{E} \, |R_1(m)| I(\Omega_K) \leq A_K \sum_{j=m}^{\infty} |\varphi_j| \sum_{i=0}^{m} \mathrm{E} \, |\epsilon_{n,i+1}| = o\big(1\big), \qquad (4.26)$$

$$\mathrm{E} \, |R_2(m)| I(\Omega_K) \leq A_K \sum_{j=0}^{m-1} |\varphi_j| \sum_{i=m-j}^{m} \mathrm{E} \, |\epsilon_{n,i+1}| = o\big(1\big), \quad (4.27)$$

$$\mathrm{E} \max_{1 \leq m \leq n} |R_3(m)| I(\Omega_K) \leq A_K \sum_{j=0}^{\infty} \mathrm{E} \, |\epsilon_{n,-j}| \sum_{i=j}^{\infty} |\varphi_i|$$

$$\leq C \sum_{i=0}^{\infty} |\varphi_i| \sum_{k=0}^{i} \mathrm{E} \, |\epsilon_{n,-k}| = o\big(1\big). \qquad (4.28)$$

Hence $(|R_1(m)| + |R_2(m)| + |R_3(m)|) I(\Omega_K) = o_{\mathrm{P}}(1)$ for each m. This proves (4.24) as $\mathrm{P}(\max_{1 \leq k \leq n} \sum_{i=1}^{d} |x_{nk}^{(i)}| > K) \to 0$ as $K \to \infty$.
Similarly, as in (4.26) and (4.27), we have

$$\max_{1 \leq m \leq n} |R_1(m)| I(\Omega_K) \leq A_K \max_{1 \leq m \leq n} \sum_{j=m}^{\infty} j \, |\varphi_j| \frac{1}{m} \sum_{i=0}^{m-1} |\epsilon_{n,i+1}|$$

$$\leq C \, A_K \max_{1 \leq m \leq n} \frac{1}{m} \sum_{i=0}^{m-1} |\epsilon_{n,i+1}| = o_{\mathrm{P}}(1),$$

$$\max_{1 \leq m \leq n} |R_2(m)| I(\Omega_K) \leq A_K \sum_{j=0}^{\infty} j \, |\varphi_j| \max_{1 \leq i < k \leq n} \frac{1}{k-i} \sum_{j=i}^{k} |\epsilon_{nj}| = o_{\mathrm{P}}(1),$$

under the additional condition (4.23). This yields

$$\max_{1 \leq m \leq n} (|R_1(m)| + |R_2(m)|) = o_{\mathrm{P}}(1)$$

due to $\mathrm{P}(\Omega_K) \to 1$ as $K \to \infty$. The result $\max_{1 \leq m \leq n} |R_3(m)| = o_{\mathrm{P}}(1)$ follows from (4.28) and $\mathrm{P}(\Omega_K) \to 1$ as $K \to \infty$. This proves (4.25) and also completes the proof of Theorem 4.7. $\qquad \square$

Remark 4.3. The decomposition (4.22) can be used together with Theorem 4.6 to provide an extension of the limit theory to more general classes of processes, which will be discussed in subsequent sections. To make the noise term $R(m)$ negligible, the conditions imposed on the ϵ_{nk} and x_{nk} are natural and close to minimal. Furthermore, if $f(x)$ is a bounded function on R, the condition $\max_{1 \leq k \leq n} \sum_{i=1}^{d} |x_{nk}^{(i)}| = O_{\mathrm{P}}(1)$ is not necessary. In other words, (4.22) remains true without any restriction on the random sequence x_{nk}.

Remark 4.4. The decomposition (4.22) can be easily extended to multivariate settings. To illustrate, let

$$u_{nk} = \sum_{j=0}^{\infty} \varphi_j \, \epsilon'_{n,k-j}, \tag{4.29}$$

where $\epsilon_{ni} = (\epsilon_{ni}^{(1)}, \cdots, \epsilon_{ni}^{(d_1)})$ and $\varphi_j = (\varphi_{j1}, \cdots, \varphi_{jd_1})$. Suppose, for each $1 \le k \le d_1$, $\sum_{j=0}^{\infty} j \, |\varphi_{jk}| < \infty$ and conditions (4.21) and (4.23) hold with $\epsilon_{ni}^{(k)}$ instead of ϵ_{ni}. Then, for any locally bounded function $f(x)$ on $\mathbb{R}^d$, we have

$$\sum_{k=1}^{m} f(x_{n,k-1}) u_{nk} = \sum_{i=1}^{d_1} \sum_{j=0}^{\infty} \varphi_{ji} \sum_{k=1}^{m} f(x_{n,k-1}) \, \epsilon_{nk}^{(i)}$$

$$+ \sum_{i=1}^{d_1} \sum_{j=0}^{m-1} \varphi_{ji} \sum_{k=0}^{m-1} [f(x_{n,k+j}) - f(x_{nk})] \epsilon_{n,k+1}^{(i)} + R(m), \tag{4.30}$$

where $\max_{1 \le m \le n} |R(m)| = o_P(1)$.

Remark 4.5. Instead of (4.22), $\sum_{k=1}^{m} f(x_{n,k-1}) u_{nk}$ can be decomposed as

$$\sum_{k=1}^{m} f(x_{n,k-1}) u_{nk} = \varphi \sum_{i=1}^{m} f(x_{n,i-1}) \epsilon_{ni}$$

$$+ \sum_{i=1}^{m} [f(x_{ni}) - f(x_{n,i-1})] \epsilon_{ni}^* - r_m, \tag{4.31}$$

where $\epsilon_{ni}^* = \sum_{j=0}^{\infty} \varphi_j^* \epsilon_{n,i-j}$, $\varphi_j^* = \sum_{s=j+1}^{\infty} \varphi_s$ and

$$r_m = f(x_{nm}) \epsilon_{nm}^* - f(x_{n0}) \epsilon_{n0}^*.$$

Indeed, applying the BN decomposition (Phillips and Solo, 1992), using summation by parts, we have

$$\sum_{i=1}^{m} f(x_{n,i-1}) u_{ni} = \sum_{i=1}^{m} f(x_{n,i-1}) \Big(\sum_{j=0}^{\infty} \varphi_j \epsilon_{n,i-j} \Big)$$

$$= \sum_{i=1}^{m} f(x_{n,i-1}) \big(\varphi \, \epsilon_{ni} + \epsilon_{n,i-1}^* - \epsilon_{ni}^* \}$$

$$= \varphi \sum_{i=1}^{m} f(x_{n,i-1}) \epsilon_{ni} - \sum_{i=1}^{m} f(x_{n,i-1}) \big(\epsilon_{ni}^* - \epsilon_{n,i-1}^* \big)$$

$$= \varphi \sum_{i=1}^{m} f(x_{n,i-1}) \epsilon_{ni} + \sum_{i=1}^{m} \{ f(x_{ni}) - f(x_{n,i-1}) \} \epsilon_{ni}^* - r_m.$$

As in the proof of Theorem 4.7, if $f(x)$ is locally bounded on $\mathbb{R}^d$, we have $\max_{1 \le m \le n} |r(m)| = o_P(1)$ provided $\max_{1 \le k \le n} \sum_{i=1}^{d} |x_{nk}^{(i)}| = O_P(1)$ and conditions (4.21) and (4.23) hold.

4.5.2   Long memory processes

For each $n \geq 1$, let

$$\epsilon_{ni} = (\epsilon_{ni}^{(1)}, \cdots, \epsilon_{ni}^{(d_1)}), \quad i \in \mathbb{Z},$$

be a vector martingale difference with respect to σ-fields $\{\mathcal{F}_{ni}\}_{i \in \mathbb{Z}}$ and

$$x_{ni} = (x_{ni}^{(1)}, \cdots, x_{ni}^{(d)}), \quad i \geq 1,$$

be a vector process adapted to the same $\{\mathcal{F}_{ni}\}_{i \geq 1}$. We consider weak convergence of $\sum_{k=1}^{[nt]} f(x_{n,k-1}) u_{nk}$, where u_{nk} is defined as in (4.29), i.e.,

$$u_{nk} = \sum_{j=0}^{\infty} \varphi_j \, \epsilon'_{n,k-j},$$

where $\varphi_j = (\varphi_{j1}, \cdots, \varphi_{jd_1})$ with $\sum_{j=0}^{\infty} j \, ||\varphi_j|| < \infty$.

Theorem 4.8. *Suppose that*

(i) $\{x_{n,[nt]}, \sum_{j=1}^{[nt]} \epsilon_{nj}\} \Rightarrow \{X(t), Y(t)\}$ *on* $D_{\mathbb{R}^d \times \mathbb{R}^{d_1}}[0, \infty)$;

(ii) $\sup_{i,j \geq 1} \frac{1}{j^2} \mathrm{E} ||x_{n,i+j} - x_{ni}||^2 = o(n^{-1})$;

(iii) for each $1 \leq k \leq d_1$, $\sup_{i \in \mathbb{Z}} \mathrm{E} (\epsilon_{ni}^{(k)})^2 = o(1)$ *and*

$$\sum_{i=1}^{n} \mathrm{E} (\epsilon_{ni}^{(k)})^2 \mathrm{I}(|\epsilon_{ni}^{(k)}| \geq \epsilon) = o(1), \quad \forall \epsilon > 0. \tag{4.32}$$

Then, for any function $f(s)$ *on* $\mathbb{R}^d$ *satisfying local Lipschitz condition*[1] *and any continuous function* $g(s)$, *we have*

$$\Big\{ x_{n,[nt]}, \, \sum_{j=1}^{[nt]} \epsilon_{nj}, \, \frac{1}{n} \sum_{k=1}^{[nt]} g(x_{nk}), \, \sum_{k=1}^{[nt]} f(x_{n,k-1}) u_{nk} \Big\}$$

$$\Rightarrow \Big\{ X(t), \, Y(t), \, \int_0^t g[X(s)] ds, \, \int_0^t f[X(s)] \, dY^*(s) \Big\}, \tag{4.33}$$

on $D_{\mathbb{R}^{d+d_1+2}}[0, 1]$, *where* $Y^*(s) = \sum_{j=0}^{\infty} \varphi_j \, Y'(s)$.

---

[1]That is, for every $K > 0$, there exists a constant C_K such that, for all $|x_j| + |y_j| \leq K, j = 1, \cdots, d$,

$$|f(x_1, ..., x_d) - f(y_1, ..., y_d)| \leq C_K \sum_{j=1}^{d} |x_j - y_j|.$$

Proof. For each $1 \leq k \leq d$, by setting $X(t) = (X_1(t), \cdots, X_d(t))$, it follows from condition (i) that

$$\max_{1 \leq i \leq n} |x_{ni}^{(k)}| \to_D \sup_{0 \leq t \leq 1} |X_k(t)| = O_P(1).$$

On the other hand, condition (iii) implies that, for each $1 \leq k \leq d_1$, $\max_{1 \leq i \leq n} |\epsilon_{ni}^{(k)}| = o_P(1)$ and

$$\sup_{j \geq 1, i \in \mathbb{Z}} \frac{1}{j} \sum_{t=1}^{j} \mathrm{E}\left(\epsilon_{n,t+i}^{(k)}\right)^2 \leq \sup_{i \in \mathbb{Z}} \mathrm{E}\,(\epsilon_{ni}^{(k)})^2 \to 0,$$

i.e., conditions (4.21) and (4.23) hold with $\epsilon_{ni}^{(k)}$ instead of ϵ_{ni}. Due to these facts, the decomposition (4.30) holds, namely, we may write

$$\sum_{k=1}^{[nt]} f(x_{n,k-1}) u_{nk} = \sum_{k=1}^{[nt]} f(x_{n,k-1}) Z_{nk} + \Delta_n(t) + R_n(t), \qquad (4.34)$$

where $\sup_{0 \leq t \leq 1} |R_n(t)| = o_P(1)$, $Z_{nk} = \sum_{j=0}^{\infty} \varphi_j \epsilon'_{nk}$ and

$$\Delta_n(t) = \sum_{i=1}^{d_1} \sum_{j=0}^{[nt]-1} \varphi_{ji} \sum_{k=0}^{[nt]-1} [f(x_{n,k+j}) - f(x_{nk})] \epsilon_{n,k+1}^{(i)}.$$

Let $y_{nk} = \sum_{j=1}^{k} Z_{nj}$. It follows from the continuous mapping theorem and condition (i) again that

$$(x_{n,[nt]}, y_{n,[nt]}) \Rightarrow \left(X(t), \sum_{j=0}^{\infty} \varphi_j Y'(s)\right).$$

By the Hölder's inequality and (4.32), we further have

$$Ey_{nn}^2 = \sum_{k=1}^{n} \mathrm{E}\, Z_{nk}^2 = \sum_{k=1}^{n} E\left(\sum_{j=0}^{\infty} \varphi_j \epsilon'_{nk}\right)^2$$

$$\leq \left(\sum_{j=0}^{\infty} j\,||\varphi_j||\right) \sum_{i=1}^{d_1} \sum_{j=0}^{\infty} j|\varphi_{ji}| \sum_{k=1}^{n} E(\epsilon_{nk}^{(i)})^2 = O(1).$$

Hence, by Theorem 4.6, (4.33) holds with Z_{nk} instead of u_{nk}. Now, to prove (4.33), it suffices to show

$$\sup_{0 \leq t \leq 1} |\Delta_n(t)| = o_P(1). \qquad (4.35)$$

For any $K > 0$, let $\Delta_n = I\big(\max_{1 \le k \le 2n} ||x_{nk}|| \le K\big) \sup_{0 \le t \le 1} |\Delta_n(t)|$. Using Hölder's inequality, condition (ii) and the fact that $f(s)$ satisfies the local Lipschitz condition, we have

$$
E\,|\Delta_n| \le C \sum_{i=1}^{d_1} \sum_{j=0}^{n} |\varphi_{ji}| \sum_{k=1}^{n} E\left(||x_{n,k+j-1} - x_{n,k-1}|| \, |\epsilon_{nk}^{(i)}| \right)
$$

$$
\le C \sum_{i=1}^{d_1} \sum_{j=0}^{n} |\varphi_{ji}| \left(\sum_{k=1}^{n} E\,||x_{n,k+j} - x_{nk}||^2 \right)^{1/2} \left(\sum_{k=1}^{n} E\,|\epsilon_{nk}^{(i)}|^2 \right)^{1/2}
$$

$$
\le C_1 \left(\sup_{i,j \ge 1} \frac{n}{j^2} E\,||x_{n,i+j} - x_{ni}||^2 \right)^{1/2} \sum_{j=0}^{\infty} j\,||\varphi_j||
$$

$$
= o(1),
$$

where we have used $\sum_{k=1}^{n} E\,|\epsilon_{nk}^{(i)}|^2 = O(1)$ due to (4.32). This implies $\Delta_n = o_P(1)$ and hence (4.35) follows since $P(\max_{1 \le k \le 2n} ||x_{nk}|| > K) \to 0$ as $n \to \infty$ first and then $K \to \infty$ due to condition (i). The proof of Theorem 4.8 now completes.   □

Example 4.6. (Long memory linear process) Let $\{\epsilon_i, \eta_i\}_{i \in \mathbb{Z}}$ be an i.i.d. sequence with zero means, unit variances and covariance $\rho = E\,\epsilon_0 \eta_0$. Standard functional limit theory shows that

$$
\left(\frac{1}{\sqrt{n}} \sum_{i=1}^{\lfloor nt \rfloor} \epsilon_i, \ \frac{1}{\sqrt{n}} \sum_{i=1}^{\lfloor nt \rfloor} \eta_i, \ \frac{1}{\sqrt{n}} \sum_{i=1}^{\lfloor nt \rfloor} \eta_{-i} \right) \Rightarrow \left(B_t, \ B_{1t}, \ B_{2t} \right)
$$

on $D_{\mathbb{R}^3}[0, \infty)$, where B_{2t} is a Brownian motion independent of (B_t, B_{1t}). The latter (B_t, B_{1t}) is a 2-dimensional Gaussian process with mean zero, stationary and independent increments, and covariance matrix:

$$
\Omega_t = \begin{pmatrix} 1 & \rho \\ \rho & 1 \end{pmatrix} t.
$$

Let $z_j = \sum_{k=0}^{\infty} \psi_k\,\eta_{j-k}$, where $\psi_k \sim k^{-\mu} h(k)$, $1/2 < \mu < 1$ and $h(k)$ is a function that is slowly varying at ∞, and define

$$
z_{nk} = \frac{1}{d_n} \sum_{j=1}^{k} z_j, \quad \text{where } d_n^2 = var\left(\sum_{j=1}^{n} z_j \right).
$$

By the continuous mapping theorem and similar arguments to those in Wang, Lin and Gullati (2003), we have

$$
\left(\frac{1}{\sqrt{n}} \sum_{i=1}^{\lfloor nt \rfloor} \epsilon_i, \ z_{n, \lfloor nt \rfloor} \right) \Rightarrow \left(B_t, B_{3/2 - \mu}(t) \right),
$$

on $D_{\mathrm{R}^2}[0,\infty)$, where $\{B_H(t)\}_{t\geq 0}$ is a fractional Brownian motion and a functional of $\{B_{1t}, B_{2t}\}_{t\geq 0}$. Furthermore, due to the stationarity of z_j and

$$d_n^2 = \mathrm{E}\,|\sum_{k=1}^{n} z_k|^2 \sim c_\mu\, n^{3-2\mu}\, h^2(n)$$

with some constant c_μ [see, e.g., Wang, Lin and Gullati (2003)], we have

$$\sup_{i,j\geq 1} \frac{1}{j^2}\mathrm{E}\,|z_{n,i+j} - z_{n,i}|^2 = \frac{1}{d_n^2}\sup_{j\geq 1}\frac{1}{j^2}\mathrm{E}\,|\sum_{k=1}^{j} z_k|^2 = o(n^{-1}).$$

It now follows from Theorem 4.8 that, for any function $f(s)$ satisfying a local Lipschitz condition and for any continuous function $g(s)$,

$$\Big\{ \frac{1}{n}\sum_{k=1}^{n} g(z_{nk}),\; \frac{1}{\sqrt{n}}\sum_{k=1}^{n} f(z_{n,k-1})\,u_k \Big\}$$

$$\Rightarrow \Big\{ \int_0^1 g[B_{3/2-\mu}(t)]dt,\; \varphi\int_0^1 f[B_{3/2-\mu}(t)]\,dB_t \Big\},$$

where $u_k = \sum_{j=0}^{\infty}\varphi_j\,\epsilon_{k-j}$ with $\varphi = \sum_{j=0}^{\infty}\varphi_j \neq 0$ and $\sum_{j=0}^{\infty} j\,|\varphi_j| < \infty$.

Remark 4.6. If $\varphi_j = (\varphi_{jst})_{1\leq s\leq m, 1\leq t\leq d_1}$ is a $m \times d_1$ matrix with $\sum_{j=0}^{\infty} j\,|\varphi_{jst}| < \infty$ for each $1 \leq s \leq m$ and $1 \leq t \leq d_1$, then

$$u_{nk} = \sum_{j=0}^{\infty}\varphi_j\epsilon'_{n,k-j}$$

is an m-dimensional vector. Using the Cramér Wold theorem, (4.33) still holds, but $Y^*(s)$ is now a vector given by

$$Y^*(s) = \Big(\sum_{j=0}^{\infty}\varphi_{j1}Y'(s),\cdots,\sum_{j=0}^{\infty}\varphi_{jm}Y'(s)\Big),$$

where $\varphi_{js} = (\varphi_{js1},\cdots,\varphi_{jsd_1}), s = 1,2,...,m$. Result (4.33) can be further extended to vector functionals such as $f(s) = (f_1(s),\cdots,f_k(s)), s \in \mathrm{R}^d$ without essential difficulties. We leave the extension to readers.

Remark 4.7. The local Lipschitz condition on $f(x)$ is a minor requirement and holds for many continuous functions. Recall $\sum_{j=1}^{[nt]}\epsilon_{nj} \Rightarrow Y(t)$ on $D_{\mathrm{R}^{d_1}}[0,\infty)$, i.e., the components ϵ_{nj} are standardized differences. It is natural therefore to assume condition (iii). As indicated in Example 4.6, condition (ii) holds for standardized sums of a long memory process such as

$$x_{nk} = \sum_{j=1}^{k}\xi_j/d_n,\quad 1 \leq k \leq n,$$

where $d_n^2 = var(\sum_{j=1}^n \xi_j) \sim C n^\alpha$ with $1 < \alpha \leq 2$. Note that in this case the standardization is $d_n = O\left(n^{\alpha/2}\right)$, which exceeds the usual $\sqrt{n}$ standardization for $I(1)$ processes. Then, $\sup_{i,j\geq 1} \frac{1}{j^2} E |x_{n,i+j} - x_{n,i}|^2 =$ $\sup_{j\geq 1} \frac{Cj^\alpha}{j^2 n^\alpha} = o\left(n^{-1}\right)$, as in condition (ii). Interestingly, however, condition (ii) excludes partial sums of a short memory process and the condition does not hold even for partial sums of i.i.d. $\left(0, \sigma^2\right)$ innovations for which it is easily seen that

$$\sup_{i,j\geq 1} \frac{1}{j^2} E |x_{n,i+j} - x_{n,i}|^2 = \sup_{j\geq 1} \frac{Cj}{j^2 n} = O\left(n^{-1}\right).$$

In next section, we remove this restriction but impose greater smoothness on $f(x)$, thereby showing that the time series structure of u_{nk} and its interaction with the properties of the nonlinear function f can have a significant effect on limit behavior.

4.5.3   *Short memory processes*

Let $\{\eta_k\}_{k\geq 1}$ and $\{\epsilon_k\}_{k\in\mathbb{Z}}$ be two random variable sequences such that

$$\left(X_{n,[nt]}, Y_{n,[nt]}\right) \Rightarrow \left(X(t), Y(t)\right), \tag{4.36}$$

on $D_{\mathbb{R}^2}[0,\infty)$, where $X_{nk} = \frac{1}{\sqrt{n}} \sum_{j=1}^k \eta_j$ and $Y_{nk} = \frac{1}{\sqrt{n}} \sum_{j=1}^k \epsilon_j, k \geq 1$. Setting $X_{n0} = 0$, we consider weak convergence of the process $S_{[nt]}$ defined by

$$S_j = \frac{1}{\sqrt{n}} \sum_{k=1}^j f(X_{n,k-1})u_k, \quad j \geq 1,$$

where $u_k = \sum_{j=0}^\infty \varphi_j \epsilon_{k-j}$ with $\varphi = \sum_{j=0}^\infty \varphi_j \neq 0$ and $\sum_{j=0}^\infty j |\varphi_j| < \infty$.

Under $\sqrt{n}$ standardization, X_{nk} is usually a partial sum of short memory process. As noted in Remark 4.7, weak convergence of $S_{[nt]}$ is more involved in this situation, particularly, the time series structure of u_k and its interaction with X_{nk} can have a significant effect on limit behavior.

To facilitate the analysis and make notation convenience, throughout the section, we assume that $\{\eta_k\}_{k\geq 1}$ is stationary and $\{\epsilon_k, \mathcal{F}_k\}_{k\in\mathbb{Z}}$ forms a martingale difference, where $\mathcal{F}_k = \sigma(\epsilon_j, \eta_j)_{j\leq k}$ (let $\eta_j = 0$ if $j \leq 0$).

Theorem 4.9. *In addition to (4.36), assume*

(i) $\sup_{i\in\mathbb{Z}} E|\epsilon_i|^3 < \infty$, $E\eta_1^2 < \infty$ *and*

$$\sup_{j\geq 1} \frac{1}{j} E\Big| \sum_{k=1}^j \eta_k \Big|^2 < \infty;$$

(ii) there exists a constant $A_0 > 0$ such that, for each $i \geq 0$,

$$\sum_{j=1}^{\infty} \varphi_j \sum_{k=1}^{j} \mathrm{E}\left(\eta_{k+i}\epsilon_{i+1} \mid \mathcal{F}_i\right) = A_0, \quad a.s.; \qquad (4.37)$$

(iii) $f'(x)$ is locally bounded and

$$|f'(x) - f'(y)| \leq C_K |x - y|^{\beta},$$

for some $0 < \beta \leq 1/3$ and $\max\{|x|, |y|\} \leq K$, where C_K is a constant depending only on K.

Then, for any continuous function $g(s)$, we have

$$\left\{X_{n,[nt]}, \ Y_{n,[nt]}, \ \frac{1}{n}\sum_{k=1}^{[nt]} g(X_{nk}), \ S_{[nt]}\right\}$$

$$\Rightarrow \left\{X(t), \ Y(t), \int_0^t g[X(s)]ds, \ S(t)\right\}, \qquad (4.38)$$

on $D_{\mathrm{R}^4}[0,1]$, where

$$S(t) = \varphi \int_0^t f[X(s)]\, dY(s) + A_0 \int_0^t f'[X(s)]ds.$$

Proof. It is readily seen that $\max_{1 \leq k \leq n} |X_{nk}| = O_P(1)$ by (4.36). Furthermore, (4.21) and (4.23) hold with $\epsilon_{ni} = \epsilon_i/\sqrt{n}$ due to condition (i). Using Theorem 4.7, we have

$$\frac{1}{\sqrt{n}}\sum_{i=1}^{[nt]} f(X_{n,i-1})u_i = \frac{\varphi}{\sqrt{n}}\sum_{i=1}^{[nt]} f(X_{n,i-1})\epsilon_i$$

$$+ \frac{1}{\sqrt{n}}\sum_{j=0}^{[nt]-1} \varphi_j \sum_{i=0}^{[nt]-1} [f(X_{n,i+j}) - f(X_{ni})]\epsilon_{i+1} + R_n(t), \qquad (4.39)$$

where $\sup_{0 \leq t \leq 1} |R_n(t)| = o_P(1)$. Noting that

$$f(X_{n,i+j}) - f(X_{ni}) = f'(X_{ni})(X_{n,i+j} - X_{n,i})$$

$$+ \int_{X_{ni}}^{X_{n,i+j}} [f'(x) - f'(X_{ni})]dx,$$

by letting $Z_{ki} = \eta_{k+i}\epsilon_{i+1}$, we may write

$$\sum_{j=0}^{[nt]-1} \varphi_j \sum_{i=0}^{[nt]-1} [f(X_{n,i+j}) - f(X_{ni})]\epsilon_{i+1}$$

$$= \frac{1}{\sqrt{n}}\sum_{j=1}^{[nt]-1} \varphi_j \sum_{k=1}^{j} \sum_{i=0}^{[nt]-1} f'(X_{ni}) Z_{ki} + R_{n1}(t)$$

$$= A_0 \sum_{i=1}^{[nt]} f'(X_{ni}) + R_{n1}(t) + R_{n2}(t) + R_{n3}(t), \qquad (4.40)$$

where

$$|R_{n1}(t)| \le \sum_{j=0}^{n-1} |\varphi_j| \sum_{i=0}^{n} |\epsilon_{i+1}| \left| \int_{X_{ni}}^{X_{n,i+j}} [f'(x) - f'(X_{ni})] dx \right|$$

$$\le \sum_{j=0}^{n-1} |\varphi_j| \sum_{i=0}^{n} |\epsilon_{i+1}| \int_{0}^{|X_{n,i+j} - X_{ni}|} |f'(x + X_{ni}) - f'(X_{ni})| dx,$$

$$R_{n2}(t) = \frac{1}{\sqrt{n}} \sum_{j=1}^{[nt]-1} \varphi_j \sum_{k=1}^{j} \sum_{i=0}^{[nt]-1} f'(X_{ni}) [Z_{ki} - \mathrm{E}(Z_{ki} \mid \mathcal{F}_i)],$$

$$|R_{n3}(t)| \le \frac{1}{\sqrt{n}} \sum_{i=0}^{[nt]-1} |f'(X_{ni})| \left| \sum_{j=1}^{[nt]-1} \varphi_j \sum_{k=1}^{j} \mathrm{E}(Z_{ki} \mid \mathcal{F}_i)] - A_0 \right|.$$

Due to (4.39) and (4.40), using Theorem 4.6, (4.38) will follow if we prove

$$\sup_{0 \le t \le 1} |R_{nj}(t)| = o_{\mathrm{P}}(n^{1/2}), \quad j = 1, 2, 3. \tag{4.41}$$

Write $\Omega_K = \{X_{ni} : \max_{1 \le i \le 2n} |X_{ni}| \le K/3\}$. Note that $|x + X_{ni}| \le K$ whenever $0 \le x \le |X_{n,i+j} - X_{ni}|$. It follows from condition (iii) and Hölder's inequality that

$$\mathrm{E} \sup_{0 \le t \le 1} |R_{n1}(t)| I(\Omega_K) \le C_K \sum_{j=0}^{n-1} |\varphi_j| \sum_{i=0}^{n} \mathrm{E}\left(|X_{n,i+j} - X_{n,i}|^{1+\beta} |\epsilon_{i+1}|\right)$$

$$\le C_K \, n^{-(1+\beta)/2} \sum_{j=0}^{n-1} |\varphi_j| \sum_{i=0}^{n} \left(\mathrm{E} \left| \sum_{k=1}^{j} \eta_{k+i} \right|^2\right)^{(1+\beta)/2} \left(\mathrm{E}|\epsilon_{i+1}|^3\right)^{1/3}$$

$$\le C n^{-(\beta-1)/2}) \left(\sup_{j \ge 1} \frac{1}{j} E \left| \sum_{k=1}^{j} \eta_k \right|^2\right)^{(1+\beta)/2} \sum_{j=0}^{n-1} j |\varphi_j|$$

$$= O(n^{-(\beta-1)/2}), \tag{4.42}$$

due to $0 < \beta \le 1/3$, the stationarity of η_j and condition (i). This implies

$$\sup_{0 \le t \le 1} |R_{n1}(t)| = O_{\mathrm{P}}(n^{(1-\beta)/2}),$$

as $\mathrm{P}(\Omega_K) \to 1$ as $K \to \infty$, i.e., (4.41) holds for $j = 1$.

To discuss $R_{n2}(t)$, we let $\widetilde{Z}_{ki} = \eta_{k+i}\epsilon_{i+1} I(|\epsilon_{i+1}| \le n^{1/4})$ and

$$W_{ki} = f'(X_{ni}) I(\max_{1 \le k \le i} |X_{nk}| \le K) [Z_{ki} - \mathrm{E}(Z_{ki} \mid \mathcal{F}_i)],$$

$$\widetilde{W}_{ki} = f'(X_{ni}) I(\max_{1 \le k \le i} |X_{nk}| \le K) [\widetilde{Z}_{ki} - \mathrm{E}(\widetilde{Z}_{ki} \mid \mathcal{F}_i)],$$

for any $K > 0$. Recalling that X_{ni} is adapted to $\mathcal{F}_i$ and $f'(x)$ is locally bounded, we have

$$\mathrm{E} \max_{1 \le m \le n-1} \Big| \sum_{i=0}^{m} \widetilde{W}_{ki} \Big|^2 \le C \sum_{i=0}^{n-1} \mathrm{E} \big(\widetilde{W}_{ki} \big)^2 \le C_1 n^{3/2} \mathrm{E} \eta_1^2. \quad (4.43)$$

On the other hand, it follows from $\sup_{i \in \mathcal{Z}} \mathrm{E} |\epsilon_i|^3 < \infty$ and $\mathrm{E} \eta_1^2 < \infty$ that

$$\mathrm{E} \sum_{i=0}^{n-1} |W_{ki} - \widetilde{W}_{ki}| \le 2 \sum_{i=0}^{n-1} \mathrm{E} |\eta_{k+i}| \epsilon_{i+1} |\mathrm{I}(|\epsilon_{i+1}| \ge n^{1/4})|$$

$$\le 2 n^{7/8} \, (\mathrm{E} \eta_1^2)^{1/2} \, (\sup_{i \in \mathbb{Z}} \mathrm{E} |\epsilon_i|^3)^{1/2} \le C n^{7/8}. \quad (4.44)$$

Write $R_{n2}^*(t) = \frac{1}{\sqrt{n}} \sum_{j=0}^{[nt]-1} \varphi_j \sum_{k=1}^{j} \sum_{i=0}^{[nt]-1} W_{ki}$. By virtue of (4.43) and (4.44), we have

$$\mathrm{E} \sup_{0 \le t \le 1} |R_{n2}^*(t)|$$

$$\le \frac{1}{\sqrt{n}} \sum_{j=1}^{n-1} |\varphi_j| \sum_{k=1}^{j} \Big(\mathrm{E} \max_{1 \le m \le n-1} \Big| \sum_{i=0}^{m} \widetilde{W}_{ki} \Big| + \mathrm{E} \sum_{i=0}^{n-1} |W_{ki} - \widetilde{W}_{ki}| \Big)$$

$$\le C n^{-1/2} (n^{3/4} + n^{7/8}) \sum_{j=0}^{\infty} j |\varphi_j| = O(n^{3/8}),$$

which implies $\sup_{0 \le t \le 1} |R_{n2}^*(t)| = o_{\mathrm{P}}(n^{1/2})$. Now, by noting

$$\sup_{0 \le t \le 1} |R_{n2}(t)| = \sup_{0 \le t \le 1} |R_{n2}^*(t)|$$

on $\Omega_K = \{X_{ni} : \max_{1 \le i \le n} |X_{ni}| \le K\}$, it follows that

$$\sup_{0 \le t \le 1} |R_{n2}(t)| = o_{\mathrm{P}}(n^{1/2})$$

due to $\mathrm{P}(\Omega_K) \to 1$ as $K \to \infty$. Hence (4.41) holds for $j = 2$.

Finally we prove (4.41) for $j = 3$. In fact, by noting that

$$\mathrm{E} |\mathrm{E} \big(Z_{ki} \mid \mathcal{F}_i \big)| \le (\mathrm{E} \eta_1^2)^{1/2} (\sup_{i \in \mathbb{Z}} \mathrm{E} |\epsilon_i|^2)^{1/2} \le C,$$

and $\frac{1}{m} \sum_{i=0}^{m} |f'(X_{ni})| \to_D \int_0^1 |f'[X(t)]| dt = O_{\mathrm{P}}(1)$, routine calculations show that

$$\sup_{0 \le t \le 1} |R_{n3}(t)|$$

$$\le \frac{O_{\mathrm{P}}(1)}{\sqrt{n}} \sum_{i=0}^{\sqrt{n}} |f'(X_{ni})| + \frac{O_{\mathrm{P}} \big(\sum_{j=\sqrt{n}}^{\infty} j |\varphi_j| \big)}{\sqrt{n}} \sum_{i=0}^{n} |f'(X_{ni})|$$

$$= o_{\mathrm{P}}(n^{1/2}),$$

as required. We now complete the proof of Theorem 4.9. $\qquad \square$

Remark 4.8. If the second order derivative of $f(x)$ exists on R, condition (iii) is trivially satisfied. Conditions (i) and (ii) usually hold for short memory processes satisfying certain stationary conditions. An illustrated example is given as follows.

Example 4.7. (Short memory linear process) Let $\{\epsilon_i, \xi_i\}_{i \in \mathbb{Z}}$ be an i.i.d. sequence with zero means, unit variances and covariance $\rho = \mathrm{E}\,\epsilon_0 \eta_0$. Let

$$z_{nk} = \frac{1}{\sqrt{n}\psi} \sum_{i=1}^{k} \eta_i, \quad k \geq 1,$$

where $\eta_j = \sum_{k=0}^{\infty} \psi_k \xi_{j-k}$ with $\sum_{k=0}^{\infty} |\psi_k| < \infty$ and $\psi = \sum_{k=0}^{\infty} \psi_k \neq 0$. Further let

$$u_k = \sum_{j=0}^{\infty} \varphi_j \,\epsilon_{k-j}$$

with $\varphi = \sum_{j=0}^{\infty} \varphi_j \neq 0$ and $\sum_{j=0}^{\infty} j\,|\varphi_j| < \infty$.

It is readily seen that, for each $i \geq 0$,

$$A_0 := \sum_{j=1}^{\infty} \varphi_j \sum_{k=1}^{j} \mathrm{E}\left(\eta_{k+i}\epsilon_{i+1} \mid \mathcal{F}_i\right) = \rho \sum_{j=0}^{\infty} \varphi_{j+1} \sum_{k=0}^{j} \psi_k,$$

where $\mathcal{F}_i = \sigma\{(\epsilon_k, \xi_k), k \leq i\}$, and standard arguments as in Wang, Lin and Gulatti (2001) show that

$$\left(\frac{1}{\sqrt{n}} \sum_{i=1}^{\lfloor nt \rfloor} \epsilon_i, \; z_{n,[nt]}\right) \Rightarrow \left(B_t, \; B_{1t}\right)$$

on $D_{\mathrm{R}^2}[0, \infty)$, where (B_t, B_{1t}) is a 2-dimensional Gaussian process with mean zero, stationary and independent increments, and covariance matrix $\Omega_t = \begin{pmatrix} 1 & \rho \\ \rho & 1 \end{pmatrix} t$.

If in addition $\mathrm{E}\,|\epsilon_0|^3 < \infty$, it now follows from Theorem 4.9 that, for any function $f(s)$ satisfying condition (iii) of Theorem 4.9 and for any continuous function $g(s)$,

$$\left\{ \frac{1}{n} \sum_{k=1}^{n} g(z_{nk}), \; \frac{1}{\sqrt{n}} \sum_{k=1}^{n} f(z_{n,k-1})\,u_k \right\}$$

$$\to_D \left\{ \int_0^1 g(B_{1t})dt, \; \varphi \int_0^1 f(B_{1t})\,dB_t + A_0 \int_0^1 f'(B_{1t})dt \right\}.$$

4.5.4 *LPWW decomposition theorem: multivariate extension*

Suppose $\epsilon_k = (\epsilon_{k1}, \cdots, \epsilon_{kd_1})$ and $\xi_k = (\xi_{k1}, \cdots, \xi_{kd}), k \in \mathbb{Z}$, are two i.i.d. random vector sequences with mean zero and finite, positive definite covariance matrixes Ω_1 and Ω_2, respectively. Let

$$\eta_k = (\eta_{k1}, \cdots, \eta_{km}) = \sum_{j=0}^{\infty} \xi_{k-j}\, \psi_j'$$

where $\psi_j = (\psi_{jst})_{1 \le s \le m, 1 \le t \le d}$ is an $m \times d$ matrix with $\sum_{j=0}^{\infty} ||\psi_j|| < \infty$, and let

$$u_k = \sum_{j=0}^{\infty} \varphi_j \epsilon_{k-j}'$$

where $\varphi_j = (\varphi_{j1}, \cdots, \varphi_{jd_1})$ with $\sum_{j=0}^{\infty} j\, ||\varphi_j|| < \infty$. Further let $f(x)$ be a real function on $\mathbb{R}^m$ and write $\mathrm{D}f(x) = \left(\frac{\partial f}{\partial x_1}, \cdots, \frac{\partial f}{\partial x_m}\right)'$. Set

$$X_{nk} = \frac{1}{\sqrt{n}} \sum_{j=1}^{k} \eta_j, \quad k \ge 1.$$

Using the Cramér–Wold theorem, standard arguments yield

$$\left(X_{n,[nt]}, \frac{1}{\sqrt{n}} \sum_{j=1}^{[nt]} \epsilon_j\right) \Rightarrow (B_1(t), B_2(t)), \tag{4.45}$$

on $D_{\mathbb{R}^{m+d_1}}[0, \infty)$, where $(B_1(t), B_2(t))_{t \ge 0}$ is a two dimensional Gaussian process with mean zero, stationary and independent increments, and covariance matrix:

$$\Omega_t = \begin{pmatrix} \Omega_2^* & \rho \\ \rho' & \Omega_1 \end{pmatrix} t,$$

where $\Omega_2^* = \sum_{j=0}^{\infty} \varphi_j \Omega_2 \varphi_j'$ and $\rho = \psi_0 \mathrm{E}\xi_1' \epsilon_1$.

We have the following result, which provides an extension of Example 4.7 to multivariate settings.

Theorem 4.10. *Suppose* $\mathrm{E}\, ||\epsilon_0||^3 < \infty$, $\mathrm{D}f(x)$ *is continuous on* $\mathbb{R}^m$ *and, on any compact set of* $\mathbb{R}^m$ *($||x|| \le K$, say),*

$$|f(x) - f(x_0) - (x - x_0)\mathrm{D}f(x_0)| \le C_K ||x - x_0||^{1+\beta}, \tag{4.46}$$

where $0 < \beta \le 1/3$ and C_K is a constant depending only on K. Then, for any continuous function $g(x)$ on R^m, we have

$$\left\{ \frac{1}{n} \sum_{k=1}^{[nt]} g(X_{nk}), \; \frac{1}{\sqrt{n}} \sum_{k=1}^{[nt]} f(X_{n,k-1}) u_k \right\}$$

$$\Rightarrow \left\{ \int_0^t g[X(s)]ds, \; \int_0^t f[B_1(s)]dB_2^*(s) + \int_0^t A_0 \, Df[B_1(s)]ds \right\}, \quad (4.47)$$

where $A_0 = \sum_{j=0}^{\infty} \sum_{i=0}^{j} \varphi_{j+1} \, \mathrm{E}\left(\epsilon_1' \xi_1\right) \psi_i'$ and $B_2^*(s) = \sum_{j=0}^{\infty} \varphi_j B_2'(s)$.

Proof.   It is readily seen that $\max_{1 \le k \le n} \|X_{nk}\| = O_\mathrm{P}(1)$, and conditions (4.21) and (4.23) hold with $\epsilon_{ik}/\sqrt{n}, 1 \le k \le d_1$, instead of ϵ_{ni}. Hence, as in (4.34), we may write

$$\frac{1}{\sqrt{n}} \sum_{k=1}^{[nt]} f(X_{n,k-1}) u_k$$

$$= \frac{1}{\sqrt{n}} \sum_{k=1}^{[nt]} f(X_{n,k-1}) \, Z_{nk} + \Delta_n(t) + R_n(t), \quad (4.48)$$

where $\sup_{0 \le t \le 1} |R_n(t)| = o_\mathrm{P}(1)$, $Z_{nk} = \sum_{j=0}^{\infty} \varphi_j \epsilon_k'$ and

$$\Delta_n(t) = \frac{1}{\sqrt{n}} \sum_{j=1}^{[nt]-1} \sum_{k=0}^{[nt]-1} [f(X_{n,k+j}) - f(X_{nk})] \varphi_j \epsilon_{k+1}'.$$

Furthermore, by letting $W_{nk}(t) = \sum_{j=1}^{[nt]-1} \varphi_j \epsilon_{k+1}' \sum_{i=1}^{j} \eta_{k+i}$, we have

$$\Delta_n(t) = \frac{1}{\sqrt{n}} \sum_{j=1}^{[nt]-1} \sum_{k=0}^{[nt]-1} (X_{n,k+j} - X_{nk}) \, Df(X_{nk}) \, \varphi_j \epsilon_{k+1}' + R_{n1}(t)$$

$$= \frac{1}{n} \sum_{k=0}^{[nt]-1} A_0 \, Df(X_{nk}) + R_{n1}(t) + R_{n2}(t) + R_{n3}(t), \quad (4.49)$$

where

$$R_{n1}(t) = \frac{1}{\sqrt{n}} \sum_{j=1}^{[nt]-1} \sum_{k=0}^{[nt]-1} [f(X_{n,k+j}) - f(X_{nk})$$

$$- (X_{n,k+j} - X_{nk}) \, Df(X_{nk})] \varphi_j \epsilon_{k+1}',$$

$$R_{n2}(t) = \frac{1}{n} \sum_{k=0}^{[nt]-1} \left[W_{nk}(t) - \mathrm{E} \, W_{nk}(t) \right] Df(X_{nk}),$$

$$R_{n3}(t) = \frac{1}{n} \sum_{k=0}^{[nt]-1} \left[\mathrm{E} \, W_{nk}(t) - A_0 \right] Df(X_{nk}).$$

Let $\Theta_K = \{X_{nk} : \max_{1 \le k \le 2n} \|X_{nk}\| \le K\}$. Recalling (4.46), simple calculations show

$$\mathrm{E} \sup_{0 \le t \le 1} |R_{n1}(t)| I(\Theta_K)$$

$$\le C_K n^{-1-\beta/2} \sum_{j=1}^{n} \sum_{k=0}^{n} \mathrm{E} \left(\left\| \sum_{i=1}^{j} \eta_{k+i} \right\|^{1+\beta} |\varphi_j \epsilon'_{k+1}| \right)$$

$$\le C\, n^{-\beta/2},$$

which yields $\sup_{0 \le t \le 1} |R_{n1}(t)| = o_{\mathrm{P}}(1)$ due to $\mathrm{P}(\Theta_K) \to 1$. Note that

$$\mathrm{E}\, W_{nk}(t) = \sum_{j=1}^{[nt]-1} \sum_{i=1}^{j} \varphi_j \mathrm{E}\left(\epsilon'_{k+1} \eta_{k+i} \right) = \sum_{j=0}^{[nt]-2} \sum_{i=0}^{j} \varphi_{j+1} \mathrm{E}\left(\epsilon'_1 \xi_1 \right) \psi'_i.$$

Similar to the proof of (4.41) with $j = 3$, we get

$$\sup_{0 \le t \le 1} |R_{n3}(t)| = o_{\mathrm{P}}(1).$$

As for $R_{n2}(t)$, by noting that, for any $k \ge 0$, $\mathrm{E}\left(W_{nk}(t) \mid \mathcal{F}_k \right) = \mathrm{E}\, W_{nk}(t)$, where $\mathcal{F}_k = \sigma(\epsilon_j, \xi_j, j \le k)$, similar arguments as in the proof of (4.41) with $j = 2$ yield $\sup_{0 \le t \le 1} |R_{n2}(t)| = o_{\mathrm{P}}(1)$. Taking these estimates into (4.48) and (4.49), we immediately have (4.47) due to (4.45), together with an application of Theorem 4.6. $\square$

Remark 4.9. As in Remark 4.6, Theorem 4.10 can be easily extended to more general multivariate setting. We leave the extension to readers.

4.5.5 *Extension to α-mixing sequences*

Let $\{u_i, v_i\}_{i \ge 1}$ be a sequence of stationary α-mixing random variables[2] with mean zeros and coefficients $\alpha(n) = O(n^{-\gamma})$ for some $\gamma > 6$, and $\mathrm{E}\,|u_1|^6 + \mathrm{E}\,|v_1|^6 < \infty$. Write

$$U_{nk} = \frac{1}{\sqrt{n}} \sum_{i=1}^{k} u_i, \quad V_{nk} = \frac{1}{\sqrt{n}} \sum_{i=1}^{k} v_i, \quad 1 \le k \le n.$$

---

[2]A sequence $\{\zeta_k, k \ge 1\}$ is said to be α-mixing if

$$\alpha(n) := \sup_{k \ge 1} \sup\{|\mathrm{P}(AB) - \mathrm{P}(A)\mathrm{P}(B)| : A \in \mathcal{F}_{n+k}^{\infty}, B \in \mathcal{F}_1^k\}$$

converges to zero as $n \to \infty$, where $\mathcal{F}_l^m = \sigma\{\zeta_l, \zeta_{l+1}, \ldots, \zeta_m\}$ denotes the σ-algebra generated by $\zeta_l, \zeta_{l+1}, \ldots, \zeta_m$ with $l < m$.

According to standard functional limit theory, for any continuous function $g(x)$, we have

$$\{U_{n,\lfloor nt \rfloor}, V_{n,\lfloor nt \rfloor}, \frac{1}{n}\sum_{k=1}^{n} g(U_{nk})\} \Rightarrow \{U(t), V(t), \int_0^1 g[U(t)]dt\}, \quad (4.50)$$

on $D_{\mathrm{R}^3}[0, \infty)$, where $(U(t), V(t))$ is a 2-dimensional Gaussian process with mean zero, stationary and independent increments, and covariance matrix:

$$\Omega_t = \begin{pmatrix} \sigma_u^2 & \sigma_{uv} \\ \sigma_{uv} & \sigma_v^2 \end{pmatrix} t,$$

where $\sigma_u^2 = \mathrm{E}\, u_1^2 + 2\sum_{i=1}^{\infty} \mathrm{E}\, u_1 u_{1+i}$ and $\sigma_v^2 = \mathrm{E}\, v_1^2 + 2\sum_{i=1}^{\infty} \mathrm{E}\, v_1 v_{1+i}$ are the long run variances of U_{nn} and V_{nn}, respectively, and $\sigma_{uv} = \mathrm{E}\, u_1 v_1 + \sum_{i=1}^{\infty}(\mathrm{E}\, u_1 v_{1+i} + \mathrm{E}\, v_1 u_{1+i})$ is the long run covariance of (U_{nn}, V_{nn}). See, de Jong and Davidson (2000a, b), for instance.

Write $\Lambda = \sum_{k=1}^{\infty} \mathrm{E}\,(u_1 v_{k+1})$ and $\Delta = \mathrm{E}\,(u_1 v_1) + \Lambda$. Regarding weak convergence of the sample covariance functional $\frac{1}{\sqrt{n}}\sum_{k=1}^{n-1} f(U_{nk})v_{k+1}$, we have the following result.

Theorem 4.11. *For any function $f(x)$ satisfying (iii) of Theorem 4.9 and for any continuous function $g(s)$, we have*

$$\left\{U_{n,\lfloor nt \rfloor}, V_{n,\lfloor nt \rfloor}, \frac{1}{n}\sum_{k=1}^{n} g(U_{nk}), \frac{1}{\sqrt{n}}\sum_{k=1}^{n-1} f(U_{nk})v_{k+1}\right\}$$
$$\Rightarrow \left\{U(t), V(t), \int_0^1 g[U(t)]dt, \int_0^1 f[U(t)]dV(t) + \Lambda \int_0^1 f'[U(t)]dt\right\}.$$
$$(4.51)$$

We also have

$$\left\{U_{n,\lfloor nt \rfloor}, V_{n,\lfloor nt \rfloor}, \frac{1}{n}\sum_{k=1}^{n} g(U_{nk}), \frac{1}{\sqrt{n}}\sum_{k=1}^{n} f(U_{nk})v_k\right\}$$
$$\Rightarrow \left\{U(t), V(t), \int_0^1 g[U(t)]dt, \int_0^1 f[U(t)]dV(t) + \Delta \int_0^1 f'[U(t)]dt\right\}.$$
$$(4.52)$$

Proof. We start with some preliminaries. Let $\mathcal{F}_t = \sigma(u_i, v_i, 1 < i < t)$, and $\mathcal{F}_s = \sigma(\phi, \Omega)$ be the trivial σ-field for $s < 0$. Put

$$z_i = \sum_{k=1}^{\infty} \mathrm{E}\,(v_{i+k}|\mathcal{F}_i) \quad \text{and} \quad \epsilon_i = \sum_{k=0}^{\infty}[\mathrm{E}\,(v_{i+k}|\mathcal{F}_i) - \mathrm{E}\,(v_{i+k}|\mathcal{F}_{i-1})].$$

Recalling $\alpha(n) = O(n^{-\gamma})$ for some $\gamma > 6$, $\mathrm{E}\,u_1 = \mathrm{E}\,v_1 = 0$ and $\mathrm{E}\,|u_1|^6 + \mathrm{E}\,|v_1|^6 < \infty$, standard arguments (see, McLeish (1975), for instance) show that $||\mathrm{E}\,(v_{i+k}|\mathcal{F}_i)||_3 \leq C\alpha(k)^{1/6}\,||v_1||_6$ and

$$||z_i||_3 \leq \sum_{k=1}^{\infty} ||\mathrm{E}\,(v_{i+k}|\mathcal{F}_i)||_3 \leq C\,||v_1||_6 \sum_{k=1}^{\infty} k^{-\gamma/6} < \infty, \qquad (4.53)$$

where $||X||_p = (\mathrm{E}\,|X|^p)^{1/p}$. We further have $\sup_{i \geq 1} \mathrm{E}\,\epsilon_i^2 < \infty$,

$$\sup_{i \geq 1} \mathrm{E}\,(|u_i|^{r_1}|z_i|^{r_2}) \leq (\mathrm{E}\,u_1^6)^{r_1/6}(\mathrm{E}\,|z_1|^3)^{r_2/3} < \infty, \qquad (4.54)$$

for any $1 \leq r_1, r_2 \leq 2$. Consequently, by letting $\lambda_k = u_k z_k - \mathrm{E}\,(u_k z_k)$, it follows that

$$\sup_{k \geq 1} \mathrm{E}\,|\mathrm{E}\,(\lambda_k \mid \mathcal{F}_{k-m})| \leq 6\alpha^{1/2}(m) \sup_{k \geq 1} ||\lambda_k||_2 \to 0, \qquad (4.55)$$

as $m \to \infty$.

We are now ready to prove Theorem 4.11. It is readily seen that $v_i = \epsilon_i + z_{i-1} - z_i$, $\{\epsilon_i, \mathcal{F}_i\}_{i \geq 1}$ forms a sequence of martingale differences, and

$$\frac{1}{\sqrt{n}} \sum_{k=1}^{n-1} f(U_{nk})v_{k+1} = \frac{1}{\sqrt{n}} \sum_{k=1}^{n-1} f(U_{nk})(\epsilon_{k+1} + z_k - z_{k+1})$$

$$= \frac{1}{\sqrt{n}} \sum_{k=1}^{n-1} f(U_{nk})\epsilon_{k+1} + \frac{1}{\sqrt{n}} \sum_{k=1}^{n-1} [f(U_{nk}) - f(U_{n,k-1})]z_k + o_P(1)$$

$$= \frac{1}{\sqrt{n}} \sum_{k=1}^{n-1} f(U_{nk})\epsilon_{k+1} + \frac{\Lambda}{n} \sum_{k=1}^{n-1} f'(U_{n,k-1}) + R_1(n) + R_2(n),$$

$$(4.56)$$

where $\Lambda = \mathrm{E}\,(u_1 z_1) = \sum_{k=1}^{\infty} \mathrm{E}\,(u_1 v_{k+1})$, and the remainder terms are

$$R_1(n) = \frac{1}{\sqrt{n}} \sum_{k=1}^{n-1} z_k \int_{U_{n,k-1}}^{U_{nk}} \left[f'(x) - f'(U_{n,k-1})\right] dx,$$

$$R_2(n) = \frac{1}{n} \sum_{k=1}^{n-1} f'(U_{n,k-1})[u_k z_k - \mathrm{E}\,(u_k z_k)] + o_P(1).$$

Write $Y_{n,\lfloor nt \rfloor} = \frac{1}{\sqrt{n}} \sum_{k=1}^{\lfloor nt \rfloor} \epsilon_k$. By virtue of Theorem 4.6 with $\epsilon_{nk} = v_k/(\sqrt{n})$ and $y_{nk} = U_{nk}$, to prove (4.51), it suffices to show that

$$\left\{U_{n,\lfloor nt \rfloor}, Y_{n,\lfloor nt \rfloor}\right\} \Rightarrow \left\{U(t), V(t)\right\}, \qquad (4.57)$$

and

$$R_i(n) = o_P(1), \quad i = 1,\, 2. \qquad (4.58)$$

The proof of (4.57) is simple. Indeed, by observing that

$$\sup_{0 \leq t \leq 1} |Y_{n,\lfloor nt \rfloor} - V_{n,\lfloor nt \rfloor}| \leq \frac{1}{\sqrt{n}} \max_{1 \leq k \leq n} |z_k|,$$

(4.57) follows from (4.50) and the fact that, for any $\eta > 0$ and $0 < \delta \leq 1$,

$$P(\max_{1 < i < n} |z_i| > \eta \sqrt{n}) \leq \sum_{i=1}^{n} P(|z_i| > \eta \sqrt{n})$$

$$\leq C n^{-1-\delta/2} \sum_{i=1}^{n} E |z_i|^{2+\delta} \to 0,$$

due to (4.53).

To prove (4.58), write $\Omega_K = \{U_{ni} : \max_{1 \leq i \leq n} |U_{ni}| \leq K\}$. As in the proof of (4.42), it follows from (iii) of Theorem 4.9 and (4.54) that

$$E |R_1(n)| I(\Omega_K) \leq \frac{C_K}{\sqrt{n}} \sum_{k=1}^{n} E\left(|U_{nk} - U_{n,k-1}|^{1+\beta} |z_k|\right)$$

$$\leq C_K \, n^{-(1+\beta/2)} \sum_{k=1}^{n} E\left(|u_k|^{1+\beta} |z_k|\right)$$

$$= O(n^{-\beta/2}). \tag{4.59}$$

This implies that $R_1(n) = O_P(n^{-\beta/2})$ due to $P(\Omega_K) \to 1$ as $K \to \infty$.

It remains to show $R_2(n) = o_P(1)$. To this end, let $m = \lfloor \log n \rfloor$ and recall $\lambda_k = u_k z_k - E(u_k z_k)$. We have

$$R_2(n) = \frac{1}{n} \sum_{k=1}^{n} f'(U_{n,k-m-1}) \lambda_k$$

$$+ \frac{1}{n} \sum_{k=1}^{n} \left[f'(U_{n,k-1}) - f'(U_{n,k-m-1}) \right] \lambda_k$$

$$= R_{21}(n) + R_{22}(n), \quad \text{say}. \tag{4.60}$$

As in the proof of (4.59), it is readily seen that

$$E |R_{22}(n)| I(\Omega_K) \leq C_K n^{-1} \sum_{k=1}^{n} E\left(|U_{n,k-1} - U_{n,k-m-1}|^{\beta} |\lambda_k|\right)$$

$$\leq C_K n^{-1-\beta/2} \sum_{k=1}^{n} \sum_{j=k-m}^{k-1} E\left(|u_j|^{\beta} |\lambda_k|\right) \leq C n^{-\beta/2} \log n,$$

since $0 < \beta \le 1/3$. Hence $R_{22}(n) = o_{\mathrm{P}}(1)$ due to $\mathrm{P}(\Omega_K) \to 1$ as $K \to \infty$. To estimate $R_{21}(n)$, write

$$IR_1(n) = \frac{1}{n} \sum_{k=1}^{n} U_k^* \left[\lambda_k - \mathrm{E}\left(\lambda_k \mid \mathcal{F}_{k-m-1} \right) \right],$$

$$IR_2(n) = \frac{1}{n} \sum_{k=1}^{n} U_k^* \, \mathbb{E}(\lambda_k \mid \mathcal{F}_{k-m-1}),$$

where $U_k^* = f'(U_{n,k-m-1}) \mathrm{I}(\max_{1 \le j \le k-m-1} |U_{n,j}| \le K)$. It is readily seen from (4.54) and (4.55) that

$$\mathrm{E}\, IR_1^2(n) \le \frac{C}{n^2} \sum_{k=1}^{n} \mathrm{E}\left[\lambda_k - \mathrm{E}\left(\lambda_k \mid \mathcal{F}_{k-m-1} \right) \right]^2 = O(n^{-1}),$$

$$\mathrm{E}\, |IR_2(n)| \le \frac{C}{n} \sum_{k=1}^{n} |\mathrm{E}\left(\lambda_k \mid \mathcal{F}_{k-m-1} \right)| = o(1),$$

which yield $IR_1(n) + IR_2(n) = o_{\mathrm{P}}(1)$. We now have $R_{21}(n) = o_{\mathrm{P}}(1)$ due to $\mathrm{P}(\Omega_K) \to 1$ as $K \to \infty$, and the fact that, on Ω_k,

$$R_{21}(n) = \frac{1}{n} \sum_{k=1}^{n} U_k^* \lambda_k = IR_1(n) + IR_2(n) = o_{\mathrm{P}}(1).$$

Combining these results proves $R_2(n) = o_{\mathrm{P}}(1)$ and also completes the proof of (4.51). The proof of (4.52) is essentially the same and the details are omitted. □

Remark 4.10. Suppose $\{u_i, v_i\}_{i \ge 1}$, where $u_i = (u_{i1}, \cdots, u_{im})$ and $v_i = (v_{i1}, \cdots, v_{id})$, are stationary α-mixing random vectors with mean zero and coefficients $\alpha(n) = O(n^{-\gamma})$ for some $\gamma > 6$. Suppose

$$\mathrm{E}\, \|u_1\|^6 + \mathrm{E}\, \|v_1\|^6 < \infty.$$

Then, for any continuous function $g : \mathrm{R}^m \to \mathrm{R}$, result (4.50) holds on $D_{\mathrm{R}^{m+d+1}}[0, \infty)$, where $(U(t), V(t))$ is an $(m+d)$-dimensional Gaussian process with mean zero, stationary and independent increments, and covariance matrix Ω_t. Furthermore, for any function $f : \mathrm{R}^m \to \mathrm{R}$ defined as in Theorem 4.10, we have

$$\left\{ \frac{1}{n} \sum_{k=1}^{n} g(U_{nk}), \ \frac{1}{\sqrt{n}} \sum_{k=1}^{n-1} f(U_{n,k}) v_{k+1} \right\}$$

$$\to_D \left\{ \int_0^1 g[U(s)] ds, \ \int_0^1 f[U(s)] dV(s) + \int_0^1 Df[U(s)] A_1 ds \right\}, \quad (4.61)$$

where $A_1 = \sum_{k=1}^{\infty} \mathrm{E}\left(u_1' v_{k+1} \right)$ and $Df(x) = \left(\frac{\partial f}{\partial x_1}, \cdots, \frac{\partial f}{\partial x_m} \right)$.

Result (4.61) provides an extension of Theorem 4.11 to multivariate settings. Its proof is similar to that of Theorem 4.10 by using the Cramér Wold theorem. We leave the details to readers.

4.6   Bibliographical Notes

Section 4.1. This section introduces the definition of stochastic integrals with respect to Brownian motion, local and semimartingales, and provides some of their basic properties. Materials in this regard is standard in advanced stochastic calculus text books such as Rogers and Williams (2000) and Prötter (2005). See, also, Liptser and Shiryaev (1989), Jacod and Shiryaev (2003).

Sections 4.2-4.4. Results on weak convergence of the sample covariance $\sum_{k=1}^{n} f(x_{nk})(y_{n,k+1} - y_{nk})$ began to emerge in the 1980s in statistics, probability and econometrics. Chan and Wei (1988), Phillips (1987, 1988a), and Strasser (1986), for example, gave results for martingale arrays. Condition UT was introduced in Stricker (1985) and later considered by Jakubowski, Mémin and Pageés (1989), where some general results were provided when $\{y_{nk}\}$ is a semimartingale. An alternative condition that is equivalent to condition UT was used in Kurtz and Protter (1991), Duffie and Protter (1992) and Kurtz and Protter (1996). Under semimartinagle assumptions, results on weak convergence of $\sum_{k=1}^{n} f(x_{nk})(y_{n,k+1} - y_{nk})$ are celebrated, and conditions imposed to establish the results are nearly necessary. For the purpose of this book, we do not collect the most general results, but rather those having easily verified sufficient conditions. Materials are mainly taken from Jakubowski, Ménin and Pageés (1989) and Kurtz and Protter (1996). For related results in this regard, we refer to Jakubowski (1996), Jacod and Shiryaev (2003).

Section 4.5. The sample covariance $\sum_{k=1}^{n} f(x_{nk})(y_{n,k+1} - y_{nk})$ arises frequently in the study of nonstationary time series, unit root testing and nonlinear cointegration regressions. In many econometric applications such as a cointegration framework, endogeneity is expected and it is therefore realistic to assume that the regressors x_{nk} are correlated with the innovations $\epsilon_{nk} = y_{n,k+1} - y_{nk}$ at some leads and/or lags. In such cases, when $f(x) = x$, Phillips (1988b) considered linear processes with i.i.d. innovations; Phillips (1987), Hansen (1992) and de Jong and Davidson (2000a, b) allowed for mixing sequences. Extensions to general $f(x)$ were investigated in an unpublished paper by de Jong (2002) and in Ibragimov and Phillips (2008). We also refer to Chang and Park (2011) and Lin and Wang (2010) for some related results. In technical part, de Jong (2002) made use of stringent uniform strong convergence condition on x_{nk} and y_{nk}. The approach adopted in Ibragimov and Phillips (2008) is to use general methods of weak convergence of discrete time semimartingales to continuous time semimartingales. The idea is conceptually elegant, offers considerable generality, and unifies convergence results for stationary and unity root cases, but the proofs are often lengthy and involve some complex derivations. Using a novel decomposition for $\sum_{k=1}^{n} f(x_{nk})(y_{n,k+1} - y_{nk})$, Liang, Phillips, Wang and Wang (2014) provided limit results for the sample covariance $\sum_{k=1}^{n} f(x_{nk})(y_{n,k+1} - y_{nk})$ when linear process, long memory and mixing variates are involved in innovations. Materials in this section are mainly taken from Liang, Phillips, Wang and Wang (2014) with certain extensions to multivariate settings.

Chapter 5

Nonlinear cointegrating regression

A typical nonlinear cointegrating regression model has the form:

$$y_t = f(x_t) + u_t, \qquad t = 1, 2, ..., n, \tag{5.1}$$

where u_t is a zero mean equilibrium error, x_t is a nonstationary regressor and $f(\cdot)$ is an unknown real function on R. In the presence of more explicit prior information, the regression function $f(\cdot)$ may be specified in parametric form as

$$f(x) = g(x, \theta_0), \tag{5.2}$$

where $g(\cdot, \theta)$ represents a parametric family of functions with unknown true parametric value $\theta_0 \in \Theta$, where $\Theta \subset \mathrm{R}^m$ for some $m \geq 1$ is a compact set.

Using the limit theorems developed in previous chapters, this chapter investigates estimation and inference theory for both nonparametric model (5.1) and parametric model (5.2).

5.1 Nonparametric estimation

Without assuming prior information on the regression function $f(x)$, an intuitive estimator of $f(x)$ can be locally weighted average of the $y_j, 1 \leq j \leq n$. The conventional Nadaraya-Watson regression estimator is given by

$$\hat{f}(x) = \sum_{t=1}^{n} y_t \, K[(x_t - x)/h] \Big/ \sum_{t=1}^{n} K[(x_t - x)/h],$$

where $K(x)$ is a real-valued function assigning weights. $K(x)$ is usually called a *kernel function* and h is a *bandwidth* which is a positive number controlling the size of the local neighborhood.

Suppose that $f(x)$ is locally approximated by a constant α. Running a local least square regression, we obtain the estimate

$$\hat{\alpha} = \text{argmin}_\alpha \sum_{k=1}^n (y_k - \alpha)^2 w_k = \sum_{k=1}^n w_k y_k \Big/ \sum_{k=1}^n w_k.$$

The Nadaraya-Watson estimator is of this form with weights $w_k = K[(x_k - x)/h]$. Similarly, if $f(x)$ can be locally approximated by

$$f(z) \approx f(x) + f'(x)(z - x) \equiv \alpha + \beta(z - x),$$

for z in a neighborhood of x, by using Taylor's expansion, an estimator $\hat{f}^L(x)$ for $f(x)$ is $\hat{\alpha}$, where $(\hat{\alpha}, \hat{\beta})$ is the minimizer of

$$\sum_{k=1}^n \big\{ y_k - [\alpha + \beta(x_k - x)] \big\}^2 K[(x_k - x)/h].$$

$\hat{f}^L(x)$ is called a *local linear estimator* of $f(x)$, which can be explicitly expressed as

$$\hat{f}^L(x) = \sum_{i=1}^n w_i(x) y_i \Big/ \sum_{i=1}^n w_i(x),$$

where $H_j(y) = y^j K(y)$, $V_{nj}(x) = \sum_{i=1}^n H_j[(x_i - x)/h], j = 1, 2$, and

$$w_i(x) = K[(x_i - x)/h]V_{n2}(x) - H_1[(x_i - x)/h]V_{n1}(x).$$

In this section, we investigate the asymptotics of $\hat{f}(x)$ and $\hat{f}^L(x)$.

5.1.1 *Nadaraya-Watson estimator*

Let x_0 be fixed. We split $\hat{f}(x_0) - f(x_0)$ as

$$\frac{\sum_{t=1}^n u_t K[(x_t - x_0)/h]}{\sum_{t=1}^n K[(x_t - x_0)/h]} + \frac{\sum_{t=1}^n [f(x_t) - f(x_0)] K[(x_t - x_0)/h]}{\sum_{t=1}^n K[(x_t - x_0)/h]}. \quad (5.3)$$

The second term of (5.3) is easy to handle with and is expected to be negligible under mild conditions. The asymptotic normality of $\hat{f}(x_0)$ is determined by the first term of (5.3), which follows from the joint convergence of the numerator and the denominator. We make use of the following assumptions for a framework on the asymptotic normality of $\hat{f}(x_0)$.

Assumption 5.1. [Regression function and Kernel function]

(a) There exists a nonnegative real function $f_1(s, x_0)$ and a constant $0 < \gamma \le 1$ such that, when $\delta > 0$ is sufficiently small,

$$|f(\delta y + x_0) - f(x_0)| \le \delta^\gamma f_1(y, x_0), \quad \text{for all } y \in \text{R};$$

(b) $K(x)$ is a non-negative bounded real function on R satisfying $0 < \int_{-\infty}^{\infty} K(x)dx < \infty$;

(c) $\int_{-\infty}^{\infty} K(s) f_1(s, x_0)ds < \infty$.

Assumption 5.2. [Regressor, error process and bandwidth] There exist constants $0 < a_n \to \infty$ such that, for each real function $g(x)$ on R,

(a) if $\int_{-\infty}^{\infty} |g(s)|ds < \infty$, then $\frac{1}{a_n} \sum_{t=1}^{n} E|g[(x_t - x_0)/h]| = O(1)$;

(b) if $|g(s)|$ is bounded and $\int_{-\infty}^{\infty} |g(s)|ds < \infty$, then

$$\left\{ \frac{1}{a_n} \sum_{t=1}^{n} g[(x_t - x_0)/h], \ \left(\frac{1}{a_n}\right)^{1/2} \sum_{t=1}^{n} u_t \, g[(x_t - x_0)/h] \right\}$$

$$\to_D \left(c_{1g} V^2, \ c_{2g} \, N V \right), \tag{5.4}$$

where c_{1g} and c_{2g} are constants depending only on $g(x)$, V is a random variable with $P(V \neq 0) = 1$ and N is a standard normal variate independent of V.

Assumption 5.1 is weak and can be simply verified for various kernels $K(x)$ and regression functions $f(x)$. For instance, if $K(x)$ is a standard normal kernel or has compact support, a wide range of regression functions $f(x)$, such as $f(x) = |x|^{\alpha}$ and $f(x) = 1/(1 + |x|^{\alpha})$ for some $\alpha > 0$, are included. Assumption 5.2(a) is mild and usually easy to be verified. Assumption 5.2(b) is a "high level" condition on x_t, u_t and h for a framework, which will be discussed in details after the following theorem.

Theorem 5.1. *Suppose Assumptions 5.1–5.2 hold. Then, for any h satisfying $a_n h^{2\gamma} \to 0$, we have*

$$a_n^{1/2} \left[\hat{f}(x_0) - f(x_0) \right] \to_D c_{2K} \, c_{1K}^{-1} V^{-1} N, \quad (5.5)$$

$$\left(\sum_{t=1}^{n} K[(x_t - x_0)/h] \right)^{1/2} \left[\hat{f}(x_0) - f(x_0) \right] \to_D c_{2K} \, c_{1K}^{-1/2} N, \quad (5.6)$$

where c_{1K} and c_{2K} are defined as in (5.4) with that $g(x)$ is replaced by $K(x)$, and N is a standard normal variate independent of V.

Proof. The proof of Theorem 5.1 is simple. Indeed, due to Assumption 5.1 and Assumption 5.2(a), we have

$$\sum_{t=1}^{n} |[f(x_t) - f(x_0)]| \, K[(x_t - x_0)/h]$$

$$\leq h^{\gamma} \sum_{t=1}^{n} K_1[(x_t - x_0)/h] = O_P(a_n \, h^{\gamma}),$$

where $K_1(y) = K(y)f_1(y, x_0)$. Since (5.4) holds true when $g(x)$ is replaced by $K(x)$, it follows from (5.3) and the continuous mapping theorem that

$$a_n^{1/2}\left[\hat{f}(x_0) - f(x_0)\right] = \frac{a_n^{-1/2}\sum_{t=1}^n u_t K[(x_t - x_0)/h]}{a_n^{-1}\sum_{t=1}^n K[(x_t - x_0)/h]}$$
$$+ \frac{O_P(a_n^{1/2}h^\gamma)}{a_n^{-1}\sum_{t=1}^n K[(x_t - x_0)/h]}$$
$$\to_D c_{2K}\, c_{1K}^{-1}\, V^{-1}\, N,$$

for any h satisfying $a_n h^{2\gamma} \to 0$. This proves (5.5). The proof of (5.6) is similar and hence the details are omitted. $\square$

It is clear from Theorem 5.1 that Assumption 5.2(b), namely, (5.4) is essential to establish the asymptotic normality of $\hat{f}(x_0)$. The "high level" condition (5.4) provides a trade off among the regressor x_t, the error process u_t and the bandwidth h. In the presence of stationary process x_t, V is usually a constant, i.e., a degenerate random variable. In this case, many classical methods are available to establish (5.4). The situation becomes very different if x_t is a nonstationary time series or an $I(1)$ process. Non-stationarity often appears in the cointegration framework. In this situation, V is usually a non-degenerate random variable, which makes the identification of Assumption 5.2(b) very challenge.

Write $\Omega = \{g : \sup_x |g(x)| < \infty, \int_{-\infty}^\infty |g(x)|dx < \infty\}$. If $g_1, g_2 \in \Omega$, then $|g_1|^\gamma \in \Omega$ for any $\gamma \geq 1$ and $\alpha g_1 + \beta g_2 \in \Omega$ for any $\alpha, \beta \in \mathrm{R}$. Using this fact and the Cramér-Wold theorem, we may prove that Assumption 5.2 is equivalent to the following Assumption 5.3.

Assumption 5.3. There exist constants $0 < a_n \to \infty$ such that, for each real function $g(x)$ on R,

(a) if $\int_{-\infty}^\infty |g(s)|ds < \infty$, then $\frac{1}{a_n}\sum_{t=1}^n E|g[(x_t - x_0)/h]| = O(1)$;
(b) if $|g(s)|$ is bounded and $\int_{-\infty}^\infty |g(s)|ds < \infty$, then

$$\frac{1}{a_n}\sum_{t=1}^n g[(x_t - x_0)/h] \to_D c_{1g}V^2 \tag{5.7}$$

and

$$\left\{\frac{1}{a_n}\sum_{t=1}^n g^2[(x_t - x_0)/h],\ \left(\frac{1}{a_n}\right)^{1/2}\sum_{t=1}^n u_t\, g[(x_t - x_0)/h]\right\}$$
$$\to_D \left(c_{1g^2}V^2,\ c_{2g}\, N\, V\right), \tag{5.8}$$

where c_{1g}, c_{1g^2} and c_{2g} are constants depending only on $g(x)$, $g^2(x)$ and $g(x)$, respectively, V is a random variable with $P(V \neq 0) = 1$ and N is a standard normal variate independent of V.

Indeed it follows from the Cramér-Wold theorem that if (5.7) holds, then

$$\left\{ \frac{1}{a_n} \sum_{t=1}^{n} g_1[(x_t - x_0)/h], \ \frac{1}{a_n} \sum_{t=1}^{n} g_2[(x_t - x_0)/h] \right\}$$
$$\to_D \left\{ c_{1g_1} V^2, \ c_{1g_2} V^2 \right\}, \tag{5.9}$$

for any $g_1, g_2 \in \Omega$. Since $g^2 \in \Omega$ whenever $g \in \Omega$, and (5.9) implies

$$\frac{\sum_{t=1}^{n} g_1[(x_t - x_0)/h]}{\sum_{t=1}^{n} g_2[(x_t - x_0)/h]} \to_P \frac{c_{1g_1}}{c_{1g_2}},$$

it is readily seen that Assumption 5.2 is equivalent to Assumption 5.3.

In Chapter 3, we have established several results in relation to Assumption 5.3 under certain conditions on x_t, u_t and h. By verifying Assumption 5.3, two illustrated corollaries of Theorem 5.1 are given as follows.

Let $\{\epsilon_j\}_{j \in \mathbb{Z}}$ be a sequence of i.i.d. random variables with that $E\epsilon_0 = 0$, $E\epsilon_0^2 = 1$ and $\lim_{|t| \to \infty} |t|^\eta |E\, e^{it\epsilon_0}| < \infty$ for some $\eta > 0$. Let $\tau \geq 0$ be a constant, $\gamma = 1 - \tau/n$, $x_0 = 0$,

$$x_k = \gamma\, x_{k-1} + \xi_k, \quad k \geq 1$$

where ξ_k is a linear process defined by $\xi_j = \sum_{k=0}^{\infty} \phi_k \epsilon_{j-k}$, with the coefficients $\phi_k, k \geq 0$, satisfying one of the following conditions:

LM. $\phi_k \sim k^{-\mu} \rho(k)$, where $1/2 < \mu < 1$ and $\rho(k)$ is a function slowly varying at ∞.

SM. $\sum_{k=0}^{\infty} |\phi_k| < \infty$ and $\phi \equiv \sum_{k=0}^{\infty} \phi_k \neq 0$.

Write $d_n^2 = E\, (\sum_{j=1}^{n} \xi_j)^2$, define $Z = \{Z_t\}_{t \geq 0}$ as in (2.82) and denote by $L_Z(t, s)$ the local time of Z.

Corollary 5.1. *Suppose that*

(i) *Assumption 5.1 holds;*

(ii) $\{u_i, \mathcal{F}_i\}_{i \geq 1}$ *is a martingale difference such that* $E\, (u_i^2 \mid \mathcal{F}_{i-1}) \to_{a.s.}$ 1, *as* $i \to \infty$, *and*

$$\sup_{i \geq 1} E\, \left[u_i^2 I(|u_i| \geq A) \mid \mathcal{F}_{i-1} \right] = o_P(1),$$

as $A \to \infty$, *where* $\mathcal{F}_i = \sigma(u_i, ..., u_1; \epsilon_{i+1}, \epsilon_i, ...)$.

Then, for any h satisfying $d_n h^{2\gamma} \to 0$ and $nh/d_n \to \infty$, we have

$$\left(\frac{nh}{d_n}\right)^{1/2} \left[\hat{f}(x_0) - f(x_0)\right] \to_D \tau \, L_Z(1,0)^{-1/2} \, N,$$

$$\left(\sum_{t=1}^{n} K[(x_t - x_0)/h]\right)^{1/2} \left[\hat{f}(x_0) - f(x_0)\right] \to_D \tau_1 \, N,$$

where $\tau = \left(\int_{-\infty}^{\infty} K^2(x)dx\right)^{1/2} / \int_{-\infty}^{\infty} K(x)dx$, $\tau_1 = \tau \left(\int_{-\infty}^{\infty} K(x)dx\right)^{1/2}$ and N is a standard normal variate independent of $L_Z(1,0)$.

Proof. By Theorem 5.1, it suffices to verify Assumption 5.3 with $V^2 = L_Z(1,0)$, $a_n = nh/d_n$, and

$$c_{1g} = \int_{-\infty}^{\infty} g(x)dx, \ c_{1g^2} = \int_{-\infty}^{\infty} g^2(x)dx, \ c_{2g} = \left(\int_{-\infty}^{\infty} g^2(x)dx\right)^{1/2}.$$

In fact, Assumption 5.3 (a) follows from Lemma 2.2(i), (5.7) follows from Theorem 2.21 and (5.8) follows from Corollary 3.5. $\qquad\square$

Corollary 5.2. *Let $\eta_i \equiv (\epsilon_i, \nu_i)', i \in \mathbb{Z}$, be a sequence of i.i.d. random vectors with $\mathrm{E}\,\eta_0 = 0$, $\mathrm{E}\,(\eta_0\eta_0') = \Sigma$ and $\mathrm{E}\,||\eta_0||^4 < \infty$. Suppose that*

(i) in addition to Assumption 5.1, $\int_{-\infty}^{\infty} |\hat{K}(t)| dt < \infty$, where $\hat{K}(t) = \int_{-\infty}^{\infty} e^{itx} K(x) dx$;

(ii) $u_k = \sum_{j=0}^{\infty} \psi_j \eta_{k-j}$, where the coefficient vector $\psi_k = (\psi_{k1}, \psi_{k2})$ satisfies $\sum_{k=0}^{\infty} k^{1/4}(|\psi_{1k}| + |\psi_{2k}|) < \infty$ and $\sum_{k=0}^{\infty} \psi_k \neq 0$.

Then, for any h satisfying $d_n h^{2\gamma} \to 0$ and $nh/d_n \to \infty$, we have

$$\left(\frac{nh}{d_n}\right)^{1/2} \left[\hat{f}(x_0) - f(x_0)\right] \to_D \tau_2 \, L_Z(1,0)^{-1/2} \, N,$$

$$\left(\sum_{t=1}^{n} K[(x_t - x_0)/h]\right)^{1/2} \left[\hat{f}(x_0) - f(x_0)\right] \to_D \tau_3 \, N,$$

where $\tau_2 = \left(\mathrm{E}\,u_0^2 \int_{-\infty}^{\infty} K^2(x)dx\right)^{1/2} / \int_{-\infty}^{\infty} K(x)dx$, $\tau_3 = \tau_2 \left(\int_{-\infty}^{\infty} K(x)dx\right)^{1/2}$ and N is a standard normal variate independent of $L_Z(1,0)$.

Proof. The proof is similar to that of Corollary 5.1, by noting that Assumption 5.3 (a) holds by Lemma 2.2(i); (5.7) holds with $c_{1g} = \int_{-\infty}^{\infty} g(x)dx$ by Theorem 2.21; and (5.8) holds with $c_{1g^2} = \int_{-\infty}^{\infty} g^2(x)dx$ and $c_{2g} = \left(\mathrm{E}\,u_0^2 \int_{-\infty}^{\infty} g^2(x)dx\right)^{1/2}$ by Theorem 3.16. $\qquad\square$

5.1.2 *Nadaraya-Watson estimator: certain extensions*

As stated in Section 5.1.1, Assumption 5.2 provides a trade off among the regressor x_t, the error process u_t and the bandwidth h, and is essential to establish the asymptotic normality of $\hat{f}(x_0)$. If u_t has certain structure, the asymptotics of $\hat{f}(x_0)$ can be investigated in less restrictive settings on x_t (e.g., Assumption 5.4 below) than that of Assumption 5.2.

To illustrate, we consider

$$u_t = \sigma(x_t)\,\nu_t, \qquad (5.10)$$

where $\sigma(x)$ is a positive real function and

$(\nu_t, \mathcal{F}_t)_{t\geq 1}$ forms a martingale difference with $\mathrm{E}\,(\nu_t^2|\mathcal{F}_{t-1}) = 1$ and $\sup_{t\geq 1} \mathrm{E}\,(|\nu_t|^q) < \infty$ a.s. for some $q > 2$.

Assumption 5.4. There exist constants $0 < a_n \to \infty$ such that, for each real function $g(x)$ on R,

(a) if $\int_{-\infty}^{\infty} |g(s)|ds < \infty$, then $\frac{1}{a_n} \sum_{t=1}^{n} \mathrm{E}|g[(x_t - x_0)/h]| = O(1)$;
(b) if $|g(s)|$ is bounded and $\int_{-\infty}^{\infty} |g(s)|ds < \infty$, then

$$\frac{1}{a_n} \sum_{t=1}^{n} g[(x_t - x_0)/h] \to_D c_{1g} V^2 \qquad (5.11)$$

where c_{1g} is a constant depending only on $g(x)$ and V is a random variable with $\mathrm{P}(V \neq 0) = 1$.

Theorem 5.2. *Suppose that*

(a) $\mathrm{E}\,(\nu_t \mid x_1, ..., x_n) = 0$ *for* $1 \leq t \leq n$, $n \geq 1$;
(b) in addition to Assumption 5.1, there exist a nonnegative real function $\sigma_1(s, x_0)$ *and a constant* $0 < \gamma_1 \leq 1$ *such that*

$$|\sigma(\delta y + x_0) - \sigma(x_0)| \leq \delta^{\gamma_1}\,\sigma_1(y, x_0), \quad \text{for all } y \in \mathrm{R}, \quad (5.12)$$

when $\delta > 0$ *is sufficiently small, and* $\int_{-\infty}^{\infty} K^2(s)\,\sigma_1^2(s, x_0)ds < \infty$;
(c) Assumption 5.4 holds.

Then, for any h *satisfying* $a_n\,h^{2\gamma} \to 0$, *we have*

$$a_n^{1/2}\left[\hat{f}(x_0) - f(x_0)\right] \to_D c_{2K}\,c_{1K}^{-1}\,V^{-1}\,N, \quad (5.13)$$

$$\Big(\sum_{t=1}^{n} K[(x_t - x_0)/h]\Big)^{1/2}\left[\hat{f}(x_0) - f(x_0)\right] \to_D c_{2K}c_{1K}^{-1/2}\,N, \qquad (5.14)$$

where $c_{2K} = \sigma(x_0)\,(c_{1K^2})^{1/2}$, c_{1K} *and* c_{1K^2} *are defined as in (5.11) with that* g *is replaced by* K *and* K^2, *respectively, and* N *is a standard normal variate independent of* V.

Proof.   Write $\Pi_n = \sum_{t=1}^{n} \left[\sigma(x_t) - \sigma(x_0) \right] \nu_t \, K[(x_t - x_0)/h]$. Note that

$$\sum_{t=1}^{n} \left[\sigma(x_t) - \sigma(x_0) \right]^2 K^2[(x_t - x_0)/h] \leq C \, h^{2\gamma_1} \sum_{t=1}^{n} K_2[(x_t - x_0)/h],$$

where $K_2(s) = K^2(s) \, \sigma_1^2(s, x_0)$, due to (5.12). It is readily seen that, under given conditions,

$$\mathrm{E}\Pi_n^2 \leq C \, h^{2\gamma_1} \sum_{t=1}^{n} \mathrm{E}K_2[(x_t - x_0)/h] = O(a_n \, h^{2\gamma_1}),$$

i.e., $a_n^{-1/2}\Pi_n = O_{\mathrm{P}}(h^{\gamma_1}) = o_{\mathrm{P}}(1)$. Now, as in the proof of Theorem 5.1, we have

$$a_n^{1/2} \left[\hat{f}(x_0) - f(x_0) \right]$$

$$= \frac{\sigma(x_0)a_n^{-1/2} \sum_{t=1}^{n} \nu_t \, K[(x_t - x_0)/h]}{V_{n0}} + \frac{a_n^{-1/2}\Pi_n}{V_{n0}} + o_{\mathrm{P}}(1)$$

$$= \frac{\sigma(x_0)a_n^{-1/2} \sum_{t=1}^{n} \nu_t \, K[(x_t - x_0)/h]}{V_{n0}} + o_{\mathrm{P}}(1)$$

where $V_{n0} = a_n^{-1} \sum_{t=1}^{n} K[(x_t - x_0)/h]$. Hence, to prove (5.13) and (5.14), it suffices to show that Assumption 5.2(b) [namely, (5.4)] holds true with that u_t is replaced by ν_t, under given conditions.

To this end, let $W_n = \sum_{k=1}^{n} \nu_k \, Z_{nk}$, where $Z_{nk} = g[(x_k - x_0)/h]/\Delta_n^{1/2}$ and

$$\Delta_n = \sum_{k=1}^{n} g^2[(x_k - x_0)/h].$$

Since, given $x_1, x_2, ..., x_n$, $(Z_{nt} \nu_t, t = 1, 2, ..., n)$ is a martingale difference, it follows from Theorem 3.9 [(3.75) there] in Hall and Heyde (1980) with $\delta = q/2 - 1$ that

$$\sup_x \left| \mathrm{P}(W_n \leq x \mid x_1, x_2, ..., x_n) - \Phi(x) \right| \leq A(\delta) \, \mathcal{L}_n^{1/(1+q)}, \quad a.s.,$$

where $A(\delta)$ is a constant depending only on δ, $\Phi(x)$ is a standard normal distribution function and

$$\mathcal{L}_n = \sum_{k=1}^{n} |Z_{nk}|^q \mathrm{E}\, |\nu_k|^q \leq C \, \Delta_n^{-\delta/2}.$$

Noting that $g^2(x)$ is bounded and integrable, we have $\Delta_n \to_{\mathrm{P}} \infty$ by Assumption 5.4(b). As a consequence, it follows that

$$\sup_x \left| P\big(W_n \leq x \mid x_1, ..., x_n\big) - \Phi(x) \right| \to_{\mathrm{P}} 0. \tag{5.15}$$

We next let $\Delta_{1n} = \sum_{k=1}^{n} g[(x_k - x_0)/h]$. Since $\beta\Delta_n + \gamma\Delta_{1n}$ is $\sigma(x_1, ...x_n)$ measurable and (5.9) holds true by (5.11), it follows from (5.15) that, for all $\alpha, \beta, \gamma \in \mathbb{R}$,

$$\left| E\left[e^{i\alpha W_n + i\beta\,\Delta_n/a_n + i\gamma\,\Delta_{1n}/a_n} \right] - e^{-\frac{1}{2}\alpha^2}\, E\left[e^{i(\beta\,c_{1g^2} + \gamma\,c_{1g})V^2} \right] \right|$$

$$\leq E\left| E\left(e^{i\alpha W_n} \,|\, x_1, ..., x_n \right) - e^{-\frac{1}{2}\alpha^2} \right|$$

$$+ e^{-\frac{1}{2}\alpha^2}\left| \mathrm{E}\, e^{i(\beta\Delta_n + \gamma\Delta_{1n})/a_n} - \mathrm{E}\, e^{i(\beta\,c_{1g^2} + \gamma\,c_{1g})V^2} \right| \to 0$$

by the dominated convergence theorem. This implies that

$$\{W_n,\; \Delta_n/a_n,\; \Delta_{1n}/a_n\} \to_D \{N,\; c_{1g^2}\,V^2,\; c_{1g}\,V^2\}$$

where N is a standard normal random variable independent of V. Hence, by the continuous mapping theorem, we obtain

$$\left\{ \frac{1}{a_n}\sum_{k=1}^{n} g[(x_k - x_0)/h],\; (\frac{1}{a_n})^{1/2}\sum_{k=1}^{n} g[(x_k - x_0)/h]\,\nu_t \right\}$$

$$= \left\{ \Delta_{1n}/a_n,\; W_n\,(\Delta_n/a_n)^{1/2} \right\} \to_D \{c_{1g}\,V^2,\; N\,(c_{1g^2}\,V^2)^{1/2}\},$$

which yields (5.4) with that u_t is replaced by ν_t. We now complete the proof of Theorem 5.2. $\qquad\square$

Remark 5.1. If ν_t is independent of x_t, it is natural to have $\mathrm{E}\,(\nu_t \mid x_1, ..., x_n) = 0$ for $1 \leq t \leq n$. If only $\mathrm{E}\,(\nu_t \mid x_1, ..., x_t) = 0$, except imposing more conditions on x_t such as those in Corollary 5.1, Assumption 5.4 [namely, (5.11)] is not sufficient to derive the results (5.13) and (5.14). Examples can be similarly constructed as that in Section 3.3.3. It is interesting to notice that Assumption 5.4 is quite mild, hosting for both stationary and nonstationary time series as special examples. Hence, under the assumption on the independence between ν_t and x_t, Theorems 5.2 provides a unified result on the asymptotic normality of $\hat{f}(x_0)$ for both stationary and nonstationary time series.

We may replace the martingale structure of ν_t in model (5.10) by a linear process, defined by

$$\nu_t = \sum_{k=0}^{\infty} \psi_k \eta_{t-k}, \text{ where } \psi := \sum_{k=0}^{\infty} \psi_k \neq 0,\; \sum_{k=0}^{\infty} |\psi_k| < \infty,$$

η_k are i.i.d. random variables with $\mathrm{E}\eta_0 = 0, \mathrm{E}\eta_0^2 = 1$ and $\mathrm{E}\,|\eta_0|^q < \infty$ for some $q > 2$.

Theorem 5.3. *Suppose that*

(a) $\mathrm{E}\,(\eta_t \mid x_1, ..., x_n) = 0$ *for* $1 \leq t \leq n$, $n \geq 1$;

(b) in addition to the conditions (b) and (c) of Theorem 5.2,

$$\frac{1}{a_n} \sum_{t=1}^{n} |g[(x_t - x_0)/h]| \, |g[(x_{t+l} - x_0)/h]| = o_P(1), \qquad (5.16)$$

for any finite $l \geq 1$.

Then, for any h satisfying $a_n h^{2\gamma} \to 0$, we have

$$a_n^{1/2} \left[\hat{f}(x_0) - f(x_0) \right] \to_D c_{2K} \, c_{1K}^{-1} \, V^{-1} \, N, \qquad (5.17)$$

$$\Big(\sum_{t=1}^{n} K[(x_t - x_0)/h] \Big)^{1/2} \left[\hat{f}(x_0) - f(x_0) \right] \to_D c_{2K} \, c_{1K}^{-1/2} \, N, \qquad (5.18)$$

where $c_{2K} = \sigma(x_0)[\gamma(0) \, c_{1K^2}]^{1/2}$ with $\gamma(k) = \sum_{j=0}^{\infty} \psi_j \psi_{j+k}$, c_{1K} and c_{1K^2} are defined as in (5.11) with that g is replaced by K and K^2, respectively, and N is a standard normal variate independent of V.

Proof. Let $\Pi_n = \sum_{t=1}^{n} \left[\sigma(x_t) - \sigma(x_0) \right] \nu_t \, K[(x_t - x_0)/h]$. Recalling $\nu_t = \sum_{k=0}^{\infty} \psi_k \eta_{t-k}$ and $\mathrm{E}\,(\eta_j | x_1, ..., x_n) = 0$ for $j \leq n$, simple calculations show that

$$\mathrm{E}\Pi_n^2 \leq \sum_{k=0}^{\infty} |\psi_k| \sum_{k=0}^{\infty} |\psi_k| \, \mathrm{E} \, \Big| \sum_{t=1}^{n} \eta_{t-k} \left[\sigma(x_t) - \sigma(x_0) \right] K[(x_t - x_0)/h] \Big|^2$$

$$\leq C \, h^{2\gamma_1} \sum_{t=1}^{n} \mathrm{E}K_2[(x_t - x_0)/h] = O(a_n h^{2\gamma_1}),$$

where $K_2(s) = K^2(s) \, \sigma_1^2(s, x_0)$, due to (5.12) and Assumption 5.4(b). Now, by using similar arguments as in the proof of Theorem 5.2, it suffices to show that if $|g(s)|$ is bounded and $\int_{-\infty}^{\infty} |g(s)| ds < \infty$, then

$$\Big\{ \frac{1}{a_n} \sum_{t=1}^{n} g[(x_t - x_0)/h], \ \Big(\frac{1}{a_n}\Big)^{1/2} \sum_{t=1}^{n} \nu_t \, g[(x_t - x_0)/h] \Big\}$$

$$\to_D \ \Big(c_{1g} \, V^2, \ N \, [\gamma(0) \, c_{1g^2} \, V^2]^{1/2} \Big), \qquad (5.19)$$

where N is a standard normal variate independent of V, i.e., Assumption 5.2(b) holds true with that u_t is replaced by ν_t.

Recalling $\nu_t = \sum_{k=-\infty}^{\infty} \psi_k \eta_{t-k}$ (set $\psi_k = 0$ if $k < 0$), we have

$$\Big(\frac{1}{a_n}\Big)^{1/2} \sum_{k=1}^{n} \nu_k \, g[(x_k - x_0)/h] = \Big(\frac{1}{a_n}\Big)^{1/2} \sum_{i=-\infty}^{\infty} \eta_i \, Z_{ni},$$

where $Z_{ni} = \sum_{j=1}^{n} \psi_{j-i} g_j$ with $g_j = g[(x_j - x_0)/h]$. Note that

$$\frac{1}{a_n} \sum_{i=-\infty}^{\infty} Z_{ni}^2 = \frac{1}{a_n} \sum_{k,l=1}^{n} g_k g_l \sum_{i=-\infty}^{\infty} \psi_{k-i} \psi_{l-i}$$

$$= \frac{1}{a_n} \sum_{k,l=1}^{n} g_k g_l \gamma(k - l)$$

$$= \frac{\gamma(0)}{a_n} \sum_{k=1}^{n} g^2[(x_k - x_0)/h] + \frac{2}{a_n} \sum_{1 \le l < k \le n} g_k g_l \gamma(k - l). \quad (5.20)$$

To prove (5.19), by using similar arguments as in the proof of Theorem 5.2 [recalling (5.15) and $E(\eta_j | x_1, ..., x_n) = 0$ for $j \le n$), it suffices to show that

$$\Lambda_n := \frac{1}{a_n} \sum_{1 \le l < k \le n} g_k g_l \gamma(k - l) = o_P(1), \quad (5.21)$$

and for any $\delta > 0$,

$$\frac{1}{a_n^{1+\delta/2}} \sum_{i=-\infty}^{\infty} |Z_{ni}|^{2+\delta} = o_P(1). \quad (5.22)$$

For any $M_0 \ge 1$, we may write

$$|\Lambda_n| \le \frac{1}{a_n} \sum_{l=1}^{n-1} \sum_{k=1}^{n-l} |g_l| |g_{k+l}| |\gamma(k)|$$

$$\le A_0 \sum_{k=1}^{M_0} \frac{1}{a_n} \sum_{l=1}^{n-1} |g_l| |g_{k+l}| + \frac{\sup_x |g(x)|}{a_n} \sum_{l=1}^{n-1} |g_l| \sum_{k=M_0}^{\infty} |\gamma(k)|, \quad (5.23)$$

where $A_0 = \sum_{k=0}^{\infty} |\gamma(k)|$. Note that $\sum_{k=0}^{\infty} |\gamma(k)| \le (\sum_{j=0}^{\infty} |\psi_j|)^2 < \infty$. result (5.21) follows from (5.23), (5.16) and Assumption 5.4(a) as $n \to \infty$ first and then $M_0 \to \infty$. Since $\sup_i |Z_{ni}| \le \sup_x |g(x)| \sum_{j=0}^{\infty} |\psi_j|$ and $a_n^{-1} \sum_{i=-\infty}^{\infty} Z_{ni}^2 = O_P(1)$ by (5.20)–(5.21) and Assumption 5.4(a), result (5.22) follows easily from the fact that $a_n \to \infty$.

This yields (5.19) and also completes the proof of Theorem 5.3. $\quad \square$

5.1.3 *Bias analysis and local linear estimator*

An explicit bias term may be incorporated into the limit theory (5.5) and (5.6) if we impose stronger smoothness conditions on f and K instead of Assumption 5.1. Furthermore, the Nadaraya-Watson estimator $\hat{f}(x)$ has the same limit distribution (to the second order including bias) as the local linear nonparametric estimator. Explicitly, we have the following theorem.

Theorem 5.4. *Suppose that*

(a) instead of Assumption 5.1, for some $p \geq 2$,

    (i) for fixed x_0, $f(x_0)$ has a continuous $(p+1)$-th order derivative in a small neighborhood of x_0;

    (ii) $K(x)$ has compact support, satisfying that $\int K(y)dy = 1$,

$$\int y^p K(y)dy \neq 0, \quad \int y^i K(y)dy = 0, \quad i = 1, 2, ..., p - 1;$$

(b) in addition to Assumption 5.2,

$$\left(\frac{1}{a_n}\right)^{1/2} \sum_{t=1}^{n} g[(x_t - x_0)/h] = O_P(1), \tag{5.24}$$

if $\int_{-\infty}^{\infty} |g(s)|ds < \infty$ and $\int_{-\infty}^{\infty} g(x)dx = 0$.

Then, for any h satisfying $a_n h^{2(p+1)} \to 0$, we have

$$a_n^{1/2} \left[\hat{f}(x_0) - f(x_0) - B_p \right] \to_D c_{2K} c_{1K}^{-1} V^{-1} N, \tag{5.25}$$

and

$$\left(h \sum_{t=1}^{n} K[(x_t - x_0)/h] \right)^{1/2} \left[\hat{f}(x_0) - f(x_0) - B_p \right]$$

$$\to_D c_{2K} c_{1K}^{-1} N. \tag{5.26}$$

where $B_p = \frac{h^p f^{(p)}(x_0)}{p!} \int_{-\infty}^{\infty} y^p K(y)dy$ and the other notation are the same as those in Theorem 5.1. Furthermore, both results (5.25) and (5.26) (with $p = 2$) hold if we replace $\hat{f}(x_0)$ by $\hat{f}^L(x_0)$.

Proof. Recalling (5.3) and Assumption 5.2, to prove (5.25) and (5.26), it suffices to show that, for any h satisfying $a_n h^{2(p+1)} \to 0$,

$$R_n := \frac{\sum_{t=1}^{n} [f(x_t) - f(x_0)] K[(x_t - x_0)/h]}{\sum_{t=1}^{n} K[(x_t - x_0)/h]}$$

$$= \frac{h^p f^{(p)}(x_0)}{p!} \int_{-\infty}^{\infty} y^p K(y)dy + o_P(a_n^{-1/2}). \tag{5.27}$$

Note that the numerator of R_n involves

$$\sum_{t=1}^{n} \{f(x_t) - f(x_0)\} K[(x_t - x_0)/h] = \sum_{j=1}^{p+1} I_j, \tag{5.28}$$

where, by letting $H_j(y) = y^j K(y)$,

$$I_j = \frac{h^j f^{(j)}(x_0)}{j!} \sum_{t=1}^{n} H_j[(x_t - x_0)/h], \quad j = 1, 2, ..., p,$$

$$I_{p+1} = \sum_{t=1}^{n} \left\{ f(x_t) - \sum_{j=0}^{p} \frac{f^{(j)}(x_0)}{j!} (x_t - x_0)^j \right\} K[(x_t - x_0)/h].$$

Since $K(x)$ has compact support (Ω, say) and $\lim_{h\to 0} \sup_{y\in\Omega} |f^{(p+1)}(yh + x_0)| \leq C$, Taylor's expansion yields

$$|I_{p+1}| \leq C \sum_{t=1}^{n} |x_t - x_0|^{p+1} K[(x_t - x_0)/h]$$

$$\leq C h^{p+1} \sum_{t=1}^{n} K[(x_t - x_0)/h]. \tag{5.29}$$

On the other hand, simple calculations from (5.24) and Assumption 5.2 [which implies (5.9)] show that

$$\frac{h^{-j} a_n^{1/2} |I_j|}{\sum_{t=1}^{n} K[(x_t - x_0)/h]} = O_P(1), \tag{5.30}$$

for $j = 1, 2, ..., p-1$, and

$$\frac{h^{-p} I_p}{\sum_{t=1}^{n} K[(x_t - x_0)/h]} \to_P \frac{f^{(p)}(x_0)}{p!} \int_{-\infty}^{\infty} y^p K(y) dy. \tag{5.31}$$

By virtue of (5.28)-(5.31), we have

$$a_n^{1/2} \left[R_n - \frac{h^p f^{(p)}(x_0)}{p!} \int_{-\infty}^{\infty} y^p K(y) dy \right]$$

$$\leq \frac{a_n^{1/2}}{\sum_{t=1}^{n} K[(x_t - x_0)/h]} \sum_{j=1}^{p-1} |I_j| + \frac{a_n^{1/2} |I_{p+1}|}{\sum_{t=1}^{n} K[(x_t - x_0)/h]}$$

$$= O_P \left[h + a_n^{1/2} h^{p+1} \right] = o_P(1),$$

whenever $a_n h^{2(p+1)} \to 0$. This proves (5.27), and also completes the proofs of (5.25) and (5.26).

We next show (5.25) and (5.26) (with $p = 2$) hold if we replace $\hat{f}(x_0)$ by $\hat{f}^L(x_0)$. We may write

$$\hat{f}^L(x) = \frac{\sum_{i=1}^{n} K[(x_i - x)/h]}{\sum_{i=1}^{n} K[(x_i - x)/h] - V_{n1}^2/V_{n2}} \, \hat{f}(x)$$

$$- \frac{(V_{n1}/V_{n2}) \sum_{i=1}^{n} H_1[(x_i - x)/h] \, y_i}{\sum_{i=1}^{n} K[(x_i - x)/h] - V_{n1}^2/V_{n2}}.$$

Note that, as in the proof of (5.30), $V_{n1}^2/V_{n2} = O_P(1)$ and $V_{n1}/V_{n2} = O_P(a_n^{-1/2})$. By a simple calculation, the claim will follow if we prove

$$R_{1n} := \frac{\sum_{i=1}^{n} H_1[(x_i - x)/h] \, y_i}{\sum_{i=1}^{n} K[(x_i - x)/h]} = o_P(1), \tag{5.32}$$

provided $a_n h^{2(p+1)} \to 0$ and $a_n \to \infty$. This is simple under given conditions. Indeed, we may split the numerator of R_{1n} as

$$f(x) \sum_{i=1}^{n} H_1[(x_i - x)/h] + \sum_{i=1}^{n} H_1[(x_i - x)/h][f(x_i) - f(x)]$$

$$+ \sum_{i=1}^{n} H_1[(x_i - x)/h] \, u_i$$

$$:= R_{2n} + R_{3n} + R_{4n}.$$

Recalling that $|H_1(x)|$ is bounded, integrable and $\int_{-\infty}^{\infty} H_1(x)dx = 0$, it follows from (5.24) and Assumption 5.2 that

$$\frac{R_{2n} + R_{4n}}{\sum_{i=1}^{n} K[(x_i - x)/h]} = O_P[a_n^{-1/2}] = o_P(1).$$

As for R_{3n}, we have

$$|R_{3n}| \leq C \sum_{i=1}^{n} |H_1[(x_i - x)/h]||x_i - x|^\gamma \leq C \, h^\gamma \sum_{i=1}^{n} K[(x_i - x)/h],$$

which implies

$$\frac{R_{3n}}{\sum_{i=1}^{n} K[(x_i - x)/h]} = O_P(h^\gamma) = o_P(1).$$

Combining all these estimates, we obtain (5.32), and also complete the proof of Theorem 5.4.   □

Remark 5.2. The local linear nonparametric estimator is popular partly because of its bias reducing properties in comparison with the Nadaraya-Watson estimator. The present theorem shows that this particular advantage is lost when the regressor x_t is nonstationary. The other main advantage of the local linear smoother is the absence of boundary effects when the distribution of x_t has bounded support. However, in the present case, x_t is recurrent with unbounded support and so this second advantage does not apply either.

5.1.4 *Uniform convergence for local linear estimator*

This section investigates uniform convergence of the local linear estimator $\hat{f}^L(x)$ in the following nonlinear cointegrating regression model:

$$y_t = f(x_t) + u_t, \quad u_t = \sigma(x_t) \nu_t, \tag{5.33}$$

where f is an unknown function to be estimated with the observed data $\{x_t, y_t\}_{t=1}^n$, $\sigma(x)$ is a heterogeneity generating function (HGF) and for a filtration $\mathcal{F}_t$ to which x_{t+1} is adapted, $\{u_t, \mathcal{F}_t\}$ forms a martingale difference. The uniform convergence of the Nadaraya-Watson estimator $\hat{f}(x)$ can be considered in a similar way, where details can be found in Chan and Wang (2014). See, also, Duffy (2014a, b).

We make use of the following assumptions in the development of our uniform asymptotics.

Assumption 5.5. $x_t = \sum_{j=1}^t \xi_j$ with $\xi_j = \sum_{k=0}^\infty \phi_k \epsilon_{j-k}$, where $\{\epsilon_j\}_{j \in \mathbb{Z}}$ is a sequence of i.i.d. random variables with $E\epsilon_0 = 0$, $E\epsilon_0^2 = 1$, $E |\epsilon_0|^r < \infty$ for some $r > 2$ and $\int_{-\infty}^\infty |E\, e^{it\epsilon_0}| dt < \infty$. Furthermore, the coefficients ϕ_k, $k \geq 0$ are assumed to satisfy one of the following conditions:

C1. $\phi_0 \neq 0$ and $\phi_k \sim k^{-\mu}$, where $1/2 < \mu < 1$.
C2. $\sum_{k=0}^\infty k|\phi_k| < \infty$ and $\phi \equiv \sum_{k=0}^\infty \phi_k \neq 0$.

Assumption 5.6. $\{u_t, \mathcal{F}_t\}_{t \geq 1}$ is a martingale difference, where $\mathcal{F}_t = \sigma(x_1, ..., x_{t+1}, u_1, ..., u_t)$, satisfying $\sup_{t \geq 1} E(|u_t|^{2p} \mid \mathcal{F}_{t-1}) < \infty$, where $p > 1/\epsilon_0$ for some $0 < \epsilon_0 \leq 1/2$.

Assumption 5.7. K has compact support, $\int_{-\infty}^\infty x K(x) dx = 0$ and $|K(x) - K(y)| \leq C|x - y|$ for all $x, y \in \mathbb{R}$.

Assumption 5.8. The first order derivative of $f(x)$ exists and there exist a $0 < \tau \leq 1$ and a real positive function $f_0(x)$ such that

$$|f'(y) - f'(x)| \leq C |y - x|^\tau f_0(x), \tag{5.34}$$

uniformly for $x \in \mathbb{R}$ and $|y - x|$ sufficiently small.

Assumption 5.9. $\inf_{x \in \mathbb{R}} \sigma(x) > 0$ and for any $|y|$ sufficiently small,

$$\sup_{x \in \mathbb{R}} \frac{|\sigma(x + y) - \sigma(x)|}{\sigma(x)} \leq C |y|, \tag{5.35}$$

where C is a positive constant.

We have the following uniform asymptotic result.

Theorem 5.5. *Suppose Assumptions 5.5–5.9 hold. Let $\epsilon_0 > 0$ be given as in Assumption 5.6, $d_n^2 = var(x_n)$ and $0 < r_n \to 0$. Then, for any h satisfying $h \to 0$ and $n^{1-\epsilon_0} h/d_n \to \infty$, we have*

$$\sup_{|x| \leq b_n} \frac{|\widehat{f}^L(x) - f(x)|}{\sigma(x)} = O_P\left\{ (nh/d_n)^{-1/2} \log^{1/2} n + h^{1+\tau} \delta_n \right\}, \tag{5.36}$$

where $\delta_n = \sup_{|x| \le b_n} [f_0(x)/\sigma(x)]$ and $b_n \le r_n d_n$.

Proof.   Let $V_n(x) = \sum_{i=1}^n w_i(x)$. We may write

$$\widehat{f}^L(x) - f(x) = \Gamma_{1n}(x) + \Gamma_{2n}(x), \tag{5.37}$$

where $\Gamma_{1n}(x) = V_n^{-1}(x) \sum_{i=1}^n w_i(x)\sigma(x_i)u_i$ and

$$\Gamma_{2n}(x) = V_n^{-1}(x) \sum_{i=1}^n w_i(x)\big[f(x_i) - f(x)\big].$$

Recall that $V_{nj}(x) = \sum_{i=0}^n H_j\big(\frac{x_i - x}{h}\big)$, where $H_j(x) = x^j K(x)$, $j = 0, 1, 2$. Due to Assumption 5.7, for each of $j = 0, 1, 2$, $H_j(x)$ satisfies Assumption 2.3(ii). Now Corollary 2.12 implies that, whenever $n^{1-\epsilon_0} h/d_n \to \infty$,

$$\sup_{x \in R} |V_{n1}(x)| = \sup_{x \in R} \big| \sum_{i=1}^n H_1\big(\frac{x_i - x}{h}\big)\big| = O_P\big[(nh/d_n)\log^{-\beta} n\big], \tag{5.38}$$

for any $\beta > 0$, $\sup_{x \in R} |V_{nj}(x)| = O_P(nh/d_n)$ and

$$\big\{ \inf_{|x| \le b_n} |V_{nj}(x)|\big\}^{-1} = O_P\big[d_n/(nh)\big],$$

for $j = 0$ and 2. It follows from these facts that

$$\big\{ \inf_{|x| \le b_n} |V_n(x)|/V_{n2}(x)\big\}^{-1} \le \big\{ \inf_{|x| \le b_n} \big|V_{n0}(x) - \frac{V_{n,1}^2(x)}{V_{n,2}(x)}\big|\big\}^{-1}$$

$$= \big\{ \inf_{|x| \le b_n} |V_{n0}(x)| - o_P(nh/d_n)\big\}^{-1}$$

$$= O_P\big[d_n/(nh)\big]$$

and $\sup_{|x| \le b_n} |V_{n1}(x)|/V_{n2}(x) = o_P(\log^{-\beta} n)$ for any $\beta > 0$. On the other hand, by taking ϵ_0 as defined in Assumption 5.6, it follows easily from Corollary 3.9 that, whenever $n^{1-\epsilon_0} h/d_n \to \infty$,

$$\sum_{i=1}^n g\big(\frac{x_i - x}{h}\big)\frac{\sigma(x_i)}{\sigma(x)}u_i = O_P\big(\frac{nh}{d_n}\big)^{1/2} \log^{1/2} n,$$

where $g(x) = K(x)$ or $H_1(x)$. Consequently, we have

$$\sup_{|x| \le b_n} \frac{|\Gamma_{1n}(x)|}{\sigma(x)} \le \big\{ \inf_{|x| \le b_n} \frac{|V_n(x)|}{V_{n2}(x)}\big\}^{-1}\big\{ \sup_{|x| \le b_n} \big| \sum_{i=1}^n K\big(\frac{x_i - x}{h}\big)\frac{\sigma(x_i)}{\sigma(x)}u_i\big|$$

$$+ \sup_{|x| \le b_n} \frac{|V_{n1}(x)|}{V_{n,2}(x)}\big| \sum_{i=1}^n H_1\big(\frac{x_i - x}{h}\big)\frac{\sigma(x_i)}{\sigma(x)}u_i\big|\big\}$$

$$= O_P\big[\big(\frac{d_n}{nh}\big)^{1/2} \log^{1/2} n\big]. \tag{5.39}$$

We next consider $\Gamma_{2n}(x)$. Note that $\sum_{i=1}^{n} w_i(x)(x_i - x) = 0$ and, by Assumption 5.8,

$$|f(y) - f(x) - f'(x)(y - x)| = \left| \int_x^y f'(s) - f'(x)ds \right|$$

$$\leq f_0(x) \int_x^y |s - x|^\tau ds = \frac{1}{\tau + 1} f_0(x)|y - x|^{\tau + 1}.$$

It follows from Assumption 5.9 that

$$\sup_{|x| \leq b_n} \frac{|\Gamma_{2n}(x)|}{\sigma(x)} = \sup_{|x| \leq b_n} \frac{\left| \sum_{i=1}^{n} w_i(x)\left[f(x_i) - f(x) - f'(x)(x_i - x) \right] \right|}{\sigma(x)V_n(x)}$$

$$\leq \sup_{|x| \leq b_n} \frac{|f_0(x)|}{2\sigma(x)} \frac{\sum_{i=1}^{n} |w_i(x)||x_i - x|^{\tau + 1}}{V_n(x)}$$

$$\leq C \, \delta_n \sup_{|x| \leq b_n} \left| \frac{\sum_{i=1}^{n} |x_i - x|^{\tau + 1} K[(x_i - x)/h]}{V_{n2}^{-1}(x)V_n(x)} + \right.$$

$$\left. \frac{[\sum_{i=1}^{n} |x_i - x|^{\tau + 2} K[(x_i - x)/h]]}{V_{n2}^{-1}(x)V_n(x)} \left\{ \frac{|V_{n1}(x)|}{V_{n2}(x)} \right\} \right|$$

$$\leq \frac{C d_n \, h^{\tau + 1} \delta_n}{nh} \sup_{|x| \leq b_n} \sum_{i=1}^{n} K\left(\frac{x_i - x}{h} \right)$$

$$\leq C \, h^{\tau + 1} \delta_n. \tag{5.40}$$

Taking (5.39) and (5.40) into (5.37), we prove (5.36). □

Remark 5.3. The convergence rate in (5.36) is sharp and probably optimal. In the situation where x_t is a stationary regressor, the sharp rate of convergence is $O_P[(nh)^{-1/2} \log^{1/2} n]$ (see, e.g., Hansen (2008)). The reason behind the difference is due to the fact that, the integrated series wanders over the entire real line but spent only $O(d_n)$ amount of sample time around any specific point, while stationary time series spent $O(n)$.

Remark 5.4. The range $|x| \leq r_n d_n$ that enables (5.36) is optimal in the situation that $b_n/d_n \to 0$ in (5.36) cannot be extended to $b_n/d_n \to C > 0$. Explanation in this regard is similar to that of Remark 2.15.

Remark 5.5. Under less restrictions on the kernel $K(x)$ and the regression function $f(x)$, a similar result can be established for the Nadaraya-Watson estimator $\hat{f}(x)$ defined by

$$\hat{f}(x) = \frac{\sum_{k=1}^{n} K\left[(x_k - x)/h \right] y_k}{\sum_{k=1}^{n} K\left[(x_k - x)/h \right]}.$$

For details, we refer to Chan and Wang (2014) and Duffy (2014a, b). The latter papers investigated the uniform asymptotics under the random optimal range for the x to be held.

Remark 5.6. Due to the definition of δ_n, when it is weighted by the HGF $\sigma(x)$, the uniform convergence rate of $\widehat{f}^L(x) - f(x)$ is improved in the tail. Note that model (5.33) can be restated as

$$y_t/\sigma(x_t) = f(x_t)/\sigma(x_t) + u_t.$$

It is not so superise to see that the limit behavior of $[\widehat{f}^L(x) - f(x)]/\sigma(x)$ is improved in the tail. This feature is first noticed in Liu, Chan and Wang (2014).

The HGF $\sigma(x)$ can be estimated by

$$\widehat{\sigma}^2(x) = \frac{\sum_{t=1}^{n}[y_t - \widehat{f}^L(x_t)]^2 K[(x_t - x)/h]}{\sum_{t=1}^{n} K[(x_t - x)/h]}.$$

The following result provides the limit behavior of $\widehat{\sigma}_n^2(x)$.

Theorem 5.6. *Suppose Assumptions 5.5, 5.7–5.9 hold, and in addition to Assumption 5.6,* $\mathrm{E}\left(u_t^2 \mid \mathcal{F}_{t-1}\right) \to 1, a.s.$ *and* $\sup_{t \geq 1} \mathrm{E}\left(|u_t|^{4p} \mid \mathcal{F}_{t-1}\right) < \infty,$ *where* $p > 1/\epsilon_0$ *for some* $0 < \epsilon_0 < 1.$ *Let* $d_n^2 = var(x_n)$ *and* $0 < r_n \to 0.$ *Then, for any* h *satisfying* $h \to 0$ *and* $n^{1-\epsilon_0}h/d_n \to \infty,$ *we have*

$$\sup_{|x| \leq b_n} \frac{\left|\widehat{\sigma}^2(x) - \sigma^2(x)\right|}{\sigma^2(x)}$$

$$= O_P\left\{h + \left(nh/d_n\right)^{-1/2} \log^{1/2} n + h^{1+\tau} \delta_n\right\}, \qquad (5.41)$$

where $\delta_n = \sup_{|x| \leq b_n} \left[f_0(x)/\sigma(x)\right]$ *and* $b_n \leq r_n d_n.$

Proof. First note that, due to Assumption 5.9 and $K(s) = 0$ if $|s| \geq A,$

$$\frac{\left|\sigma^i(x_k) - \sigma^i(x)\right|}{\sigma^i(x)} K[(x_k - x)/h] \leq C h K[(x_k - x)/h],$$

for $i = 1, 2,$ all $x \in \mathrm{R}$ and h sufficiently small. Similarly, by recalling (5.36), $\inf_{x \in \mathrm{R}} \sigma(x) > 0$ and Assumption 5.8, we have

$$\left|\widehat{f}^L(x) - f(x_k)\right|^i K[(x_k - x)/h]/\sigma^i(x)$$

$$\leq 2\left\{\sup_{|x| \leq b_n} \frac{\left|\widehat{f}^L(x) - f(x)\right|^i}{\sigma^i(x)} + \frac{|f(x_n) - f(x)|^i}{\sigma^i(x)}\right\} K[(x_k - x)/h]$$

$$\leq C \Delta_n K[(x_k - x)/h],$$

for $i = 1, 2$, $|x| \le b_n$ and h sufficiently small, where

$$\Delta_n = h + (nh/d_n)^{-1/2} \log^{1/2} n + h^{1+\tau} \delta_n.$$

By virtue of these estimates, simple calculations show that (recalling $V_{n0}(x) = \sum_{k=1}^n K[(x_k - x)/h]$)

$$\frac{|\hat{\sigma}^2(x) - \sigma^2(x)|}{\sigma^2(x)} \le V_{n0}^{-1}(x) \Big| \sum_{k=1}^n K\Big(\frac{x_k - x}{h}\Big)(u_k^2 - 1)\Big|$$

$$+ V_{n0}^{-1}(x) \sum_{k=1}^n K\Big(\frac{x_k - x}{h}\Big) \Big\{ \frac{\big|\hat{f}^L(x) - f(x_k)\big|^2}{\sigma^2(x)}$$

$$+ \frac{|\sigma^2(x_k) - \sigma^2(x)|}{\sigma^2(x)} u_k^2 + 2 \frac{\big|\hat{f}^L(x) - f(x_k)\big|}{\sigma(x)} \frac{\sigma(x_k)}{\sigma(x)} |u_k| \Big\}$$

$$\le V_{n0}^{-1}(x) \Big| \sum_{k=1}^n K\Big(\frac{x_k - x}{h}\Big)(u_k^2 - 1)\Big|$$

$$+ V_{n0}^{-1}(x)\Delta_n \sum_{k=1}^n \Big[1 + \frac{\sigma^2(x_k)}{\sigma^2(x)}\Big](1 + u_k^2)] K\Big(\frac{x_k - x}{h}\Big),$$

for $|x| \le b_n$ and h sufficiently small. Now, by using similar arguments as that in the proof of Theorem 5.5, we get

$$\sup_{|x| \le b_n} \frac{|\hat{\sigma}^2(x) - \sigma^2(x)|}{\sigma^2(x)} = O_P\Big\{ h + (nh/d_n)^{-1/2} \log^{1/2} n + h^{1+\tau} \delta_n \Big\},$$

as required. $\qquad\square$

5.1.5 *Multivariate contingrating regression*

It is important in econometrics application to consider model (5.1) with multivariate regressor $x_t = (x_{1t}, x_{2t}, ..., x_{mt})$, where $m \ge 2$. Namely it is of interests to investigate the multivariate nonlinear regression model:

$$y_t = f_1(x_{1t}, x_{2t}, ..., x_{mt}) + u_t, \quad t = 1, 2, ..., n. \tag{5.42}$$

As in one-dimension situation, the kernel estimator $\hat{f}_1(x_1, x_2, ..., x_m)$ of $f_1(x_1, x_2, ..., x_m)$ can be similarly defined by

$$\hat{f}_1(x_1, x_2, ..., x_m) = \frac{\sum_{t=1}^n y_t \Pi_{j=1}^m K_j[(x_{jt} - x_j)/h_j]}{\sum_{t=1}^n \Pi_{j=1}^m K_j[(x_{jt} - x_j)/h_j]}, \tag{5.43}$$

where $K_j(x)$ are non-negative kernel functions and the bandwidth $h_j \equiv h_{jn} \to 0$ for $j = 1, ..., m$.

Set $D_n = \sum_{t=1}^n \Pi_{j=1}^m K_j[(x_{jt} - x_j)/h_j]$. In order to investigate the asymptotics of $\hat{f}_1(x_1, x_2, ..., x_m)$, the denominator D_n in (5.43) is naturally expected to have certain rate in diverging to infinity. When $m \geq 2$, the diverging rate required on D_n is attainable only if one component of x_t is $I(1)$ and all others are $I(0)$ processes.

Example 5.1. Let $\epsilon_{jk}, j = 1, ..., m', k \geq 1$, where $m' \leq m$, be a sequence of i.i.d. $N(0, 1)$ random variables, and write $x_{jt} = \sum_{k=1}^t \epsilon_{jk}, j = 1, ..., m'$. It is readily seen that

$$\mathrm{E}D_n = \sum_{t=1}^n \Pi_{j=1}^m \mathrm{E}K_j[(x_{jt} - x_j)/h_j] \leq C \Pi_{j=1}^{m'} h_j \sum_{t=1}^n t^{-m'/2},$$

i.e, $D_n = O_P(1)$ if $m' \geq 3$ and $D_n = O_P(h_1 h_2 \log n)$ if $m' = 2$.

It follows from Example 5.1 that the denominator D_n cannot diverge to infinity if at least three components of x_t are $I(1)$ processes. Even in the situation that two components of x_t are I(1) processes, the $\log n$ rate on the denominator D_n is too restrictive for the asymptotics of $\hat{f}_1(x_1, x_2, ..., x_m)$.

We next consider the asymptotics of $\hat{f}_1(x_1, x_2, ..., x_m)$ by assuming that x_{1t} is nonstationary and $x_{jt}, 2 \leq j \leq m$ are stationary time series. We further assume $m = 2$ for the sake of notation convenience. The extension to $m \geq 2$ is straightforward. We make use of the following assumptions.

Assumption 5.10. [Regression function and Kernel function]

(a) The kernels $K_1(x)$ and $K_2(x)$ have a common compact support Ω satisfying $\int_\Omega K_1(x)dx = \int_\Omega K_2(x)dx = 1$ and $\int_{-\infty}^\infty |\hat{K}_1(t)|dt < \infty$, where $\hat{K}_1(t) = \int_{-\infty}^\infty e^{itx} K_1(x)dx$;

(b) When (x, y) is in a compact set, we have

$$|f_1(x + \delta_1, y + \delta_2) - f_1(x, y)| \leq C (|\delta_1| + |\delta_2|), \qquad (5.44)$$

whenever δ_1 and δ_2 are sufficiently small.

Assumption 5.11. [Regressors]

(a) $x_{1t} = \sum_{j=1}^t \xi_j$ with $\xi_j = \sum_{k=0}^\infty \phi_k \epsilon_{j-k}$, where $\sum_{k=0}^\infty |\phi_k| < \infty$, $\phi = \sum_{k=0}^\infty \phi_k \neq 0$, and $\{\epsilon_j\}_{j \in \mathbb{Z}}$ is a sequence of i.i.d. random variables with $\mathrm{E}\epsilon_0 = 0$, $\mathrm{E}\epsilon_0^2 = 1$ and $\lim_{|t| \to \infty} |t|^\eta |\mathrm{E}\,e^{it\epsilon_0}| < \infty$ for some $\eta > 0$.

(b) $x_{2t} = \Gamma(\eta_t, ..., \eta_{t-m_0+1})$ for some $m_0 > 0$, where $\eta_t = (\epsilon_t, \nu_t), t \geq 1$, is a sequence of i.i.d. random vectors ($\eta_t = 0$ if $t \leq 0$), $\Gamma(.)$ is a real measurable function of its components and x_{2m_0} has a continuous density function $p(x)$.

Assumption 5.12. [Error processes]

(a) $\{u_i, \mathcal{F}_i\}_{i \geq 1}$ is a martingale difference such that, as $i \to \infty$, $\mathrm{E}\,(u_i^2 \mid \mathcal{F}_{i-1}) \to_{a.s.} 1$, and, as $A \to \infty$,

$$\sup_{i \geq 1} \mathrm{E}\,\left[u_i^2 \mathrm{I}(|u_i| \geq A) \mid \mathcal{F}_{i-1}\right] = o_\mathrm{P}(1).$$

(b) (x_{1t}, x_{2t}) is adapted to $\mathcal{F}_{t-1}$.

Theorem 5.7. *Under Assumptions 5.10-5.12, for any h_1 and h_2 satisfying $nh_1^2 h_2^2 \to \infty$ and $n(h_1 + h_2)^4 h_1^2 h_2^2 \to 0$, we have*

$$D_n^{1/2}\left[\hat{f}_1(x_1, x_2) - f_1(x_1, x_2)\right] \to_D N(0, \sigma^2), \qquad (5.45)$$

where $\sigma^2 = \int_\Omega K_1^2(x)dx \int_\Omega K_2^2(x)dx$.

Proof. Let $V_t = K_1[(x_{1t} - x_1)/h_1]\, K_2[(x_{2t} - x_2)/h_2]$. We may write

$$D_n^{1/2}\left[\hat{f}_1(x_1, x_2) - f_1(x_1, x_2)\right] = D_n^{-1/2} \sum_{t=1}^n u_t V_t + R_n. \qquad (5.46)$$

where, by Assumption 5.10,

$$|R_n| = D_n^{-1/2} \sum_{t=1}^n \left|f_1(x_{1t}, x_{2t}) - f_1(x_1, x_2)\right| V_t$$
$$\leq C(|h_1| + |h_2|)\, D_n^{1/2}.$$

Since, by Assumption 5.12(b),

$$EK_2^\alpha[(x_{2t} - x_2)/h_2] = h_2\left\{p(x_2) \int_\Omega K_2^\alpha(x)dx + o(1)\right\},$$

for all $t \geq m_0$ and any $\alpha > 0$, it follows from some similar arguments as in the proofs of Lemma 2.2 (i) and (ii) that, for $i = 1, 2$,

$$\frac{1}{\sqrt{n}\, h_1\, h_2} \sum_{t=1}^n V_t^i = \frac{p(x_2) \int_\Omega K_2^i(y)dy}{\sqrt{n}\, h_1} \sum_{t=1}^n K_1^i[(x_{1t} - x_1)/h_1]$$
$$+ o_\mathrm{P}(1); \qquad (5.47)$$

$$\mathrm{E}\,|R_n| \leq C(|h_1| + |h_2|)\,(ED_n)^{1/2}$$
$$\leq C\sqrt{n}\, h_1\, h_2(|h_1| + |h_2|). \qquad (5.48)$$

Recall $nh_1^2h_2^2 \to \infty$ and $n(h_1 + h_2)^4 h_1^2 h_2^2 \to 0$. (5.48) implies $R_n = o_P(1)$. Due to (5.47), as in Theorem 2.21, we have

$$\left\{ \frac{\sum_{j=1}^{[nt]} \epsilon_j}{\sqrt{n}}, \ \frac{1}{\sqrt{n}h_1 h_2} \sum_{t=1}^n V_t, \ \frac{1}{\sqrt{n}h_1 h_2} \sum_{t=1}^n V_t^2 \right\}$$

$$\Rightarrow \left\{ B_t, \ p(x_2) L_B(1,0), \ \tau L_B(1,0) \right\},$$

on $D_{\mathrm{R}^3}[0,\infty)$, where $B = \{B_t\}_{t\geq 0}$ is a standard Brown motion and

$$\tau = p(x_2) \int_\Omega K_1^2(x)dx \int_\Omega K_2^2(x)dx.$$

Now, by using Corollary 3.5, we obtain

$$\left\{ \frac{1}{\sqrt{n}h_1 h_2} \sum_{t=1}^n V_t, \ \left(\frac{1}{\sqrt{n}h_1 h_2} \right)^{1/2} \sum_{t=1}^n u_t V_t \right\}$$

$$\Rightarrow \left\{ p(x_2) L_B(1,0), \ \tau^{1/2} N L_B^{1/2}(1,0) \right\}, \qquad (5.49)$$

where N is a standard normal variate independent of $L_B(1,0)$. (5.49) implies that $D_n^{-1/2} \sum_{t=1}^n u_t V_t \to_D N(0,\sigma^2)$. Taking these facts into (5.46), we prove (5.45). The proof of Theorem 5.7 is complete. $\qquad\square$

5.2   Parametric estimation

For the parametric model (5.2), namely $y_t = g(x_t, \theta_0) + u_t$, the least squares estimator $\widehat{\theta}_n$ of the true unknown value θ_0 is defined as

$$\widehat{\theta}_n = \arg\min_{\theta\in\Theta} \sum_{t=1}^n \left[y_t - g(x_t, \theta) \right]^2. \qquad (5.50)$$

In this section, we investigate weak consistency and asymptotic normality of $\widehat{\theta}_n$. Recall that Θ is a compact subset of R^m for some finite m, and we always assume the true θ_0 is an interior point of Θ.

5.2.1   *Weak consistency*

We make use of the following assumptions for a framework on the weak consistency of $\widehat{\theta}_n$. Write $T(x) = \sup_{\theta\in\Theta} |g(x,\theta) - g(x,\theta_0)|$.

Assumption 5.13. For each $\theta_1, \theta_2 \in \Theta$, we have

$$\left| g(x,\theta_1) - g(x,\theta_2) \right| \leq h(\|\theta_1 - \theta_2\|) T(x), \qquad (5.51)$$

where $h(x)$ is a bounded real function such that $h(x) \downarrow h(0) = 0$, as $x \downarrow 0$.

Assumption 5.14. For an increasing sequence $0 < \kappa_n \to \infty$, we have

(i) $\kappa_n^{-2} \sum_{t=1}^{n} T(x_t) \left[1 + |u_t| + T(x_t) \right] = O_P(1)$;

(ii) $\kappa_n^{-2} \sum_{t=1}^{n} [g(x_t, \theta) - g(x_t, \theta_0)] u_t = o_P(1)$ for each $\theta \in \Theta$.

Assumption 5.15. For any $\eta > 0$ and $\theta \neq \theta_0$, where $\theta, \theta_0 \in \Theta$, there exist $n_0 > 0$ and $M > 0$ such that

$$P\left(\kappa_n^{-2} \sum_{t=1}^{n} \left[g(x_t, \theta) - g(x_t, \theta_0) \right]^2 \geq 1/M \right) \geq 1 - \eta, \qquad (5.52)$$

for all $n > n_0$, where $0 < \kappa_n \to \infty$ is given in Assumption 5.14.

Theorem 5.8. *Under Assumptions 5.13–5.15, $\hat{\theta}_n$ is a consistent estimator of θ_0, i.e. $\hat{\theta}_n \to_P \theta_0$.*

Proof. Let $Q_n(\theta) = \sum_{t=1}^{n} \left[y_t - g(x_t, \theta) \right]^2$ and

$$D_n(\theta, \theta_0) = Q_n(\theta) - Q_n(\theta_0).$$

Write $\mathcal{A}_\delta = \{ \theta : \|\theta - \theta_0\| < \delta \}$, an open subset containing θ_0. If $\hat{\theta}_n \to_P \theta_0$ is not true, then there exist $\delta_0 > 0$ and a subsequence n_k of integers such that $\mathcal{A}_{\delta_0} \subset \Theta$ and

$$\lim_{k \to \infty} P(\|\widehat{\theta}_{n_k} - \theta_0\| \geq \delta_0) > 0.$$

Since $\widehat{\theta}_n$ minimizes $Q_n(\theta)$ over $\theta \in \Theta$, this implies that

$$P\left(\inf_{\theta \in \Theta \cap \mathcal{A}_{\delta_0}^c} D_{n_k}(\theta, \theta_0) \leq 0 \right) > 0,$$

for all k sufficiently large, where $A_{\delta_0}^c$ denotes the complementary set of A_{δ_0}. Hence, to prove $\hat{\theta}_n \to_P \theta_0$, it suffices to show that, for any $\eta > 0$ and $\delta > 0$, there exist $n_0 > 0$ and $M > 0$ such that, for all $n > n_0$,

$$P\left(\kappa_n^{-2} \inf_{\theta \in \Theta \cap \mathcal{A}_\delta^c} D_n(\theta, \theta_0) \geq 1/M \right) \geq 1 - \eta. \qquad (5.53)$$

Denote $\mathcal{N}_\lambda(\theta_1) = \{ \theta : \|\theta - \theta_1\| < \lambda \}$. Since Θ is compact, by the finite covering property of compact set, (5.53) will follow if we prove:

(i) for any fixed $\theta_1 \in \Theta$,

$$I_n(\lambda, \theta_1) := \sup_{\theta \in \mathcal{N}_\lambda(\theta_1)} \kappa_n^{-2} \left| D_n(\theta, \theta_0) - D_n(\theta_1, \theta_0) \right| \to_P 0, \qquad (5.54)$$

as $n \to \infty$ first and then $\lambda \to 0$, and

(ii) for any $\eta > 0$ and $\theta_1 \neq \theta_0$, there exist $M_0 > 0$ and $n_0 > 0$ such that for all $n \geq n_0$ and $M \geq M_0$,

$$P\left(D_n(\theta_1, \theta_0) \geq \kappa_n^2/M\right) \geq 1 - \eta/2. \tag{5.55}$$

Indeed, due to (5.54), for any $\eta > 0$ and $M > 0$, there exist $n_0 > 0$ and $\lambda_0 > 0$ such that, for all $n > n_0$,

$$P(\max_{1 \leq j \leq m_0} I_n(\lambda_0, \theta_j) \geq 1/2M) \leq \eta/2,$$

where m_0 and $\theta_j, 1 \leq j \leq m_0$, are chosen so that

$$\theta_j \neq \theta_0 \quad \text{and} \quad \Theta \subset \bigcup_{j=1}^{m_0} \mathcal{N}_{\lambda_0}(\theta_j).$$

Consequently, by taking $M \geq M_0/(2m_0)$, it follows from (5.55) that

$$P\left(\inf_{\theta \in \Theta \cap \mathcal{A}_\delta^c} D_n(\theta, \theta_0) \geq \kappa_n^2/M\right)$$

$$\geq P\left(\inf_{1 \leq j \leq m_0} D_n(\theta_j, \theta_0) \geq \kappa_n^2/(2M)\right) - \eta/2$$

$$\geq \inf_{1 \leq j \leq m_0} P\left(D_n(\theta_j, \theta_0) \geq \kappa_n^2/(2m_0 M)\right) - \eta/2 \geq 1 - \eta,$$

which yields (5.53).

The proofs of (5.54) and (5.55) are simple under given assumptions. Indeed, by Assumption 5.13, we have

$$I_n(\lambda, \theta_1) \leq \sup_{\theta \in \mathcal{N}_\lambda(\theta_1)} \kappa_n^{-2} \sum_{t=1}^n \left[g(x_t, \theta) - g(x_t, \theta_1)\right]^2$$

$$+ 2 \sup_{\theta \in \mathcal{N}_\lambda(\theta_1)} \kappa_n^{-2} \sum_{t=1}^n |g(x_t, \theta) - g(x_t, \theta_1)|$$

$$\left(|u_t| + |g(x_t, \theta) - g(x_t, \theta_0)|\right)$$

$$\leq C \, h(\lambda) \, \kappa_n^{-2} \sum_{t=1}^n T(x_t) \left[|u_t| + T(x_t)\right].$$

The result (5.54) follows immediately from Assumption 5.14 (i). By noting

$$D_n(\theta_1, \theta_0) = \sum_{t=1}^n \left[g(x_t, \theta_1) - g(x_t, \theta_0)\right]^2 - 2\sum_{t=1}^n \left[g(x_t, \theta_1) - g(x_t, \theta_0)\right] u_t,$$

we have (5.55) due to Assumption 5.14 (ii) and Assumption 5.15. □

Remark 5.7. Assumption 5.13 provides certain restriction on the nonlinear regression function $g(x, \theta)$ which is a standard condition used in nonlinear least squares estimation theory. See, for instance, Wu (1981), Lai (1994) and Skouras (2000).

Remark 5.8. Assumption 5.15 is mild and holds if, for any $\theta \neq \theta_0$,

$$\kappa_n^{-2} \sum_{t=1}^{n} \left[g(x_t, \theta) - g(x_t, \theta_0) \right]^2 \to_D Z, \tag{5.56}$$

where Z is a random variable satisfying $P(Z > 0) = 1$.

If $T(x)$ is bounded and integrable (typical examples include

$$f(x, \theta_1, \theta_2) = \theta_1 |x|^{\theta_2} I(x \in [a, b]), \quad e^{-\theta |x|}, \quad e^{\theta |x|} / (1 + e^{\theta |x|}),$$

etc, where a and b are finite constants), some sufficient conditions on x_t that enable (5.56) are given in Theorems 2.19 and 2.22. As an illustration, suppose that $\{x_k / d_n\}_{k \geq 1. n \geq 1}$, where $d_n^2 = var(x_n)$, is a strong smooth array (see Definition 2.5) and there exists a stochastic process $X = \{X_t\}_{t \geq 0}$ having a continuous local time $L_X(t, s)$ such that $x_{[nt]} / d_n \Rightarrow X_t$ on $D[0, \infty)$. An immediate application of Theorem 2.19 yields that if $d_n \to \infty$ and $d_n / n \to 0$, then

$$\frac{d_n}{n} \sum_{t=1}^{n} \left[g(x_t, \theta) - g(x_t, \theta_0) \right]^2 \to_D \tau L_X(1, 0),$$

where $\tau = \int_{-\infty}^{\infty} \left[g(x, \theta) - g(x, \theta_0) \right]^2 dx$. Since $P(L_X(1, 0) > 0) = 1$, (5.56) is satisfied if $\tau \neq 0$ for any $\theta \neq \theta_0$.

If $T(x)$ is non-integrable, certain smooth conditions on $g(x, \theta)$ are required to establish (5.56).

Assumption 5.16. There exist a locally Riemann integrable function $H(x, \theta, \theta_0)$ and a positive real function $\nu(\lambda)$ such that, for any $\theta, \theta_0 \in \Theta$,

$$g(\lambda x, \theta) - g(\lambda x, \theta_0) = \nu(\lambda) H(x, \theta, \theta_0) + R(\lambda, x, \theta, \theta_0), \tag{5.57}$$

where (i) $\sup_{\theta, \theta_0 \in \Theta} |R(\lambda, x, \theta, \theta_0)| \leq a(\lambda) T_1(x)$, where $\lim_{\lambda \to \infty} a(\lambda) / \nu(\lambda) = 0$ and $T_1(x)$ is locally bounded; or (ii)

$$\sup_{\theta, \theta_0 \in \Theta} |R(\lambda, x, \theta, \theta_0)| \leq b(\lambda) T_2(\lambda x),$$

where $\lim_{\lambda \to \infty} b(\lambda) / \nu(\lambda) < \infty$ and $T_2(x)$ is bounded satisfying $T_2(x) \to 0$ as $x \to \infty$.

Typical examples satisfying Assumption 5.16 include

$$g(x, \theta) = (x + \theta)^2, \ \theta e^x / (1 + e^x), \ \theta \log |x|, \ \theta |x|^\alpha$$

(α is fixed) and $\theta_0 + \theta_1 |x| + \ldots + \theta_k |x|^k$.

Suppose there exists a stochastic process $X = \{X_t\}_{t \geq 0}$ having a continuous local time $L_X(t, s)$ such that $x_{[nt]}/d_n \Rightarrow X_t$ on $D[0, \infty)$, where $d_n^2 = var(x_n) \to \infty$. It follows from Theorem 2.15 that, under Assumption 5.16,

$$\frac{1}{n\nu^2(d_n)} \sum_{t=1}^n \left[g(x_t, \theta) - g(x_t, \theta_0)\right]^2 \to_D \int_0^1 H^2(X_s, \theta, \theta_0) ds.$$

Hence (5.56) is satisfied if $\int_{|x| \leq \delta} H^2(x, \theta, \theta_0) dx \neq 0$ for any $\theta \neq \theta_0$ and $\delta > 0$.

Remark 5.9. Assumption 5.14 provides a trade off between the error process u_t and the regressor x_t. The constants sequence κ_n is usually taken to enable (5.56). Suppose $\{u_t, \mathcal{F}_t\}_{t \geq 1}$ is a martingale difference sequence such that

   (i) $\sup_{t \geq 1} E\left(|u_t|^2 | \mathcal{F}_{t-1}\right) < \infty$ and
   (ii) x_t is adapted to $\mathcal{F}_{t-1}$.

A simple sufficient condition that enables Assumption 5.14 is

$$\frac{1}{\kappa_n^2} \sum_{t=1}^n \left[T(x_t) + T^2(x_t)\right] = O_P(1). \tag{5.58}$$

See, e.g., Lemma 2 of Lai and Wei (1982). As in Remark 5.8, if $T(x)$ is bounded and integrable, then (5.58) is satisfied with $\kappa_n^2 = n/d_n$, provided that $\{x_k/d_n\}_{k \geq 1. n \geq 1}$, where $d_n^2 = var(x_n)$, is a strong smooth array. If Assumption 5.16 holds, then (5.58) is satisfied with $\kappa_n^2 = n\nu^2(d_n)$, provided that $x_{[nt]}/d_n \Rightarrow X_t$ on $D[0, \infty)$, where $d_n^2 = var(x_n)$ and X_t is a process having a continuous local time.

Remark 5.10. Assumptions 5.13 and 5.14 provide a general framework, allowing for the regressor x_t to be deterministic, stationary, nonstationary and/or multivariate. If endogeneity and/or multivariate regressor x_t appear, verifications of Assumptions 5.13 and 5.14 are complicated, which requires detailed analysis. See Section 3.5 for some results in this regard.

5.2.2 Asymptotic distribution

This section considers limit distribution of $\hat{\theta}_n$. As in Section 5.2.1, write $Q_n(\theta) = \sum_{t=1}^n \left[y_t - g(x_t, \theta) \right]^2$. Let $\dot{Q}_n$ and $\ddot{Q}_n$ be the first and second order derivatives of $Q_n(\theta)$ so that $\dot{Q}_n = \partial Q_n / \partial \theta$ and $\ddot{Q}_n = \partial^2 Q_n / \partial \theta \partial \theta'$. Similar definitions are used for $\dot{g}$ and $\ddot{g}$. We assume these quantities exist whenever they are introduced. We have the following framework.

Theorem 5.9. *Suppose θ_0 is an interior point of Θ. Suppose there exist a sequence of constants $\{k_n\}_{n \geq 1}$ and a sequence of $m \times m$ nonrandom nonsingular matrices D_n satisfying $k_n \to \infty$ and $k_n \parallel D_n^{-1} \parallel \to 0$ such that the following conditions hold: for some $1 \leq \delta_n \leq k_n^{1-\epsilon_0}$ where $0 < \epsilon_0 < 1$,*

(i) $\displaystyle \sup_{\theta : \parallel D_n(\theta - \theta_0) \parallel \leq k_n} \parallel (D_n^{-1})' \sum_{t=1}^n \left[\dot{g}(x_t, \theta) \dot{g}(x_t, \theta)' \right.$

$$\left. - \dot{g}(x_t, \theta_0) \dot{g}(x_t, \theta_0)' \right] D_n^{-1} \parallel = o_P(\delta_n^{-2}),$$

(ii) $\displaystyle \sup_{\theta : \parallel D_n(\theta - \theta_0) \parallel \leq k_n} \parallel (D_n^{-1})' \sum_{t=1}^n \ddot{g}(x_t, \theta) \left[g(x_t, \theta) - g(x_t, \theta_0) \right] D_n^{-1} \parallel$

$$= o_P(\delta_n^{-2}),$$

(iii) $\displaystyle \sup_{\theta : \parallel D_n(\theta - \theta_0) \parallel \leq k_n} \parallel (D_n^{-1})' \sum_{t=1}^n \ddot{g}(x_t, \theta) \, u_t \, D_n^{-1} \parallel = o_P(\delta_n^{-2}),$

(iv) $\displaystyle Y_n := (D_n^{-1})' \sum_{t=1}^n \dot{g}(x_t, \theta_0) \dot{g}(x_t, \theta_0)' D_n^{-1} \to_D M, \text{ where } M > 0, \text{ a.s.,}$

and

$$Z_n := (D_n^{-1})' \sum_{t=1}^n \dot{g}(x_t, \theta_0) \, u_t = O_P(\delta_n).$$

Then there exists a sequence of estimators $\{\hat{\theta}_n\}_{n \geq 1}$ satisfying $\lim_{n \to \infty} P\big(\dot{Q}_n(\hat{\theta}_n) = 0 \big) = 1$ and

$$D_n(\hat{\theta}_n - \theta_0) = Y_n^{-1} Z_n + o_P(1). \tag{5.59}$$

If we replace (iv) by the following condition $(iv)'$, then

$$D_n(\hat{\theta}_n - \theta_0) \to_D M^{-1} Z. \tag{5.60}$$

$(iv)'$ for any $\alpha_i' = (\alpha_{i1}, ..., \alpha_{im}) \in \mathbb{R}^m, i = 1, 2, 3,$

$$(\alpha_1' Y_n \alpha_2, \ \alpha_3' Z_n) \to_D (\alpha_1' M \alpha_2, \ \alpha_3' Z), \tag{5.61}$$

where $M > 0$, a.s. Y_n and Z_n are defined as in (iv) and Z is an a.s. finite random variable.

Proof.   The proof follows similar argument as that of Lemma 1 in Andrews and Sun (2004). Let

$$\Theta_0 = \{\theta \in \Theta : \parallel D_n(\theta - \theta_0) \parallel \le k_n, \parallel \theta - \theta_0 \parallel \le \delta\}$$

for some $\delta > 0$ so that $\{\theta \in \Theta : \parallel \theta - \theta_0 \parallel \le \delta\} \subset \Theta$ and $Q_n(\theta)$ is twice differentiable on $\theta \in \{\theta \in \Theta : \parallel \theta - \theta_0 \parallel \le \delta\}$. Note that

$$\frac{1}{2}\dot{Q}_n(\theta_0) = -\sum_{t=1}^{n} \dot{g}(x_t, \theta_0)\big[y_t - g(x_t, \theta_0)\big] = -\sum_{t=1}^{n} \dot{g}(x_t, \theta_0)u_t,$$

$$\frac{1}{2}\ddot{Q}_n(\theta) = \sum_{t=1}^{n} \dot{g}(x_t, \theta)\dot{g}(x_t, \theta)' + \Delta_n,$$

where $\Delta_n = -\sum_{t=1}^{n} \ddot{g}(x_t, \theta)u_t + \sum_{t=1}^{n} \ddot{g}(x_t, \theta)\big[g(x_t, \theta) - g(x_t, \theta_0)\big]$, and recall the definitions of Y_n and Z_n. It follows from the Taylor's expansion that

$$\begin{aligned}
& Q_n(\theta) - Q_n(\theta_0) \\
&= \dot{Q}_n(\theta_0)'(\theta - \theta_0) + \frac{1}{2}(\theta - \theta_0)'\ddot{Q}_n(\theta_0)(\theta - \theta_0) + R_n(\theta, \theta_0) \\
&= \big[D_n(\theta - \theta_0) - Y_n^{-1}Z_n\big]'Y_n\big[D_n(\theta - \theta_0) - Y_n^{-1}Z_n\big] \\
&\quad + 2Z_n'Y_n^{-1}Z_n + R_{1n}(\theta, \theta_0),
\end{aligned} \tag{5.62}$$

for all $\theta \in \Theta_0$, where

$$|R_n(\theta, \theta_0)| \le \sup_{\theta_1 \in \Theta_0} |(\theta - \theta_0)'\big[\ddot{Q}_n(\theta_1) - \ddot{Q}_n(\theta_0)\big](\theta - \theta_0)| \quad \text{and}$$

$$R_{1n}(\theta, \theta_0) = R_n(\theta, \theta_0) + (\theta - \theta_0)'\Delta_n(\theta - \theta_0).$$

In view of conditions (i)–(iii), simple calculations show that, for all $\theta \in \Theta_0$,

$$\begin{aligned}
& |R_{1n}(\theta, \theta_0)| \\
&\le \parallel D_n(\theta - \theta_0) \parallel^2 \Big\{ \sup_{\theta_1 \in \Theta_0} \parallel (D_n^{-1})'\big[\ddot{Q}_n(\theta_1) - \ddot{Q}_n(\theta_0)\big]D_n^{-1} \parallel \\
&\qquad\qquad\qquad + \sup_{\theta_1 \in \Theta_0} \parallel (D_n^{-1})'\Delta_n D_n^{-1} \parallel \Big\} \\
&= o_P(\delta_n^{-2}) \parallel D_n(\theta - \theta_0) \parallel^2 .
\end{aligned} \tag{5.63}$$

Let $\widetilde{\theta}_n = \theta_0 - D_n^{-1}Y_n^{-1}Z_n$. It follows from (iv) and $k_n/\delta_n \to \infty$ that

$$P(\widetilde{\theta}_n \notin \Theta_0) \le P(\|Y_n^{-1}Z_n\| \ge k_n) + P(\|D_n^{-1}Y_n^{-1}Z_n\| \ge \delta) \to 0.$$

This, together with (5.62) and (5.63), yields

$$Q_n(\widetilde{\theta}_n) - Q_n(\theta_0) = 2Z_n'Y_n^{-1}Z_n + R_{1n}(\widetilde{\theta}_n, \theta_0),$$

where $R_{1n}(\widetilde{\theta}_n, \theta_0) = o_P(1)$. For any $\epsilon > 0$ and $n \geq 1$, let

$$\Theta_n(\epsilon) = \{\theta \in \Theta : ||D_n(\theta - \theta_0) - Y_n^{-1} Z_n|| \leq \epsilon\}.$$

Recall $||Y_n^{-1} Z_n|| = O_P(\delta_n) = o_P(k_n)$ and $k_n ||D_n^{-1}|| = o(1)$. It follows that $P[\Theta_n(\epsilon) \subset \Theta_0] \to 1$, as $n \to \infty$ and $\sup_{\theta \in \Theta_n(\epsilon)} |R_{1n}(\theta, \theta_0)| = o_P(1)$. Consequently, for any $\theta \in \partial\Theta_n(\epsilon)$, where $\partial\Theta_n(\epsilon)$ denotes the boundary of $\Theta_n(\epsilon)$, we have

$$Q_n(\theta) - Q_n(\widetilde{\theta}_n) = 2\nu_n' Y_n \nu_n + o_P(1),$$

where ν_n is a vector with $||\nu_n|| = \epsilon > 0$. Since $Y_n \to_D M > 0$, a.s., we have $P(\frac{1}{2}\nu_n' Y_n \nu_n > 0) \to 1$ as $n \to \infty$. Hence, for each $\epsilon > 0$, the event that the minimum of $Q_n(\theta)$ over $\Theta_n(\epsilon)$ is in the interior of $\Theta_n(\epsilon)$ has probability that goes to one as $n \to \infty$. In particular, for each $\epsilon > 0$, there exists a point $\hat{\theta}_n(\epsilon) \in \Theta_n(\epsilon)$ (not necessary unique) so that $P(\dot{Q}_n[\hat{\theta}_n(\epsilon)] = 0) \to 1$, as $n \to \infty$. In consequence, there exists a sequence of $\hat{\theta}_n = \hat{\theta}_n(1/J_n) \in \Theta_n(1/J_n)$ where $J_n \to \infty$ so that $P(\dot{Q}_n(\hat{\theta}_n) = 0) \to 1$, as $n \to \infty$, and (5.59) holds.

Finally, by noting (iv)' implies that (5.59) holds with $\delta_n = 1$, the asymptotic distribution (5.60) follows immediately from (5.59), (5.61) and the continuous mapping theorem. □

Remark 5.11. If $g(x, \theta)$ is linear on θ, conditions (i)-(iii) holds automatically. Limit distribution of $\hat{\theta}_n$ is determined by condition (iv) which is discussed in Chapters 3 and 4. See the following illustrated example. For related results in general nonlinear settings on $g(x, \theta)$, we refer to Park and Phillips (2001), Chan and Wang (2015), and Wang and Phillips (2015).

Example 5.2. Let $g(x, \theta) = \theta_1 + \theta_2 g(x)$, namely, we have the model:

$$y_t = \theta_1 + \theta_2 g(x_t) + u_t. \tag{5.64}$$

It is readily seen that conditions (i)-(iii) holds and $\dot{g}(x, \theta) = (1, g(x))'$ for any $\theta = (\theta_1, \theta_2) \in \mathbb{R}^2$. By Theorem 5.9, there exists an estimator $\hat{\theta}_n$ of $\theta = (\theta_1, \theta_2)$ such that the limit distribution of $\hat{\theta}_n - \theta$ (under suitable standardization) is determined by (5.61), which depends on the choice of $g(x), x_t$ and u_t.

In nonlinear cointegrating regression, $g(x)$ usually satisfies one of the following conditions:

A. $g(x)$ is bounded and integrable;

B. $g(x)$ is asymptotically homogeneous, i.e.,

$$g(\lambda x) = \nu(\lambda) H(x) + R(\lambda, x),$$

where $H(x)$ is locally Riemann integrable, $\nu(\lambda)$ is positive and $R(\lambda, x)$ is negligible when $\lambda \to \infty$ or $\lambda x \to \infty$ (see Assumption 5.16).

If condition A holds, Theorem 3.14 (see also Corollary 3.5) can be used to establish (5.61). Let $x_t = \gamma x_{t-1} + \xi_t$, where $\gamma = 1 - \tau/n$ for some $\tau \geq 0$. If ξ_t and u_t satisfy the conditions of Corollary 5.1, it follows from Corollary 3.5 with $h = 1$ that

$$\left(\left(\frac{d_n}{n}\right)^{1/2} \sum_{k=1}^{n} g(x_k)u_k, \ \frac{d_n}{n} \sum_{k=1}^{n} g^2(x_k) \right)$$
$$\Rightarrow \left(V N, \ V^2 \right), \tag{5.65}$$

where N is a normal variate independent of $V^2 = \int_{-\infty}^{\infty} g^2(x)dx L_Z(1,0)$. Set $D_n = \text{diag}\{\sqrt{n}, (n/d_n)^{1/2}\}$. Using (5.65), some routine calculations show

$$Y_n = D_n^{-1} \sum_{k=1}^{n} \dot{g}(x_k, \theta)\dot{g}(x_k, \theta_0)' D_n^{-1}$$
$$= \begin{pmatrix} 1 & 0 \\ 0 & \frac{d_n}{n} \sum_{k=1}^{n} g^2(x_k) \end{pmatrix} + o_P(1) \to_D \begin{pmatrix} 1 & 0 \\ 0 & V^2 \end{pmatrix},$$
$$Z_n = D_n^{-1} \sum_{k=1}^{n} \dot{g}(x_k, \theta)u_k = O_P(1).$$

Now it follows easily from (5.59) that

$$\sqrt{n}(\hat{\theta}_{1n} - \theta_1) = \frac{1}{\sqrt{n}} \sum_{k=1}^{n} u_k + o_P(1) \to_D N(0,1),$$
$$\left(\frac{n}{d_n}\right)^{1/2}(\hat{\theta}_{2n} - \theta_2) = \left[\frac{d_n}{n} \sum_{k=1}^{n} g^2(x_k)\right]^{-1} \left(\frac{d_n}{n}\right)^{1/2} \sum_{k=1}^{n} g(x_k)u_k + o_P(1)$$
$$\to_D V^{-1}N.$$

If condition B holds, Theorems 4.6, 4.8-4.11 can be used to establish (5.61). As an illustration, let $x_t = \sum_{j=1}^{t} \eta_j$ and $u_t = \sum_{j=0}^{\infty} \varphi_j \epsilon_{t-j}$, where $\varphi = \sum_{j=0}^{\infty} \varphi_j \neq 0$ and $\sum_{j=0}^{\infty} j|\varphi_j| < \infty$. Suppose that

(i) $\left(\frac{1}{\sqrt{n}} \sum_{j=1}^{[nt]} \eta_j, \ \frac{1}{\sqrt{n}} \sum_{j=1}^{[nt]} \epsilon_j \right) \Rightarrow (X(t), Y(t))$ on $D_{\mathbb{R}^2}[0, \infty)$;

(ii) $\sup_{i \in \mathcal{Z}} E|\epsilon_i|^3 < \infty$, $E\eta_1^2 < \infty$ and

$$\sup_{j \geq 1} \frac{1}{j} E \Big| \sum_{k=1}^{j} \eta_k \Big|^2 < \infty;$$

(iii) there exists a constant $A_0 > 0$ such that, for each $i \geq 0$,

$$\sum_{j=1}^{\infty} \varphi_j \sum_{k=1}^{j} E\left(\eta_{k+i}\epsilon_{i+1} \mid \mathcal{F}_i\right) = A_0, \quad a.s.;$$

(iv) $H'(x)$ is locally bounded and

$$|H'(x) - H'(y)| \leq C_K |x - y|^{\beta},$$

for some $0 < \beta \leq 1/3$ and $\max\{|x|, |y|\} \leq K$, where C_K is a constant depending only on K.

Then, by letting $D_n = \text{diag}(\sqrt{n}, \sqrt{n}\nu(\sqrt{n}))$, it follows from Theorem 4.9 with minor modification that

$$Y_n = D_n^{-1} \sum_{k=1}^{n} \dot{g}(x_k, \theta) \dot{g}(x_k, \theta_0)' D_n^{-1}$$

$$\to_D M = \begin{pmatrix} 1 & \int_0^1 H[X(t)]dt \\ \int_0^1 H[X(t)]dt & \int_0^1 H^2[X(t)]dt \end{pmatrix},$$

$$Z_n = D_n^{-1} \sum_{k=1}^{n} \dot{g}(x_k, \theta) u_k$$

$$\to_D Z = \begin{pmatrix} \varphi Y(1) \\ \varphi \int_0^1 H[X(s)]dY(s) + A_0 \int_0^1 H'[X(s)]ds \end{pmatrix},$$

and (5.61) holds. Now (5.60) yields

$$D_n^{-1}(\hat{\theta}_n - \theta) = Y_n^{-1} Z_n + o_P(1) \to_D M^{-1} Z.$$

5.3   Model specification testing

Nonparametric function estimation is often the first step in analyzing data when there is no prior information on functional form. As is apparent from Section 5.1, nonparametric estimation has the merit of simplicity in terms of both practical implementation and asymptotics. In comparison to parametric counterparts in Section 5.2, nonparametric estimators typically deliver slower convergence rates. Parametric estimation can therefore be attractive in practical work, whilst allowing for some potential functional misspecification, making it desirable to perform a test of parametric specification.

In view of model (5.1), this section considers testing a specific parametric null hypothesis such as

$$H_0: \quad g(x) = g(x, \theta), \quad \theta \in \Omega_0, \tag{5.66}$$

for $x \in \mathbb{R}$, where $g(x, \theta)$ is a given real function indexed by a vector θ of unknown parameters which lie in the compact parameter space $\Omega_0 \subset \mathbb{R}^d$ for some $d \geq 1$. To test H_0 we make use of the following test statistic

$$T_n = \int_{-\infty}^{\infty} \left\{ \sum_{k=1}^{n} K[(x_k - x)/h][y_k - g(x_k, \hat{\theta}_n)] \right\}^2 \pi(x) dx,$$

involving the parametric regression residuals $y_t - g(x_t, \hat{\theta}_n)$, where $\pi(x)$ is a positive integrable weight function, $K(x)$ is a non-negative kernel function, h is a bandwidth satisfying $h \equiv h_n \to 0$ as the sample size $n \to \infty$ and $\hat{\theta}_n$ is a parametric estimator of θ under the null H_0, which is consistent whenever $\theta \in \Omega_0$.

The statistic T_n is a modification of the test statistic discussed by Härdle and Mammen (1993) for the random sample case. Under the null H_0, we may split T_n as

$$T_n = \int_{-\infty}^{\infty} \left\{ \sum_{k=1}^{n} K[(x_k - x)/h][g(x_k, \hat{\theta}_n) - g(x_k, \theta) + u_k] \right\}^2 \pi(x) dx$$

$$= T_{1n} + T_{2n} + T_{3n}, \tag{5.67}$$

where $T_{1n} = \int_{-\infty}^{\infty} \left\{ \sum_{k=1}^{n} K[(x_k - x)/h]u_k \right\}^2 \pi(x) dx$,

$$T_{2n} = \int_{-\infty}^{\infty} \left\{ \sum_{k=1}^{n} K[(x_k - x)/h][g(x_k, \hat{\theta}_n) - g(x_k, \theta)] \right\}^2 \pi(x) dx,$$

and $|T_{3n}|^2 \leq 4T_{1n}T_{2n}$ by Hölder's inequality. It is expected that T_{1n} dominates the asymptotics of T_n when Theorems 3.15 and 3.17 can apply, and terms T_{2n} and T_{3n} are negligible in comparison with T_{1n} under certain conditions on x_t, u_t, $g(x, \theta)$ and $\hat{\theta}_n$, as summarized in the following assumptions.

Assumption 5.17. (i) $\{\epsilon_j\}_{j \in \mathbb{Z}}$ is a sequence of i.i.d. random variables with that $\mathrm{E}\epsilon_0 = 0$, $\mathrm{E}\epsilon_0^2 = 1$ and $\lim_{|t| \to \infty} |t|^\eta |\mathrm{E}\, e^{it\epsilon_0}| < \infty$ for some $\eta > 0$. (ii)

$$x_k = \gamma\, x_{k-1} + \xi_k, \quad k \geq 1,$$

where $x_0 = 0$, $\tau \geq 0$ is a constant, $\gamma = 1 - \tau/n$ and $\xi_j = \sum_{k=0}^{\infty} \phi_k \, \epsilon_{j-k}$ with $\sum_{k=0}^{\infty} |\phi_k| < \infty$ and $\phi \equiv \sum_{k=0}^{\infty} \phi_k \neq 0$.

Assumption 5.18. $\{u_i, \mathcal{F}_i\}_{i \geq 1}$, where $\mathcal{F}_i = \sigma(u_i, ..., u_1; \epsilon_{i+1}, \epsilon_i, ...)$, is a martingale difference such that $\mathrm{E}\,(u_i^2|\mathcal{F}_{i-1}) \to_{a.s.} \sigma^2 > 0$ and, as $K \to \infty$,

$$\sup_{i \geq 1} \mathrm{E}\,(|u_i|^2 \mathrm{I}(|u_i| \geq K) \mid \mathcal{F}_{i-1}) = o_{\mathrm{P}}(1).$$

Assumption 5.19. $K(x)$ has compact support, $\int_{-\infty}^{\infty} K(x)dx = 1$ and $|K(x) - K(y)| \leq C|x - y|$ whenever $|x - y|$ is sufficiently small.

Assumption 5.20. (i) There exist $g_1(x)$ and $g_2(x)$ such that

$$|g(x,\theta) - g(x,\theta_0)| \leq C \,||\theta - \theta_0||\, g_1(x),$$

for each $\theta, \theta_0 \in \Omega_0$, and for some $0 < \beta \leq 1$,

$$|g_1(x + y) - g_1(x)| \leq C\,|y|^{\beta}\, g_2(x),$$

whenever y is sufficiently small. (ii) π is bounded and $\int_{-\infty}^{\infty}[1 + g_1^2(x) + g_2^2(x)]\,\pi(x)\,dx < \infty$.

Assumption 5.21. Under H_0, $||\hat{\theta}_n - \theta|| = o_{\mathrm{P}}[(nh^2)^{-1/2}]$.

Assumption 5.20 covers a wide class of functionals $g(x,\theta)$ and weight functions $\pi(x)$, including $g(x,\theta) = (x + \theta)^2$, $\theta e^x/(1 + e^x)$, $\theta \log|x|$, $\theta|x|^{\alpha}$ (α is fixed) and $\theta_0 + \theta_1|x| + ... + \theta_k|x|^k$ when $\pi(x) = e^{-x^2/2}$ or $\pi(x)$ has compact support. At this level of generality for $g(x,\theta)$, the condition on $\pi(x)$ is close to being necessary. Assumption 5.19 is slightly stronger than being necessary, and can be weakened to include the normal kernel function if more restrictions are imposed on the weight function $\pi(x)$. As $h \to 0$, Assumption 5.21 can be achieved as in Section 5.2. It is possible to extend Assumption 5.17 so that long memory processes are covered. See, e.g., Wang and Phillips (2015). Assumption 5.18 is a standard martingale structure required for an error process u_k, which can be replaced by the following Assumption 5.22. The latter allows for endogeneity in the model.

Assumption 5.22. (i) $\eta_i = (\epsilon_i, \nu_i)'$, $i \in \mathbb{Z}$, is a sequence of i.i.d. random vectors with $\mathrm{E}\,\eta_0 = 0$, $\mathrm{E}\,(\eta_0\eta_0') = \Sigma$ and $\mathrm{E}\,||\eta_0||^4 < \infty$, where ϵ_i is the same as in Assumption 5.17. (ii) $u_k = \sum_{j=0}^{\infty} \psi_j \, \eta_{k-j}$, where the coefficient vector $\psi_k = (\psi_{k1}, \psi_{k2})$ satisfies $\sum_{k=0}^{\infty} k^{1/4}(|\psi_{1k}| + |\psi_{2k}|) < \infty$ and $\sum_{k=0}^{\infty} \psi_k \neq 0$.

We have the following result for the limit distribution of T_n.

Theorem 5.10. *Suppose Assumptions 5.17-5.21 hold. Then, under H_0,*

$$\frac{T_n}{\sqrt{nh}} \to_D \tau_0 \, L_Z(1,0), \qquad (5.68)$$

for any h satisfying $nh^4 \log n \to 0$ and $n^{1/2-\delta_0} h \to \infty$, where

$$\tau_0 = \phi^{-1} \sigma^2 \int_{-\infty}^{\infty} K^2(s)ds \int_{-\infty}^{\infty} \pi(x)dx,$$

δ_0 can be as small as required and $Z = \{Z_t\}_{t \geq 0}$ is an O-U process defined by $Z_t = \int_0^t e^{\tau(t-s)} dB_s$.

If Assumption 5.18 is replaced by Assumption 5.22, (5.68) still holds if τ_0 is replaced by

$$\widetilde{\tau}_0 = \phi^{-1} \operatorname{E} u_0^2 \int_{-\infty}^{\infty} K^2(s)ds \int_{-\infty}^{\infty} \pi(x)dx.$$

Proof. Using Assumptions 5.20–5.21 and Corollary 2.5, we have

$$T_{2n} \leq C \, \|\hat{\theta}_n - \theta_0\|^2 \int_{-\infty}^{\infty} \left\{ \sum_{k=1}^{n} K[(x_k - x)/h] g_1(x_k) \right\}^2 \pi(x)dx$$

$$= o_P(\sqrt{nh}).$$

The result (5.68) follows from (5.67) and, under Assumptions 5.17–5.19, $\frac{T_{1n}}{\sqrt{nh}} \to_D \tau_0 \, L_Z(1,0)$ by Theorem 3.15.

If Assumption 5.18 is replaced by Assumption 5.22, it follows from Theorem 3.17 that $\frac{T_{1n}}{\sqrt{nh}} \to_D \widetilde{\tau}_0 \, L_Z(1,0)$. Hence (5.68) still holds when τ_0 is replaced by $\widetilde{\tau}_0$. □

To investigate asymptotic power, let the compact parametric space Ω_0 be convex, and the true parameter θ_0 be an interior point of Ω_0. We consider the following local alternative models

$$H_1 : \quad g(x) = g(x, \theta_0) + \rho_n \, m(x), \qquad (5.69)$$

where ρ_n is a sequence of constants measuring local deviations from the null hypothesis and $m(x)$ is a real function. Local alternatives of the form (5.69) are commonly used in the theory of nonparametric inference involving stationary data. See, for instance, Horowitz and Spokoiny (2001). We impose the following smoothness conditions on $m(x)$ and the consistency rate condition on $\hat{\theta}_n$ under H_1 to aid the asymptotic development here. It should be mentioned that, since $m(x)$ is assumed to be free of θ, the majority of common estimates for θ such as the LS estimates discussed in Section 5.2 share the same convergence rates under H_0 and H_1.

Assumption 5.23. (i) There exist $m_1(x)$ and $\gamma \in (0, 1]$ such that $|m(x + y) - m(x)| \le C\,|y|^\gamma\, m_1(x)$ for any $|y|$ sufficiently small. (ii) $\int_{-\infty}^{\infty} m^2(x)\,\pi(x)\,dx > 0$ and $\int_{-\infty}^{\infty}[1 + m^2(x) + m_1^2(x)]\,\pi(x)\,dx < \infty$.

Assumption 5.24. Under H_1, $\|\hat{\theta}_n - \theta_0\| = o_P[(nh^2)^{-1/2}]$.

Theorem 5.11. *Suppose Assumptions 5.17–5.20, 5.23 and 5.24 hold. Then, under H_1, we have*

$$\lim_{n \to \infty} \mathrm{P}\Big(\frac{T_n}{\sqrt{nh}} \ge t_0\Big) = 1,$$

for any $t_0 > 0$, any $h \to 0$ satisfying $n^{1/2-\delta_0}h \to \infty$ where δ_0 can be as small as required, and any ρ_n satisfying $nh^2\rho_n^4 \to \infty$. Result still holds if Assumption 5.18 is replaced by Assumption 5.22.

Proof. Under the alternative H_1, the test statistic T_n can be written as

$$T_n = \int_{-\infty}^{\infty} \Big\{\sum_{k=1}^{n} K[(x_k - x)/h]\big[u_k^* + \rho_n m(x_k)\big]\Big\}^2 \pi(x)dx$$

$$= T_{1n} + T_{2n} + T_{3n} + 2\rho_n T_{4n} + \rho_n^2 T_{5n},$$

where $u_k^* = u_k + g(x_k, \hat{\theta}_n) - g(x_k, \theta_0)$, $T_{jn}, j = 1, 2, 3$, are defined as in (5.67),

$$T_{5n} = \int_{-\infty}^{\infty} \Big\{\sum_{k=1}^{n} K[(x_k - x)/h]\, m(x_k)\Big\}^2 \pi(x)dx,$$

and $|T_{4n}| \le [(T_{1n} + T_{2n} + T_{3n})]^{1/2}\,(T_{5n})^{1/2}$ by Hölder's inequality. As seen in the proof of Theorem 5.10, $T_{1n} + T_{2n} + T_{3n} = O_P(\sqrt{nh})$. On the other hand, it follows from Corollary 2.5 that

$$\frac{1}{nh^2} T_{5n} \to_D \phi^{-2} \int_{-\infty}^{\infty} m^2(x)\pi(x)dx\, L_Z^2(1, 0).$$

Now, by letting $\epsilon_n = \sqrt{nh}\rho_n^2$, we have $\epsilon_n \to \infty$,

$$\Delta_n := \frac{1}{\sqrt{nh}}\,|T_{1n} + T_{2n} + T_{3n} + 2\rho_n T_{4n}|$$

$$= O_P(1) + O_P(\epsilon_n^{1/2}),$$

and for any $t_0 > 0$, as $n \to \infty$,

$$\mathrm{P}\Big(\frac{T_n}{\sqrt{nh}} \ge t_0\Big) \ge \mathrm{P}\Big(\frac{T_n}{\sqrt{nh}} \ge \epsilon_n^{3/4}\Big) = P\Big[\big(\frac{1}{nh^2}T_{5n} \ge \epsilon_n^{-1/4} - \epsilon_n^{-1}\Delta_n\big]$$

$$\ge P\Big[\big(\frac{1}{nh^2}T_{5n} \ge \epsilon_n^{-1/4}/2\big] \to 1,$$

which proves Theorem 5.11. $\qquad\square$

Remark 5.12. According to Theorem 5.11 the T_n test has nontrivial power against local alternatives of the form (5.69) whenever $\rho_n \to 0$ at a rate that is slower than $(nh^2)^{-1/2}$, as $nh^2 \to \infty$. This result differs from the stationary situation where a test generally has nontrivial power only if $\rho_n \to 0$ at a rate slower than $n^{-1/2}$. Moreover, the rate condition here is only related to the bandwidth h, not to the magnitude of $m(x)$, since the weight function $\pi(x)$ in the test offsets direct impact of the magnitude of $m(x)$ under the alternative.

Remark 5.13. The error variance $\mathrm{E}\,u_0^2$ (σ^2, respectively) that appears in the definition of $\tilde{\tau}_0$ (τ_0, respectively) can be estimated by

$$\hat{\sigma}_n^2 = \frac{\sum_{t=1}^n [y_t - g(x_t, \hat{\theta}_n)]^2 K[(x_t - x)/h]}{\sum_{t=1}^n K[(x_t - x)/h]},$$

based on a localized version of the usual residual sum of squares. Furthermore, ϕ can be estimated by standard HAC methods. Alternative HAR sieve methods (Phillips, 2005; Sun, 2011; Chen et al., 2014) or fixed-b kernel methods (Sun, 2014) may be used after some changes to the limit theory to address the random limit theory involved in the estimation of ϕ but these methods will not be explored here. These facts imply that, in the short memory case, the test T_n is applicable under endogeneity for practical implementation in specification testing.

Remark 5.14. To test H_0, we may use the following kernel-smoothed test statistic

$$S_n = \sum_{\substack{s,t=1 \\ s \neq t}}^n \hat{u}_s \hat{u}_t \, K[(x_t - x_s)/h],$$

where $\hat{u}_t = y_t - g(x_t, \hat{\theta}_n)$. The statistic S_n has commonly been applied to test parametric specifications in stationary time series regression. In nonlinear cointegrating regression, the kernel weights focus attention in the statistic on those components in the sum where the nonstationary regressor x_t nearly intersects itself. This smoothing scheme gives prominence to product components $\hat{u}_{t+1}\hat{u}_{s+1}$ in the sum where s and t may differ considerably but for which the corresponding regressor process takes similar values (that is, $x_t, x_s \simeq x$ for some x), thereby enabling a test of H_0. The involvement of the kernel weights $K[(x_t - x_s)/h]$, however, makes the asymptotics of S_n very complicated and difficult. We refer to Wang and Phillips (2012) for some results in this regard.

5.4 Bibliographical Notes

Section 5.1. Early contributions to the study of nonparametric estimation for a nonlinear cointegrating regression model include Phillips and Park (1998), Karlsen and Thostheim (2001), Guerre (2004) and Karlsen, et al. (2007). The latter developed an asymptotic theory for the Nadaraya-Watson estimators by using the framework of recurrent Markov chains. This section offers an alternative approach to the asymptotic theory. For the limit behaviors of Nadaraya-Watson estimators, Theorem 5.1 provides a general framework, allowing for both stationary and nonstationary time series as special examples and allowing for endogeneity in the model. In the nonlinear cointegration framework, local time theory and extended martingale limit theorems developed in Chapters 2 and 3 can be applied to derive the asymptotics, which is essentially different from that of Karlsen, et al. (2007). Theorem 5.1 and its corollaries come from, but essentially improve the related results by Wang and Phillips (2009a, b, 2015).

The materials for the bias of the Nadaraya-Watson estimators and for the asymptotics of local nonlinear estimators are taken from Wang and Phillips (2011) and Wang (2014). It is interesting to notice that the linear term is eliminated from the asymptotic bias. In consequence and in contrast to the stationary case, the Nadaraya-Watson estimator has the same limit distribution (to the second order including bias) as the local nonlinear estimator in nonlinear cointegrating regression. The result for multivariate cointegrating regression seems to be new to literature. As for the uniform convergence of local linear estimator, we refer to Liu, Chan and Wang (2015). See, also, Wang and Wang (2013), Chan and Wang (2014), Duffy (2014a, b). The result for recurrent Markov chain can be found in Gao, et al. (2014).

Section 5.2. Parametric estimation for nonlinear regression with integrated progresses was first considered in Park and Phillips (2001). Using a different routine, Chan and Wang (2015) currently established some similar, but more general results than those of Park and Phillips (2001). Moreover, the techniques used in Chan and Wang (2015) is related to joint distributional convergence of a martingale under consideration and its conditional variance (see Chapter 3), rather than using classical martingale limit theorem which requires establishing the convergence in probability for the conditional variance. In nonlinear cointegrating regression, there are some advantages for the methodology developed in Chan and Wang (2015) since it is usually difficult to establish the convergence in probability for the conditional variance. The main results of this section are partially taken from Chan and Wang (2015). The framework for weak consistency in Theorem 5.8 is particular useful in nonlinear cointegrating regression, where the set of sufficient conditions is easy to apply for various nonstationary regressors, including partial sums of linear processes and recurrent Markov chains. For related consistency results in nonlinear stochastic regression model, we refer to Wu (1981), Lai (1994) and Skouras (2000).

Section 5.3. The main part of this section is taken from Wang and Phillips (2015). For other related results, we refer to Gao, et al. (2009a, b), Wang and Phillips (2012).

Complementary notes. The past decade has witnessed great progress in the development of nonlinear cointegrating regression. This chapter provides estimation and inference theory for some simple nonlinear cointegrating regression models, illustrating the applications of the limit theory developed in previous chapters. There are numerous extensions in literature, which are partially summarized below.

Additive nonlinear regression model. Chang et al. (2001) extended the estimation theory for model (5.2), developed by Park and Phillips (2001), to more realistic models having multivariate integrated regressors that are additive nonlinear, as well as other stationary regressors and deterministic trends. Chang and Park (2003) investigated the index-type models such as the neural network and the smooth transition regression. More currently, Chang, et al. (2012) considered nonstationary regression with logistic transition.

Semiparametric cointegrating regression model. A typical semiparametric cointegrating regression model has the form:

$$y_t = \alpha + \beta\,x_t + g(x_t) + u_t$$

where α, β are unknown parameters and $g(x)$ is an unknown real function. We may estimate (α, β) and $g(x)$ in two steps. In the first step, we estimate (α, β) by the nonlinear least squares from the regression

$$y_t = \alpha + \beta\,x_t + e_t, \quad \text{where } e_t = g(x_t) + u_t,$$

namely, we obtain the estimator of (α, β):

$$(\hat{\alpha}_n, \hat{\beta}_n) = \arg\min_{\alpha,\beta} \sum_{t=1}^{n}(y_t - \alpha - \beta\,x_t)^2.$$

In the second step, we estimate the unknown $g(x)$ from

$$\hat{y}_t = g(x_t) + u_t$$

where $\hat{y}_t = y_t - \hat{\alpha}_n - \hat{\beta}_n x_t$. The conventional kernel estimator of $g(x)$ is given by

$$\hat{g}_n(x) = \sum_{t=1}^{n} \hat{y}_t \, K[(x_t - x)/h] \Big/ \sum_{t=1}^{n} K[(x_t - x)/h],$$

where $K(x)$ is a kernel function and h is a bandwidth. The asymptotics of $(\hat{\alpha}_n, \hat{\beta}_n)$ and $\hat{g}_n(x)$ can be established in a similar way as in Sections 5.1 and 5.2. The details under more general settings can be found in Gao and Phillips (2013), and Kim and Kim (2012).

Functional coefficient regression model. A varying coefficient regression model has the form:

$$y_t = \beta(z_t)' x_t + u_t,$$

where y_t, z_t and u_t are scalar, $x_t = (x_{t1}, ..., x_{td})'$ is a dimensional vector and $\beta(x) = (\beta_1(x), ..., \beta_d(x))'$ is an unknown $d \times 1$ column vector function. We may estimate $\beta(x)$ by using local linear fitting from the observation y_t, z_t and x_t, namely, $\beta(x)$ is estimated by

$$\hat{\beta}(x) = \arg\min_{\beta} \sum_{t=1}^{n} \left[y_t - \beta' x_t\right]^2 K[(z_t - x)/h],$$

giving the estimator error

$$\hat{\beta}(x) - \beta(x) = \sum_{t=1}^{n} x_t \, u_t \, K[(z_t - x)/h] \Big/ \sum_{t=1}^{n} x_t' x_t \, K[(z_t - x)/h].$$

If one of z_t and x_t is stationary, we may establish the asymptotics of $\hat{\beta}(x)$ by using some modified arguments as in Section 5.1. If both x_t and z_t are integrated processes, it is more involved to investigating the limit behaviors of $\hat{\beta}(x)$, requiring some new limit theorems that are not discussed in this book. We refer to Cai, Li and Park (2009), Xiao (2009), Sun, Cai and Li (2013) and Gao and Phillips (2015) for some related results under very restrictive conditions.

Nonstationary volatility model. A volatility model has the form:

$$y_t = \sigma_t \, \epsilon_t,$$

where $(\epsilon_t, \mathcal{F}_t)_{t \geq 1}$ is a martingale difference so that $\sigma_t^2 = \mathrm{E}\,(y_t^2 \mid \mathcal{F}_{t-1})$. The popular ARCH-GARCH model is referred to the cases that

$$\sigma_t^2 = \alpha_0 + \sum_{j=1}^{p} \alpha_j \sigma_{t-j}^2 + \sum_{k=1}^{q} \beta_k\, y_{k-1}^2,$$

where $\alpha_j, \beta_k, 0 \leq j \leq p, 1 \leq k \leq q$, are unknown parameters. Park (2002) introduced a new volatility process σ_t^2 defined by

$$\sigma_t^2 = m(x_t),$$

where x_t is an integrated time series and $m(x)$ is a smooth but unknown function. We refer to this model as the *nonstationary volatility model*. Note that

$$y_t^2 = m^2(x_t) + m^2(x_t)(\epsilon_t^2 - 1).$$

The unknown function $m(x)$ can be similarly estimated as in Section 5.1. We refer to Han and Zhang (2012), Wang and Wang (2013), and Han and Kristensen (2014) for more details.

For other related works, we refer to Shi and Phillips (2011), and Kasparis, et al. (2014) for weakly identified cointegrating regression; Kasparis, et al. (2015) for nonparametric predictive regression; Linton and Wang (2015) for nonparametric transformation regression; Marmer (2008) for nonlinearity, nonstationarity and spurious forecasts; and Choi and Saikonnen (2004, 2010), Hong and Phillips (2010) and Kasparis and Phillips (2012) for specification tests for nonlinear regression models with nonstationarity.

Appendix A

Concepts of stochastic processes

Definition of stochastic process

Let $\{\Omega, \mathcal{F}, \mathrm{P}\}$ be a probability space. A *(stochastic) process* is a family $X = \{X_t\}_{t \in I}$ of mappings from Ω into R^d. For a fixed time t, X_t is a random variable (vector if $d > 1$). For fixed $\omega \in \Omega$, each mapping $t \to X_t$ is called a *path*, or a *trajectory*, of the process X.

A process X is continuous, right continuous, increasing, etc., if all its paths have the same properties. We say X a *càdlàg* (resp. *càglàd*) process if X is right continuous and admits left-hand limit (resp. left continuous and admits right-hand limit).

A process X is *(square) integrable* if $(\mathrm{E}X_t^2 < \infty) \ \mathrm{E}\,|X_t| < \infty, \forall t \in I$; is *uniformly integrable* if $\lim_{\alpha \to \infty} \sup_{t \in I} \mathrm{E}\,|X_t| \mathrm{I}(|X_t| \geq \alpha) = 0$.

Let $\pi_n(I)$ be a sequence of partitions of I with $\lim_{n \to \infty} \mathrm{mesh}[\pi_n(I)] = 0$. A process X is said to be of *finite variation* (FV) if

$$\lim_{n \to \infty} \sum_{t_i \in \pi_n([a,b])} |X_{t_i} - X_{t_{i-1}}| < \infty, \quad a.s,$$

for each compact interval $[a, b]$ of I. $[X, X]$ denotes a *quadratic variation* process of X, where

$$[X, X]_t = \lim_{n \to \infty} \sum_{t_i \in \pi_n([0,t])} (X_{t_i} - X_{t_{i-1}})^2.$$

The *covariation* of X and Y is defined by

$$[X, Y] = \frac{1}{4}([X + Y, X + Y] - [X - Y, X - Y]).$$

Two processes X and Y are *modifications* if $X_t = Y_t, a.s.$ for each t; are *indistinguishable* if $X_t = Y_t, a.s.$ for all t. If Y is a modification of X, and suppose that both processes have a.s. right continuous sample paths, then X and Y are indistinguishable.

Stochastic basis and predictable process

Let $\{\Omega, \mathcal{F}, P\}$ be a probability space. A filtration $\mathbf{F}$ is a family $(\mathcal{F}_t)_{t \geq 0}$ of sub-σ fields of $\mathcal{F}$ such that $\mathcal{F}_s \subseteq \mathcal{F}_t$ for all $s \leq t$. A *stochastic basis* $\{\Omega, \mathcal{F}, \mathbf{F}, P\}$ is a probability space $\{\Omega, \mathcal{F}, P\}$ equipped with a filtration $\mathbf{F}$ satisfying: (i) $\mathcal{F}_0$ contains all P-null sets of $\mathcal{F}$ and (ii) $\mathcal{F}_t = \cap_{s > t} \mathcal{F}_s$, i.e., the filtration $\mathbf{F}$ is right continuous.

A *stopping time* (with respect to $\mathbf{F}$) is a mapping $T : \Omega \to \bar{R}_+$ such that $\{T \leq t\} \in \mathcal{F}_t$ for all $t \geq 0$. If $\{\Omega, \mathcal{F}, \mathbf{F}, P\}$ is a stochastic basis, then T is a stopping time if and only if $\{T < t\} \in \mathcal{F}_t$ for all $t \geq 0$.

A process X is *adapted* to $\mathbf{F}$ if X_t is $\mathcal{F}_t$-measurable for each $t \geq 0$.

A process X is *predictable* (with respect to $\mathbf{F}$) if the mapping $X_t(\omega) : (t, \omega) \to R^d$ is $\mathcal{P}$-measurable, where $\mathcal{P}$ is a σ-field on $\Omega \times R_+$, generated by the subsets: $A \times \{0\}$, $A \in \mathcal{F}_0$ and $A \times (s, t], A \in \mathcal{F}_s$ for $s < t$.

For each cádlág, adapted FV process A, there exists a pair (B, C) of increasing processes such that $A = B - C$ and the total variation $\mathrm{var}(A)$ is given by

$$\mathrm{var}(A)_t \equiv \int_0^t |dA_s| = B_t + C_t. \tag{A.1}$$

Moreover, if A is predictable, then B, C and $var(A)$ are also predictable. If H is a continuous (bounded) process, for each ω, the Riemann (Lebesgue)-Stieltjes integral

$$(H \cdot A)_t \equiv \int_0^t H_s dA_s$$

is well defined. If A and H are predictable, then $H \cdot A$ is predictable.

Martingale, local martingale and semimartingale

Let $\{\Omega, \mathcal{F}, \mathbf{F}, P\}$ be a stochastic basis. A real-valued cádlág, adapted process X is called a *martingale* (resp. *supermartingale, submartingale*) with respect to $\mathbf{F}$ if

(i) $\mathrm{E} |X_t| < \infty$ for all $t < \infty$,
(ii) $\mathrm{E}[X_t | \mathcal{F}_s] = X_s, a.s.$ (resp. $\mathrm{E}[X_t | \mathcal{F}_s] \leq X_s$, $\mathrm{E}[X_t | \mathcal{F}_s] \geq X_s$) for all $0 \leq s \leq t$.

A real-valued cádlág, adapted process X is called a *local martingale* (resp. *local supermartingale, local submartingale*) with respect to $\mathbf{F}$ if there exists a sequence of increasing stopping times T_n with $\lim_{n \to \infty} T_n = \infty, a.s.$ such that X^{T_n} is a martingale (resp. supermartingale, submartingale) for each $n \geq 1$, where $X^M = \{X_{t \wedge M}\}_{t \geq 0}$.

A real-valued cádlág, adapted process is called *semimartingale* if it can be decomposed as

$$X_t = X_0 + M_t + A_t, \tag{A.2}$$

where $M_0 = A_0 = 0$, M is a local martingale and A is a cádlág, adapted FV process. A semimartingale is called *special* if A is a predictable FV process.

The following properties are well-known. See, e.g., Protter (2005).

(1) If X is a uniformly integrable martingale, then $Y = \lim_{t \to \infty} X_t, a.s.$ exists, $E |Y| < \infty$ and $X_t = E(Y \mid \mathcal{F}_t), t \geq 0$.

(2) Let X be a martingale, T be a finite valued stopping time (i.e., $P(T < \infty) = 1$) and $\varphi(x)$ be convex such that $\varphi(X_t)$ is integrable. Then X^T is still a martingale and $\varphi(X)$ is a submartingale. In particular, X^+, $|X|$ and X^2 (if $E X_t^2 < \infty$) are submartingales.

(3) If X is a supermartingale with $\sup_{t \geq 0} E |X_t| < \infty$, then $Y = \lim_{t \to \infty} X_t, a.s.$ exists, and $E |Y| < \infty$.

(4) (**Doob's inequality**) Let X be a submartingale taking non-negative values. For any $\epsilon > 0$ and $p \geq 1$, we have

$$P(\sup_{0 \leq s \leq t} X_s \geq \epsilon) \leq \epsilon^{-p} E X_t^p.$$

Let $S_t = \sup_{0 \leq s \leq t} X_s$ and $\| Z \|_p = (E |Z|^p)^{1/p}$. We have

$$\| S_t \| \leq \frac{e}{e - 1} (1+ \| X_t \log X_t \|),$$

and for $p > 1$,

$$\| X_t \|_p \leq \| S_t \|_p \leq \frac{p}{p - 1} \| X_t \|_p .$$

(5) (**Doob-Meyer theorem**) Any submartingale X has a unique decomposition

$$X_t = X_0 + M_t + A_t,$$

where M is a local martingale and A is a cádlág, predictable increasing process with $M_0 = A_0 = 0$. Furthermore, if in addition X is non-negative, then M is a martingale and A is integrable. If in addition X is of Class D, i.e., the collection

$\{X_\tau : \tau$ is a finite valued stopping time$\}$ is uniformly integrable, then M is a uniformly integrable martingale and A_∞ is integrable.

(6) Any local, sub-, super-martingale is a semimartingale. Let X be a semimartingale with $X_0 = 0$. X is a special semimartingale if and only if the process J or X^* is locally integrable, where $J_t = \sup_{s \le t} |\Delta X_s|$ with $\Delta X_s = X_s - X_{s-}$ and $X_t^* = \sup_{0 \le s \le t} |X_s|$. For a special semimartingale, its decomposition (A.2) with A being predictable is unique, which is usually called the *canonical decomposition*.

(7) For each semimartingale X, its quadratic variation $[X, X]$ is a cádlág, increasing, adapted process satisfying

$$[X, X]_t = [X, X]_t^c + X_0^2 + \sum_{0 < s \le t} (\Delta X_s)^2,$$

where $[X, X]^c$ denotes the path-by-path continuous part of $[X, X]$. Furthermore, there exists a unique decomposition

$$[X, X]_t = L_t + \langle X, X \rangle_t,$$

where L is local martingale and $\langle X, X \rangle$ is a cádlág, predictable FV process. $\langle X, X \rangle$ is usually called the angle bracket of X or *compensator* of $[X, X]$.

If X is continuous, then $\langle X, X \rangle = [X, X]$. If X is a martingale with $\mathbb{E}X_t^2 < \infty$, then $X^2 - [X, X]$ is a local martingale and there exists a unique local martingale L such that $X^2 = \langle X, X \rangle + L$.

(8) Any semi- or local martingale X has an a.s. unique decomposition

$$X_t = X_0 + X_t^c + X_t^d,$$

where X^c is a continuous local martingale with $X_0^c = 0$ and X^d is a purely discontinuous semi- or local martingale.

For each semimartingale X and given $\beta > 0$, we may write X as

$$X_t = X_0 + A_t + M_t,$$

where $M_0 = A_0 = 0$, A is a cádlág, adapted FV process and M is one of the following processes:

(i) M is locally square integrable local martingale;
(ii) M is local martingale with jumps bounded by β;
(iii) with M^* is a local martingale,

$$M_t = M_t^* + \sum_{0 < s \le t} \Delta X_s I_{(|\Delta X_s| > \beta)}.$$

(9) For each semimartingale X, if $f(x)$ is a convex function, then $f(X)$ is a semimartingale. If X and Y are semimartingales, so is $XY = \{X_t Y_t\}_{t \geq 0}$.

Martingale difference array

Let $\{X_{ni}\}_{n \geq 1, i \geq 1}$ be an array of random variables. Let $\{\mathcal{F}_{ni}\}_{i \geq 1}$ be a sequence of σ-fields such that, for each $n \geq 1$,

$$\mathcal{F}_{ni} \subseteq \mathcal{F}_{n,i+1}, \quad i \geq 1,$$

and $\{X_{ni}\}_{i \geq 1}$ is adapted to $\{\mathcal{F}_{ni}\}_{i \geq 1}$. $\{X_{ni}, \mathcal{F}_{ni}\}_{n \geq 1, i \geq 1}$ is called an array of martingale differences if, for each $n \geq 1$, $\mathrm{E}(X_{ni} \mid \mathcal{F}_{n,i-1}) = 0$ for all $i \geq 1$, where $\mathcal{F}_{n0} = \{\phi, \Omega\}$. Let τ_1 and τ_2 be two stopping times such that $\tau_1 \leq \tau_2$. We have, for any $p \geq 1$,

$$\mathrm{E}\left(\sum_{j=\tau_1+1}^{\tau_2} X_{nj} \mid \mathcal{F}_{n,\tau_1} \right) = 0,$$

$$\mathrm{E}\left[\left(\sum_{j=\tau_1+1}^{\tau_2} X_{nj} \right)^2 \mid \mathcal{F}_{n,\tau_1} \right] = \mathrm{E}\left(\sum_{j=\tau_1+1}^{\tau_2} X_{nj}^2 \mid \mathcal{F}_{n,\tau_1} \right),$$

$$\mathrm{P}\left(\max_{\tau_1 < j \leq \tau_2 \wedge n} |X_{nj}| \geq \epsilon \right) \leq \frac{1}{\epsilon^p} \mathrm{E} \left| \sum_{j=\tau_1+1}^{\tau_2 \wedge n} X_{nj} \right|^p$$

Gaussian process, (fractional) Brownian motion and Lévy process

Let $X = \{X_t\}_{t \in I}$ be a stochastic process on $\{\Omega, \mathcal{F}, \mathrm{P}\}$. We say X is *stationary* if, for all $t, t + h \in I$,

$$\{X_{t+h}\}_{t \in I} =_D \{X_t\}_{t \in I},$$

where $=_D$ denotes the equality of the finite-dimensional distributions; X is *self-similar* with index $H > 0$ if, for any $a > 0$ and $at \in I$,

$$\{X_{at}\}_{t \in I} =_D \{a^H X_t\}_{t \in I};$$

X has *stationary increments* if, for all $s + h, t + h \in I$,

$$\{X_{t+h} - X_{s+h}\}_{s,t \in I} =_D \{X_t - X_s\}_{s,t \in I};$$

X has *independent increments* if, for all $t_1 < t_2 < \dots < t_n \in I$ and all $n \geq 1$,

$$X_{t_2} - X_{t_1}, X_{t_3} - X_{t_2}, \cdots, X_{t_n} - X_{t_{n-1}}$$

are independent random variables.

We call X a *Gaussian process* if all its finite-dimensional distributions are multivariate normal. A d-dimensional Gaussian process $X = \{X_t\}_{t\geq 0}$, where $X_t = (X_{1t}, \cdots, X_{dt})$, is determined by its expectation vector $m_t = EX_t$ and its covariance matrix

$$\rho(s,t) = E(X_s - m_s)'(X_t - m_t), \quad s, t \geq 0.$$

If X has independent and stationary increments, then $\rho(s,t) = \rho(s \wedge t, s \wedge t)$.

We call $B = \{B_t\}_{t\geq 0}$ a (one-dimensional)*Brownian motion* if

(i) B is a one-dimensional Gaussian process;
(ii) $EB_s = 0$ and $EB_s B_t = s \wedge t = \min\{s, t\}$;
(iii) the sample paths of B are continuous with probability one.

It can be proved that $B_0 = 0$, $[B, B]_t = t$, B has stationary and independent increments and the sample paths of B are of unbounded variation on any interval (nowhere differentiable). A d-dimensional Brownian motion B is a process

$$B_t = (B_{1t}, \cdots, B_{dt}),$$

where $B_1, \cdots, B_d$ are independent, one-dimensional Brownian motions.

A Gaussian process $B_H = \{B_H(t)\}_{t\geq 0}$ is called a *fractional Brownian motion* with index $0 < H < 1$ if it is self-similar with index $0 < H < 1$, has stationary increments and $EB_H(1)^2 = 1$. It follows from the self-similar property that $B_H(0) = 0$, $EB_H(t) = 0$ and the covariance function of B_H is

$$r_H(s,t) = E\left[B_H(s)B_H(t)\right] = \frac{1}{2}\left(|s|^{2H} + |t|^{2H} - |t - s|^{2H}\right).$$

Furthermore, B_H can be represented as

$$B_H(t) = \kappa_H \int_{-\infty}^{t} \left[((t - s)_+)^{H-1/2} - ((-s)_+)^{H-1/2}\right]dB_s,$$

where $t_+ = \max\{t, 0\}$, $B = \{B_t\}_{t\geq 0}$ is a Brownian motion, $B^* = \{B_{-t}\}_{t\geq 0}$ is an independent copy of B, and $\kappa_H > 0$ is a constant such that $EB_H^2(1) = 1$, i.e.,

$$\kappa_H = \left(\int_0^{\infty} \left[(1 + s)^{H-1/2} - s^{H-1/2}\right]^2 ds + \frac{1}{2H}\right)^{-1/2}.$$

$H \in (0, 1)$ is also called the Hurst index. When $H = 1/2$, B_H is a Brownian motion. When $H \neq 1/2$, $B_H(t)$ is not a semimartingale and does not have independent increments.

We call $X = \{X_t\}_{t \geq 0}$ an *Ornstein-Uhlenbeck* (O-U) process if

$$X_t = \alpha \int_0^t e^{-\beta(t-s)} dB_s,$$

where $\alpha \in \mathbb{R}$, $\beta > 0$ and B is a Brownian motion. An Ornstein-Uhlenbeck process is a Gaussian process with mean zero and covariance:

$$\text{cov}(X_s, X_t) = \frac{\alpha^2}{2\beta} \left(e^{-\beta|t-s|} - e^{-\beta(t+s)} \right).$$

We call $X = \{X_t\}_{t \geq 0}$ a *Lévy process* if

(i) X is cádlág with $X_0 = 0, a.s.$;
(ii) X has stationary and independent increments;
(iii) X is stochastically continuous, i.e., for all $a > 0$ and for all $s \geq 0$,

$$\lim_{t \to s} P\left(|X_t - X_s| > a \right) = 0$$

or equivalently $X_t \to_p X_s$ as $t \to s$.

[**Lévy-Khintchine formula**] If X is a Lévy process, then its characteristic function $E\, e^{i\theta X_t} = e^{-t\psi(\theta)}$, where

$$\psi(\theta) = -ia\theta + \frac{1}{2}\sigma^2\theta^2 - \int_{-\infty}^{\infty} \left(e^{i\theta y} - 1 - i\theta y \mathbf{I}_{(|y|<1)} \right) \nu(dy).$$

Here ν is a measure on $\mathbb{R}$ satisfying $\int_{-\infty}^{\infty}(1 \wedge x^2)\nu(dx) < \infty$ and $\nu(\{0\}) = 0$, which is called a Lévy measure. The triplet (a, σ^2, ν) is called the characteristics of X, which uniformly determines the law of X.

(i) If $\nu = 0$, then X has continuous paths and X must be Brownian motion with shift. That is, $X_t = \sigma B_t + at$.
(ii) If $a = \sigma = 0$ and $\nu(R) < \infty$, then X is a *compound Poisson process* with intensity $\lambda = \nu(R)$. That is,

$$P(X_t = n) = \frac{e^{-\lambda t}(\lambda t)^n}{n!}, \quad n = 0, 1, 2, \ldots$$

(iii) If $\sigma = 0$ and $v(dx) = |x|^{-1-\beta} dx$ with $0 < \beta < 2$, then X is a stable Lévy process with the index β.

A *stable Lévy process* with index $0 < \alpha \leq 2$ is a Lévy process in which each increment $X_t - X_s$ is a α-stable random variable. That is, for $s < t$,

$$E\, e^{i\theta(X_t - X_s)} = \exp\left\{ -b(t-s)|\theta|^\alpha (1 + i\beta\omega(\theta, \alpha)) \right\},$$

where $b > 0$ is a scale parameter, $|\beta| \le 1$ and

$$\omega(\theta, \alpha) = \begin{cases} \mathrm{sgn}(\lambda)\tan(\pi\alpha/2), & \alpha \ne 1, \\ \frac{2}{\pi}\mathrm{sgn}(\lambda)\log|\lambda|, & \alpha = 1. \end{cases}$$

With the sole exception of the Brownian motion with drift ($\alpha = 2$), the random variables of a stable Lévy process with $0 < \alpha < 2$ all have infinite variance, and if $\alpha \le 1$, they have infinite mean. In particular, a stable Lévy process with index $\alpha = 1$ and $\beta = 0$ is a Cauchy process where X_t has the density $f_t(x) = \frac{t}{\pi(x^2+t^2)}$.

For any Borel set A, let

$$N(t, A) = \#\{0 \le s \le t : \Delta X_s \in A\}, \quad \widetilde{N}(t, A) = N(t, A) - t\nu(A),$$

where $\Delta X_s = X_s - X_{s-}$.

[**Lévy-Itô decomposition**] Given a Lévy triplet (a, σ^2, ν), there exists three independent Lévy processes $X^{(1)}, X^{(2)}, X^{(3)}$, which lie in the same probability space, such that:

$$X = X^{(1)} + X^{(2)} + X^{(3)},$$

where $X_t^{(1)} = at + \sigma B_t$ is a Brownian motion with drift, the pure continuous part of X; $X_t^{(2)} = \int_{|x| \ge 1} x N(t, dx)$, a compound Poisson process (the jumps of X with size being greater than 1); $X_t^{(3)} = \int_{|x| < 1} x \widetilde{N}(t, dx)$, a square integrable pure jump martingale. Note that $X = M + A$, where M is the martingale $\sigma B + X^{(3)}$ and A is the finite variation process $A_t = at + X_t^{(2)}$. X is a semimartingale having the continuous part $X^c = \sigma B$.

Appendix B

Metric space

A metric space is a pair (S, d) where S is a set and d is a metric on S, i.e., a mapping $d : S \times S \to \mathrm{R}$ such that, for any $x, y, z \in S$,

(i) $d(x, y) \geq 0$;
(ii) $d(x, y) = 0$ if and only if $x = y$;
(iii) $d(x, y) = d(y, x)$;
(iv) $d(x, z) \leq d(x, y) + d(y, z)$.

Given a metric space (S, d), we define an open ball of radius $r > 0$ about x as the set

$$B(x; r) = \{y \in S : d(x, y) < r\}.$$

A set $M \subset S$ is called *open* if for every $x \in M$ there exists an $r > 0$ such that $B(x; r)$ is contained in M. The complement of an open set is called *closed*. A set $M \subset S$ is called *compact* if each open cover of M contains a finite subcover. Two metrics d_1 and d_2 on S are said to be *equivalent* if for each x and ϵ there is a δ with $B_1(x, \delta) \subset B_2(x, \epsilon)$ and $B_2(x, \delta) \subset B_1(x, \epsilon)$, where $B_i(x; r) = \{y \in S : d_i(x, y) < r\}, i = 1, 2$.

Given a metric space (S, d), a sequence $x_n \in S$ is said to converge to the limit $x \in S$ if and only if $d(x_n, x) \to 0$, as $n \to \infty$. A subset $M \subset S$ is closed if and only if every sequence in M that converges to a limit in S has its limit in M. A compact set M is closed and bounded.

A space S is *separable* if it contains a countable dense subset. A metric space (S, d) is *complete* if every Cauchy sequence x_n (i.e., $d(x_m, x_n) \to 0$, as $m, n \to \infty$) in S has a limit lying in S.

Example B.3. Let $S = \mathrm{R}^d$ and $d(x, y) = \sqrt{\sum_{j=1}^{d}(x_j - y_j)^2}$. (S, d) is a separable and complete metric space.

Example B.4. Suppose X is a measure space with a positive measure μ. Let $S = L^2$ be the space of all measurable functions f on X satisfying $\int_X f^2 d\mu < \infty$, i.e.,

$$L^2 = \{f : \int_X f^2 d\mu < \infty\}.$$

Define $d(f,g) = \int_X (f-g)^2 d\mu$. (S,d) is a complete metric space (which is usually called L^2-space).

Example B.5. Let $S = C_{\mathrm{R}^m}[0,1]$ or $C_{\mathrm{R}^m}[0,\infty)$ be the space of continuous functions from $[0,1]$ to R^m or from $[0,\infty)$ to R^m. Define

$$d(x,y) = \sup_{0 \le t \le 1} |x(t) - y(t)| \quad \text{or} \quad d(x,y) = \sup_{0 \le t < \infty} |x(t) - y(t)|.$$

(S,d) is a separable and complete metric space. The metric $d(x,y)$ is called uniform metric (*uniform topology*).

Example B.6. Let $S = D_{\mathrm{R}^m}[0,M]$, where $M > 0$, be the space of cádlág (right continuous with left limit) functions from $[0,M]$ to R^m.

Let Λ denote the class of strictly increasing, continuous mapping $\lambda : [0,M] \to [0,M]$ [we have $\lambda(0) = 0$ and $\lambda(M) = M$]. Let $||x|| = \sup_{0 \le t \le M} |x(t)|$ and $I(t) = t$. Define,

$$d(x,y) = \inf_{\lambda \in \Lambda} \{||\lambda - I|| \vee ||x - y\lambda||\},$$

where $y\lambda(t) = y(\lambda(t))$, and

$$d_0(x,y) = \inf_{\lambda \in \Lambda} \{||\lambda||_0 \vee ||x - y\lambda||\},$$

where $||\lambda||_0 = \sup_{0 \le s < t \le M} \left| \log \frac{\lambda(t) - \lambda(s)}{t - s} \right|$.

Both d and d_0 are metrics and they are equivalent. The topology generated by d or d_0 is called the *Skorokhod topology*. The space $D_{\mathrm{R}^m}[0,M]$ is separable under d and d_0, and is complete under d_0 (not under d). See, e.g, Billingsley (1995).

Elements x_n of $D_{\mathrm{R}^m}[0,M]$ converges to a limit x in the Skorokhod topology (i.e., $d(x_n, x) \to 0$ or $d_0(x_n, x) \to 0$) if and only if there exists a sequence $\lambda_n \subset \Lambda$ such that

$$\sup_{0 \le s \le M} \left\{ |\lambda_n(s) - s| + |x_n[\lambda_n(s)] - x(s)| \right\} \to 0, \quad \text{as } n \to \infty.$$

Let $x_n(s) = \mathrm{I}_{[1-1/n,2)}(s)$ and $x(s) = \mathrm{I}_{[1,2)}(s)$. It is readily seen that $d(x_n, x) \to 0$, but $d(x_n - x, 0) \not\to 0$ since, by letting

$$y_n(s) = x_n(s) - x(s) = \mathrm{I}_{[1-1/n,1)}(s),$$

we have $\sup_{s \leq 1} |y_n[\lambda_n(s)]| = 1$ for any $\lambda_n(s)$ satisfying $\sup_s |\lambda_n(s) - s| \to 0$. Thus the fact that x_n and y_n converge to x and y, respectively, does not imply $x_n + y_n$ converges to $x + y$. As a consequence, $D_{\mathbf{R}^m}[0, M]$ with the Skorokhod topology is not a topological space.

Example B.7. Let $D_{\mathbf{R}^m}[0, \infty)$ be the space of cádlág (right continuous and left-hand limit exists) functions from $[0, \infty)$ to $\mathbf{R}^m$. Write

$$g_M(t) = \begin{cases} 1 & \text{if } t \leq M - 1, \\ M - t & \text{if } M - 1 \leq t \leq M, \\ 0 & \text{if } t \geq M, \end{cases}$$

and, for $x \in D_{\mathbf{R}^m}[0, \infty)$, let

$$x^m(t) = g_M(t)x(t), \quad t \geq 0.$$

Define

$$d_\infty(x, y) = \sum_{M=1}^{\infty} 2^{-M} (1 \wedge d_0(x^M, y^M))$$

where $d_0(x, y)$ is defined as in Example B.6. d_∞ is a metric on $D_{\mathbf{R}^m}[0, \infty)$ and defines the Skorokhod topology.

$(D_{\mathbf{R}^m}[0, \infty), d_\infty)$ is a separate and complete metric space.

Let Λ_∞ be the set of continuous, increasing maps of $[0, \infty)$ onto itself. Elements x_n of $D_{\mathbf{R}^m}[0, \infty)$ converges to a limit x in the Skorokhod topology (i.e., $d_\infty(x_n, x) \to 0$) if and only if there exists a sequence $\lambda_n \subset \Lambda_\infty$ such that

$$\sup_{0 \leq s < \infty} |\lambda_n(s) - s| \to 0 \qquad \text{and}$$

$$\sup_{s \leq N} |x_n[\lambda_n(s)] - x(s)| \to 0, \quad \text{for each } N \geq 1.$$

As in Example B.6, we may have $d_\infty(x_n, x) \to 0$, but $d_\infty(x_n - x, 0) \not\to 0$, indicating that $D_{\mathbf{R}^m}[0, \infty)$ is not a topological space.

Example B.8. (Product space). Let (S_1, φ_1) and (S_2, φ_2) be two metric spaces with metrics d_1 and d_2, respectively. Let

$$S = S_1 \times S_2 = \{(\omega_1, \omega_2) : \omega_1 \in S_1, \omega_2 \in S_2\}.$$

If S_1 and S_2 are separate, then S is separate. Consequently, $\varphi = \varphi_1 \times \varphi_2$, where φ is the Borel σ-field on S. Furthermore we may define the product metric d on $S = S_1 \times S_2$ by

$$d(z_1, z_2) = \left[d_1(x_1, x_2)^2 + d_2(y_1, y_2)^2\right]^{1/2}, \qquad \text{or}$$

$$d(z_1, z_2) = \max\{d_1(x_1, x_2), d_2(y_1, y_2)\},$$

where $z_i = (x_i, y_i), i = 1, 2$.

We may prove $C_R[0,1] \times C_R[0,1] = C_{R^2}[0,1]$ in the uniform topology. Indeed, $z = (x,y) \in C_R[0,1] \times C_R[0,1]$ if and only if $x(t)$ and $y(s)$ are continuous on $[0,1]$ if and only if $(x(t), y(t))$ is continuous on $[0,1]$ if and only if $z = (x,y) \in C_{R^2}[0,1]$. Furthermore, under uniform topology, x_n and y_n converge to x and y, respectively, on $C_R[0,1]$ if and only if (x_n, y_n) converges to (x,y) on $C_{R^2}[0,1]$.

However, $D_R[0,1] \times D_R[0,1] \neq D_{R^2}[0,1]$ in the Skorokhod topology. By definition, (x_n, y_n) converges to (x,y) on $D_R[0,1] \times D_R[0,1]$ if and only if x_n and y_n converge to x and y, respectively, on $D_R[0,1]$. But (x_n, y_n) converges to (x,y) on $D_{R^2}[0,1]$ if and only if $\alpha x_n + \beta y_n$ converges to $\alpha x + \beta y$ on $D_R[0,1]$ for any $\alpha, \beta \in R$. See, e.g., Lemma (A.28) of Holley and Stroock (1979). Now, to see $D_R[0,1] \times D_R[0,1] \neq D_{R^2}[0,1]$, let

$$x_n(s) = I_{[1/2-1/n,1)}(s),$$
$$y_n(s) = x(s) = y(s) = I_{[1/2,1)}(s).$$

As in Example B.6, x_n and y_n converge to x and y, respectively, but $x_n - y_n$ fails to converge to 0, on $D_R[0,1]$. That is, (x_n, y_n) converges to (x,y) on $D_R[0,1] \times D_R[0,1]$ by definition, but (x_n, y_n) fails to converge to (x,y) on $D_{R^2}[0,1]$.

The convergence of (x_n, y_n) can be considered either in $D_R[0,1] \times D_R[0,1]$ or $D_{R^2}[0,1]$. The latter convergence is stronger as we require one sequence λ_n of changes of time such that $(x_n[\lambda_n(t)], y_n[\lambda_n(t)])$ converges uniformly to $(x(t), y(t))$ on $t \in [0,1]$. In $D_R[0,1] \times D_R[0,1]$, there are two changes of time, λ_{1n} and λ_{2n}, such that $(x_n[\lambda_{1n}(t)], y_n[\lambda_{2n}(t)])$ converges uniformly to $(x(t), y(t))$ on $t \in [0,1]$.

Appendix C

Convergence of probability measure

Given a separate and complete metric space (S, d), let φ be the σ-field generated by the open sets (Borel σ-field). A *probability measure* on (S, φ) is a nonnegative, countably additive set function P with $P(S) = 1$.

For any set A, let ∂A denote the *boundary* of A. Let P, P_n be a sequence of probability measure on (S, φ). We say that P_n converges weakly to P and write $P_n \Rightarrow P$ if

$$\lim_n P_n(A) = P(A), \quad \text{for each } A \in \varphi \text{ with } P(\partial A) = 0.$$

The following conditions are equivalent.

(i) $P_n \Rightarrow P$;

(ii) $\lim_n \int f dP_n = \int f dP$, for all bounded, uniformly continuous real f on S;

(iii) $\limsup_n P_n(F) \le P(F)$ for all closed F;

(iv) $\liminf_n P_n(G) \ge P(G)$ for all open G;

(v) Each subsequence $\{P_{n_k}\}$ contains a further subsequence $\{P_{n_{k_m}}\}$ converging weakly to P.

Let h map (S, d) to another metric space (S', d') and φ' be the Borel σ-field associated with S'. Let Ph^{-1} be the probability measure on (S', φ') generated by P, i.e., $Ph^{-1}(A) = P[h^{-1}(A)]$ for any $A \in \varphi'$. Suppose h is measurable and denote by D_h the set of its discontinuities. The following mapping theorem holds.

- If $P_n \Rightarrow P$ and $P(D_h) = 0$, then $P_n h^{-1} \Rightarrow Ph^{-1}$.

Tightness and relative compactness

A probability measure P on (S, φ) is *tight* if for each $\epsilon > 0$ there exists a compact set K such that $P(K) > 1 - \epsilon$. A family of probability measures

$\prod$ on (S, φ) is *tight* if for each $\epsilon > 0$ there exists a compact set K such that $\inf_{P \in \prod} P(K) > 1 - \epsilon$. We call $\prod$ *relative compact* if every sequence of $\prod$ contains a weakly convergent subsequence.

The following facts are well-known.

(i) If (S, φ) is separate and complete, then each probability measure on (S, φ) is tight.

(ii) If $\prod$ is tight on (S, φ) and if h is a continuous mapping from S to S', then $\prod' = \{Ph^{-1} : P \in \prod\}$ is tight on (S', φ'), where $Ph^{-1}(A) = P(B : h(B) \in A)$ for $A \in \varphi'$.

(iii) (**Prohorov theorem**) If $\prod$ is tight, then it is relatively compact. If S is separate and complete, $\prod$ is tight if and only if it is relatively compact.

Random elements

Let X be a mapping from a probability space $(\Omega, \mathcal{F}, \mathrm{P})$ into (S, φ). If, for any $A \in \varphi$, $X^{-1}(A) \in \mathcal{F}$, we say X a random element of S.

- Random variable X: if $S = \mathrm{R}$;
- Random vector X: if $S = \mathrm{R}^k$;
- Random function X: if $S = D[0, 1]$ or other function space.

The distribution of X is the probability measure $P = \mathrm{P}X^{-1}$ on (S, φ):

$$P(A) = \mathrm{P}(X^{-1}A) = \mathrm{P}(\omega : X(\omega) \in A) = \mathrm{P}(X \in A), \quad A \in \varphi.$$

A set $A \in \varphi$ is called X-*continuous* if $\mathrm{P}(X \in \partial A) = 0$, where ∂A denotes the boundary of A.

Convergence of random elements

Let X_n, X be random elements of S and d be a metric on S so that (S, d) is separable. We say $X_n \Rightarrow X$ (*convergence in distribution*) if $P_n \Rightarrow P$; $X_n \to_\mathrm{P} X$ (*convergence in probability*) if

$$\mathrm{P}(d(X_n, X) \geq \epsilon) \to 0, \quad \text{for each } \epsilon > 0;$$

$X_n \to_{a.s.} X$ (*convergence almost surely*) if

$$\mathrm{P}(\lim_{n \to \infty} d(X_n, X) = 0) = 1.$$

The following statements are equivalent.

(i) $X_n \Rightarrow X$;

(ii) $\lim_n \mathrm{E}f(X_n) = \mathrm{E}f(X)$, for all bounded, uniformly continuous real f on S;

(iii) $\limsup_n \mathrm{P}(X_n \in F) \le \mathrm{P}(X \in F)$ for all closed F;

(iv) $\liminf_n \mathrm{P}(X_n \in F) \ge \mathrm{P}(X \in G)$ for all open G.

(v) $\lim_n \mathrm{P}(X_n \in A) = \mathrm{P}(X \in A)$, for each $A \in \mathcal{F}$ with $\mathrm{P}(\partial A) = 0$.

Let (X_n, Y_n) be a random element of $S \times S$. Then $d(x, y)$ maps $S \times S$ continuously to R, and so $d(X_n, Y_n)$ is a random variable. The following results are of fundamental importance.

(i) If $X_n \Rightarrow X$ and $d(X_n, Y_n) \Rightarrow 0$, then $Y_n \Rightarrow X$. Thus if $X_n \to_{\mathrm{P}} X$, then $X_n \Rightarrow X$;

(ii) Let (X_{un}, Y_n) be a random element of $S \times S$. If $X_{un} \Rightarrow X_u$ for each u, $X_u \Rightarrow X$ and

$$\lim_{u \to \infty} \lim_{n \to \infty} \mathrm{P}(d(X_{un}, Y_n) \ge \epsilon) = 0,$$

for $\epsilon > 0$, then $Y_n \Rightarrow X$.

(iii) If $X_n \Rightarrow X$ and X_n are uniformly integrable, i.e.,

$$\lim_{A \to \infty} \sup_n \int_{|X_n| \ge A} |X_n| dP = 0,$$

then X is integrable and $\mathrm{E}X_n \to EX$.

Suppose h maps (S, d) to another metric space (S', d'). Suppose h is measurable and $\mathrm{P}(X \in D_h) = 0$, where D_h is the set of discontinuities for h. The following *continuous mapping theorem* holds.

(i) If $X_n \to_{a.s.} X$, then $h(X_n) \to_{a.s.} h(X)$;

(ii) If $X_n \to_{\mathrm{P}} X$, then $h(X_n) \to_{\mathrm{P}} h(X)$;

(iii) If $X_n \Rightarrow X$, then $h(X_n) \Rightarrow h(X)$.

Skorokhod representation theorem

Let (S, d) be a separate metric space and assume X_n, X are random elements of S. Then $X_n \Rightarrow X$ if and only if there exists a probability space supporting random elements X_n' of S such that

(i). $X_n =_D X_n'$, for each $n \ge 1$;

(ii). $X =_D X'$;

(iii). $d(X_n', X') \to 0$, a.s. as $n \to \infty$.

Skorokhod representation theorem is a powerful tool to prove the continuous mapping theorem and other related result. See, e.g., Billingsley (1995). Skorokhod representation theorem implies that $X_n =_D X'_n$, for each $n \geq 1$. But in the construction, the relation among X_n are not carried over to the X'_n, that is, the distribution of (X'_n, X'_m) has no connection with that of (X_n, X_m).

Convergence of stochastic processes on $D_{\mathrm{R}^m}[0, M]$, $D_{\mathrm{R}^m}[0, \infty)$

Let $D_{\mathrm{R}^m}[0, M]$ be the space of cádlág functions from $[0, M]$ to R^m equipped with the Skorokhod topology. As in Example B.6, $(D_{\mathrm{R}^m}[0, M], d_0)$ is a separate and complete metric space. Similarly, by Example B.7, $(D_{\mathrm{R}^m}[0, M], d_\infty)$ is a separate and complete metric space.

Let $X_n(t), X(t), t \in [0, M]$, be a sequence of d-dimensional cádlág stochastic processes defined on a probability space $(\Omega, \mathcal{F}, \mathrm{P})$. Write $X_n = \{X_n(t)\}_{0 \leq t \leq M}$ and $X = \{X(t)\}_{0 \leq t \leq M}$. X_n, X are random elements (functions) of $D_{\mathrm{R}^m}[0, M]$ since they map Ω into $D_{\mathrm{R}^m}[0, M]$. If $X_n \Rightarrow X$, we say $X_n(t)$ weakly converges to $X(t)$, and write

$$X_n(t) \Rightarrow X(t), \quad \text{on } D_{\mathrm{R}^m}[0, M].$$

Let P_n be the distribution of X_n. If $\{P_n\}$ is tight and

$$(X_n(t_1), \cdots, X_n(t_k)) \to_D (X(t_1), \cdots, X(t_k)),$$

for any $0 \leq t_1, t_2, \cdots, t_k \leq M, k \geq 1$, then $X_n \Rightarrow X$ or $X_n(t) \Rightarrow X(t)$ on $D_{\mathrm{R}^m}[0, M]$.

Similarly, if $X_n(t), X(t), t \geq 0$, is a sequence of d-dimensional cádlág stochastic processes, $X_n = \{X_n(t)\}_{t \geq 0}$ and $X = \{X(t)\}_{t \geq 0}$ are random elements (functions) of $D_{\mathrm{R}^m}[0, \infty)$. We have $X_n \Rightarrow X$ or

$$X_n(t) \Rightarrow X(t), \quad \text{on } D_{\mathrm{R}^m}[0, \infty)$$

if and only if, for any $M > 0$,

$$X_n(t) \Rightarrow X(t), \quad \text{on } D_{\mathrm{R}^m}[0, M].$$

Bibliography

Akonom, J. (1993). Comportement asymptotique du temps d'occupation du processus des sommes partielles. *Ann. Inst. H. Poincaré Probab. Statist.*, **29**, 57–81.

Aldous, D. J. (1986). Self-intersections of 1-dimensional random walks. *Probab. Theory Relat. Fields*, **72**, 559–587.

Aldous, D.J. and Eagleson, G. K.(1978). On mixing and stability of limit theorems. *Ann. Probab.*, **2**, 325–331.

Andrews, D. W. K. (1995). Nonparametric kernel estimation for semiparametric models. *Econometric Theory*, **11**, 560–596.

Andrews, D. W. K. and Sun, Y. (2004). Adaptive local polynomial Whittle estimation of long-range dependence. *Econometrica*, **72**, 569–614.

Ango Nze, P. and Doukhan P. (2004). Weak dependence: Models and applications to econometrics. *Econometric Theory*, **20**, 995–1045.

Barlow, M. T. (1988). Necessary and sufficient conditions for the continuity of local time of Lévy processes. *Ann. Probab.*, **16**, 1389–1427.

Berkes, I. and Horváth, L. (2006). Convergence of integral functionals of stochastic processes. *Econometric Theory*, **22**, 304–322.

Berman, S. M. (1969). Local times and sample function properties of stationary Gaussian processes. *Transactions of the American Mathematical Society* **137**, 277–299.

Berman, S. M. (1973). Local nondeterminism and local times of Gaussian processes. *Indiana Univ. Math. J.*, **23**, 69–94.

Berman, S. M. (1978). Gaussian processes with biconvex covariances. *Journal of Multivariate Analysis*, **8**, 30–44.

Berman, S. M. (1987). Spectral conditions for local nondeterminism. *Stochastic Process. Appl.* **27**, 73–84.

Berman, S. M. (1991). Self-intersections and local nondeterminism of Gaussian processes. *Ann. Probab.*, **19**, 160–191.

Billingsley, P. (1968, 1995). *Convergence of probability measures.* John Wiley & Sons, Inc., New York.

Billingsley, P. (1974). Conditional distributions and tightness. *Annals of Probability,* **2**, 480–485.

Borodin, A. N.; Ibragimov, I. A. (1995). *Limit theorems for functionals of random walks.* Proc. Steklov Inst. Math., no. 2.

Buchmann, B. and Chan, N. H. (2007). Asymptotic theory of least squares estimators for nearly unstable processes under strong dependence. *Ann. Statist.*, **35**, 2001–2017.

Bosq, D. (1998). *Nonparametric Statistics for Stochastic Processes: Estimation and Prediction, 2nd ed. Lecture Notes in Statistics 110*, Springer-Verlag.

Cai, Z., Li, Q. and Park, J. Y. (2009). Functional-coefficient models for nonstationary time series data. *Journal of Econometrics*, **148**, 101–113.

Chan, N. and Wang, Q. (2014). Uniform convergence for non-parametric estimators with nonstationary data. *Econometric Theory*, **30**, 1110–1133.

Chan, N. and Wang, Q. (2015). Nonlinear regression with nonstationary time series. *Journal of Econometrics*, **185**, 182–195.

Chan, N. H. and Wei, C. Z. (1987). Asymptotic inference for nearly nonstationary $AR(1)$ process. *Annals of Statistics,* **15**, 1050–1063.

Chan, N. H. and Wei, C. Z. (1988). Limit distributions of least-squares estimates of unstable auto-regressive processes. *Annals of Statistics,* **16**, 367–401.

Chang, Y. and Park, J. Y. (2003). Index models with integrated time series. *Journal of Econometrics*, **114**, 73–106.

Chang, Y. and Park, J. Y. (2011). Endogeneity in nonlinear regressions with integrated time series. *Econometric Reviews*, **30**, 51–87.

Chang, Y., Park, J. Y. and Phillips, P. C. B. (2001). Nonlinear econometric models with cointegrated and deterministically trending regressors. *Econometrics Journal*, **4**, 1–36.

Chang, Y. Jiang, B. and Park, J. Y. (2012). Nonstationary rergession with logistic transition. *Econometrics Journals*, **15**, 255–287.

Chen, X. (2000). On the limit laws of the second order for additive functionals of Harris recurrent Markov chains. *Probability Theory And Related Fields* **116**, 89–123.

Chen, J., Li, D. and Zhang, L. (2009). Robust estimation in nonlinear cointegrating model. *Journal of Multivariate Analysis,* **101**, 707–717.

Chen, X., Liao, Z., and Sun, Y. (2014). Sieve inference on possibly misspecified semi-nonparametric time series models. *Journal of Econometrics*, **178**, 639–658.

Choi, I. and Saikkonen, P. (2004). Testing linearity in cointegrating smooth transition regressions. *Econom. J.*, **7**, 341–365.

Choi, I. and Saikkonen, P. (2010). Tests for nonlinear cointegration. *Econometric Theory*, **26**, 682–709.

Christopeit, N. (2009). Weak convergence of nonlinear transformations of integrated processes: The multivariate case. *Econometric Theory*, **25**, 1180–1207.

Csörgö, M. and Révész, P. (1981). *Strong approximations in probability and statistics*. Probability and Mathematical Statistics. Academic Press, Inc., New York-London

Cuzick, J. (1978). Local nondeterminism and the zeros of Gaussian processes. *Ann. Probab.*, **6**, 72–84. [Correction: 15, 1229 (1987).]

Diaconis, P. and Freedman, D. (1999). Iterated random functions. *SIAM Rev.* **41**, 41–76.

Durrett, R. and Resnick, S. (1978). Functional limit theorems for dependent variables. *Ann. Probability* **6**, 829–346.

Duffie, D. and Protter, P. (1992). From discrete to continuous time finance: weak convergence of the financial gain process. *Mathematical Finance*, **2**, 1–15.

Dozzi, M. (2013). Occupation density and sample path properties of N-parameter processes. Topics in spatial stochastic processes (Martina Franca, 2001), 127–166, *Lecture Notes in Math.*, 1802, Springer, Berlin, 2003.

Duffy, J. (2014 a). A Uniform law for the convergence to local time. *Unpublished manuscript*, Yale University.

Duffy, J. (2014 b). Uniform in bandwidth converbgence rates, on a maximal domain, in structure nonparametric cointegrating regression. *Unpublished manuscript*, Yale University.

Eagleson, G. K.(1976). Some simple conditions for limit theorems to be mixing, *Theo. Verojatnost. i Primenen*, **21**, 653–660.

Fan, J. and Gijbels, I. (1996). *Local polynomial modelling and its applications*, Chapman & Hall/CRC.

Feller, W. (1971). *An Introduction to Probability Theory and Its Applications*, JOhn Wiley & Sons, Inc. New York.

Formanov, S. K. (1975). Invariance principles for homogeneous Markov chains. *Dokl. Akad. Nauk SSSR* **221**, 42–44.

Gaquet, C. and Witomski, P. (1999). *Fourier Analysis and Applications*. Springer, New York.

Gao, J., Li, D. and Tjøstheim, D. (2011). Uniform consistency for nonparametric estimates in null recurrent time series. *Working paper series No. 0085*, The university of Adelaide, School of Economics.

Gao, J., Kanaya, S., Li, D. and Tjøstheim, D. (2014). Uniform consistency of nonparametric estimators in null recurrent time series. *Econometric Theory*, **30**, 1–42.

Gao, J., Maxwell, K., Lu, Z., Tjøstheim, D. (2009a). Nonparametric specification testing for nonlinear time series with nonstationarity. *Econometric Theory*, **25**, 1869–1892.

Gao, J., Maxwell, K., Lu, Z., Tjøstheim, D. (2009b). Specification testing in nonlinear and nonstationary time series autoregression. *The Annals of Statistics*, **37**, 3893–3928.

Gao, J. and Phillips, P. C. B. (2013). Semiparametric estimation in triangular system equations with nonstationarity. *Journal of Econometrics*, **176**, 59–79.

Gao, J. and Phillips, P. C. B. (2013). Functional coefficient nonstationary regression, Cowles fundation discussion paper, No. 1911.

Geman, D. (1976). A note on the continuity of local times. *Proc. of the Amer. Math. Society*, **57**, 321–326.

Geman, D. and Horowitz, J. (1980). Occupation densities. *Ann. Probab.*, **8**, 1–67.

Götze, F. and Hipp, C. (1983). Asymptotic expansions for sums of weakly dependent random vectors. *Z. Wahrsch. Verw. Gebiete*, **64**, 211–239.

Guerre, E. (2004). Design-Adaptive pointwise nonparametric regression estimation for recurrent Markov time series. *Unpublished manuscript*.

Hall, P. and Heyde, C. C. (1980). *Martingale limit theory and its application*, Probability and Mathematical Statistics. Academic Press, Inc.

Han, H. and Zhang, S. (2012). Nonstationary noparametric volatility model, *Econometrics Journal*, **15**, 204–225.

Han, H. and Kristensen, D. (2014). Asymptotic theory for the QMLE's in GARCH-X models with stationary and nonstationary covariates. *Journal of Business and Economic Statistics*, **32**, 416–429.

Hansen, B. E. (1992). Covergence to stochastic integrals for dependent heterogeneous processes. *Econometric Theory*, **8**, 489–500.

Hansen, B. E. (2008). Uniform convergence rates for kernel estimation with dependent data. *Econometric Theory* **24**, 726–748.

Härdle, W. (1990). *Applied nonparametric regression*, Cambridge University Press.

Härdle, W. and Mammen, E. (1993). Comparing nonparametric versus parametric regression fits. *Annals of Statistics.* **21**, 1926–1947.

Hawkes, J. (1986). Local times as stationary processes. In From Local Times to Global Geometry (K. D. Ellworthy, ed.) 111–120. Longman, Chicago.

Helland, I. G. (1982). Central limit theorems for martingale with discreate or continuous time. *Scand J. Statist.* **9**, 79–94.

van der Hofstad, R. and Wolfgang, K. (2001). A survey of one dimensional random polymers. *J. Statist. Phys.* **103**, 915–944.

van der Hofstad, R., den Hollander, F. and Wolfgang, K. (1997). Central limit theorems for the Edwards model. *Ann. Probab.* **25**, 573–597.

van der Hofstad, R., den Hollander, F. and Wolfgang, K. (2003). Weak interaction limits for one-dimensional random polymers. *Probab. Theory Related Fields* **125**, 483–521.

Holley, R. and Stroock, D. W. (1979). Central limit phenomena of various interacting systems. *Ann. Math.* **110**, 333–393.

Hong, S. H. and Phillips, P. C. B. (2010). Testing Linearity in Cointegrating Relations with an Application to Purchasing Power Parity. *Journal of Business and Economic Statistics,* **28**, 96–114.

Horowitz, J. L. and Spokoiny, V. G. (2001). An adaptive rate-optimal test of a parametric mean-regression model against a nonparametric alternative. *Econometrica,* **69**, 599–631.

Ibragimov, R. and Phillips, P. C. B. (2008). Regression asymptotics using martingale convergence methods. *Econometric Theory,* **24**, 888–947.

Itô, K. (1944). Stochastic integral, *Proc. Imp. Acad. Tokyo,* **20**, 519–524.

Jacod, J. and Shiryaev, A. N. (2003). *Limit theorems for stochastic processes.* Second edition. Springer-Verlag, Berlin.

Jakubowski, A. (1996). Convergence in various topologies for stochastic integrals driven by semimartingales. *Annals of Probability,* **24**, 2141–2153.

Jakubowski, A., Ménin, J. and Pageés, G. (1989). Convergence en loi des suites d'integrales stochastiques sur l'espace D^1 de Skorokhod, *Probab. Theory Rel. Fields,* **81**, 111–137.

Jeganathan, P. (1995). Some aspects of asymptotic theory with applications to time series models. *Econometric Theory,* **11**, 818–867.

Jeganathan, P. (2004). Convergence of functionals of sums of r.v.s to local times of fractional stable motions. *Ann. Probab.* **32** , no. 3A, 1771–1795.

Jeganathan, P. (2008). Limit theorems for functional sums that converge to fractional Brownian and stable motions. Cowles Foundation Discussion Paper No. 1649, Cowles Foundation for Research in Economics, Yale University.

de Jong, R. (2002). Nonlinear Regression with Integrated Regressors but Without Exogeneity, Mimeograph, Department of Economics, Michigan State University.

de Jong, R. (2004). Addendum to: "Asymptotics for nonlinear transformations of integrated time series". *Econometric Theory*, **20**, 627–635.

de Jong, R. and Davidson J. (2000 a). The functional central limit theorem and weak convergence to stochastic integral I: weak dependent processes. *Econometric Theory*, **16**, 621–642.

de Jong, R. and Davidson J. (2000 b). The functional central limit theorem and weak convergence to stochastic integral II: fractionally integrated processes. *Econometric Theory*, **16**, 643–666.

de Jong, R. and Wang, Chien-Ho (2005). Further results on the asymptotics for nonlinear transformations of integrated time series. *Econometric Theory*, **21**, 413–430.

Karlsen, H. A. and Tjøstheim, D. (2001). Nonparametric estimation in null recurrent time series. *Annal of Statistics* **29**, no. 2, 372–416.

Karlsen, H. A., Myklebust, T. and Tjøstheim, D. (2007). Nonparametric estimation in a nonlinear cointegration model. *Annal of Statistics* **35**, 252–299.

Kasparis, I. and Phillips, P. C. B. (2009). Dynamic misspecification in nonparametric cointegrating regression. Cowles foundation discussion paper No. 1700.

Kasparis, I. and Phillips, P. C. B. (2012). Dynamic misspecification in nonparametric cointegrating regression. *Journal of Econometrics*, **168(2)**, 270–284.

Kasparis, I., Andreou, E. and Phillips, P. C. B. (2014). Nonparametric predictive regression. *Journal of Econometrics*, in Press.

Kasparis, I., Phillips, P. C. B. and Magdalinos, T. (2014). Nonlinearity induced weak instrumentation. *Econometric Review*, **33**, 676–712

Khoshnevisan, D., Xiao, Y. and Zhong, Y. (2003a). Local times of additive Lévy processes. *Stoch. Process. Appl.* **104**, 193–216.

Khoshnevisan, D., Xiao, Y., Zhong, Y. (2003b). Measuring the range of an additive Lévy processes. *Ann. Probab.* **31**, 1097–1141.

Kesten, H. (1969). Hitting probabilities of single points for processes with stionary independent increments, *Mem. Aner. Math. Soc.* **93**.

Kristensen, D. (2009). Uniform convergence rates for kernel estimators with heterogeneous dependent data. *Econometric Theory*, **25**, 1433–1445.

Kim, S. K. and Kim, I. M. (2012). Partial parametric estimation for nonstationary nonlinear regressions. *Journal of Econometrics*, **167**, 448–457.

Kurtz, T. G., and Protter, P. (1991). Weak limit theorems for stochastic integrals and stochastic differential equations. *Annals of Probability*, **19**, 1035–1070.

Kurtz, T. G., and Protter, P. (1996). Weak convergence of stochastic integrals and differential equations, *Probabilistic models for nonlinear partial differential*, Springer.

Lai, T. L. (1994). Asymptotic theory of nonlinear least square estimations. *The Annals of Statistics*, **22**, 1917–1930.

Lai, T. L. and Wei, C. Z. (1982). least square estimates in stochastic regression models with applications to identification and control of dynamic systems. *The Annals of Statistics*, **10**, 154–166.

Liang, H., Phillips, P. C. B. and Wang, H. and Wang, Q. (2014). Weak convergence to stochastic integrals for econometric applications. *Cowles foundation discussion paper*, CFDP 1971.

Liebscher, E. (1996). Strong convergence of sums of α-mixing random variables with applications to density estimation. *Stochastic Processes and Their Applications*. **65**, 69–80.

Liero, H. (1989). Strong uniform consistency of nonparametric regression function estimates. *Probability Theory and Related Fields*. **82**, no. 4, 587–614.

Lin, Z. and Wang, H. (2010). On Convergence to Stochastic Integrals. arXiv:1006.4693

Linton, O. and Wang, Q. (2015). Nonparametric transformation regression with nonstationary data. *Econometric Theory*, in Press.

Liptser, R. S. and Shiryaev, A. N. (1989). *Theory of Martingale,* Kluwer Academic Publication.

Liu, W. Chan, N. and Wang, Q. (2015). Uniform approximation to local time with applications in nonlinear co-integrating regression, manuscript.

Lukács, E. (1970). *Characteristic functions.* Hafner Publishing Co., New York.

Marmer, V. (2008). Nonlinearity, nonstationarity, and spurious forecasts. *Journal of Econometrics*, **142**, 1–27.

Masry, E. (1996). Multivariate local polynomial regression for time series: Uniform strong consistency and rates. *Journal of Time Series Analysis*. **17**, 571–599.

McLeish, D. L. (1975). A maximal inequality and dependent strong laws. *Annals of Probability,* **3**, 829–839.

Mykland, P. A. and Zhang L. (2006). ANOVA for diffusions and Itô processes. *Annals of Statistics*, 34, 1931–1963.

Mykland, P. A. and Zhang, L. (2012). The econometrics of high-frequency data. In *Statistical Methods for stochastic differential equations*, Edited by Kessler,

Lindneer and Sorensen. Monographs on statistics and applied probability 124.

Nolan, J. (1989). Local nondeterminism and local times for stable processes. *Probab. Th. Rel. Fields* 82, 387–410.

Nummelin, E. (1984). General Irreducible Markov Chains and Non-negative Operaors. *Cambridge University Press,* Cambridge, England.

Park, J. Y. (2002), Nonstationary nonlinear heteroskedasticity. *Journal of Econometrics,* **110**, 383–415.

Park, J. Y., Phillips, P. C. B. (1988). Statistical Inference in Regressions With Integrated Processes: Part 1. *Econometric Theory* 4, 468–497.

Park, J. Y. and Phillips, P. C. B. (1989). Statistical Inference in Regressions With Integrated Processes: Part 2. *Econometric Theory* 5, 95–131.

Park, J. Y. and Phillips, P. C. B. (1999). Asymptotics for nonlinear transformation of integrated time series. *Econometric Theory,* **15**, 269–298.

Park, J. Y., Phillips, P. C. B. (2000). Nonstationary binary choice. *Econometrica,* 68, 1249–1280.

Park, J. Y. and Phillips, P. C. B. (2001). Nonlinear regressions with integrated time series. *Econometrica,* **69**, 117–161.

Peligrad, M. (1991). Properties of uniform consistency of the kernel estimators of density and of regression functions under dependence conditions. *Stochastics and Stochastic Reports,* **40**, 147–168.

de La Pena, V. H. (1999). A general class of exponential inequalities for martingales and ratios. *Annal of Probability,* **27**, 537–564.

Petrov, V. V. (1995). *Limit Theorems of Probability Theory: Sequences of Independent Random Variables.* Oxford Studies in Probability.

Phillips, P. C. B. (1987). Time Series Regression with a Unit Root. *Econometrica,* **55**, 277–302.

Phillips, P. C. B. (1988a). Multiple regression with integrated processes. In N. U. Prabhu, (ed.), *Statistical Inference from Stochastic Processes, Contemporary Mathematics,* **80**, 79–106.

Phillips, P. C. B. (1988b). Weak convergence to sample covariance matrices to stochastic integrals via martingale approximation. *Econometric Theory,* **4**, 528–533.

Phillips, P. C. B. (1989). Partially identified econometric models. *Econometric Theory* 5, 181–240.

Phillips, P. C. B. (1991). Optimal Inference in Cointegrated Systems. *Econometrica* 59, 283–306.

Phillips, P. C. B. (2005). HAC Estimation by Automated Regression. *Econometric Theory*, **21**, 116–142.

Phillips, P. C. B.(2009). Local Limit Theory and Spurious Nonparametric Regression. *Econometric Theory*, **25**, 1466–1497.

Phillips, P. C. B. and Hansen B. E. (1990). Statistical inference in instrumental variables regression with $I(1)$ processes. *Review of Economic Studies* **57**, 99–125.

Phillips, P. C. B . and Solo, V. (1992). Asymptotics for Linear Processes. *Ann. Statist.* **20**, 971–1001.

Phillips, P. C. B and Outliaris, S. (1990). Asymptotic Properties of Residual Based Tests for Cointegration. *Econometrica*, **58**, 165–193.

Phillips, P. C. B. and Park, J. Y. (1998). Nonstationary Density Estimation and Kernel Autoregression. *Cowles Foundation discuss paper No. 1181.*

Pitt, L. D. (1978). Local times for Gaussian vector fields. *Indiana Univ. Math. J.* **27**, 309–330.

Podolskij, M. and Vetter, M. (2010). Understanding limit theorems for semimartigale: a short survey. *Statistica Neerlandia*, **64**, 329–351.

Pötscher, B. M. (2004). Nonlinear functions and convergence to Brownian motion: beyond the continuous mapping theorem. *Econometric Theory*, **20**, 1–22.

Protter, P. E. (2005). *Stochastic integration and differential equations,* Second edition, Springer.

Rényi, A. (1963). On stable sequences of events. *Sankhya Ser. A,* **25**, 293–302.

Revuz, D. and Yor, M. (1994). *Continuous Martingales and Brownian Motion.* Fundamental Principles of Mathematical Sciences 293. Springer-Verlag

Rogers, L. C. G. and Williams, D. (2000). *Diffusion, Markov processes and martingales,* Cambridge University Press.

Rootzen, H. (1983). Central limit theory for martingales via random change of time, *In Probab and Math Statist.* Uppsala University, Uppsala, 154–189.

Shi, X. and Phillips, P. C. B.(2012). Nonlinear cointegrating regression under weak identification. *Econometric Theory*, **28**, 509–547.

Skouras, K. (2000). Strong consistency in nonlinear stochastic regression models. *The Annals of Statistics*, **28**, 871–879.

Slominski, L. (1989). Stability of strong solutions of stochastic differential equations, *Stoch. Proc. and Applic.* **31**, 173–202.

Strasser, H. (1986). Martingale difference arrays and stochastic integrals. *Probabity Theory and Related Fields*, **72**, 83–89.

Stricker, C. (1985). Lois de semimartingales et criteres de compacite, *Lecture Notes in Math.* 1123, Springer-Verlag, Berling-Heidelberg-New York.

Sun, Y. (2011). Robust trend inference with series variance estimator and testing-optimal smoothing parameter. *Journal of Econometrics,* **164**, 345–366.

Sun, Y. (2014). Let's fix it: Fixed-b asymptotics versus small-b asymptotics in heteroskedasticity and autocorrelation robust inference. *Journal of Econometrics,* **178**, 659–677.

Sun, Y. Cai, Z. and Li, Q. (2013). Semiparamtric functional coefficient models with integrated covariates. *Econometric Theory,* on line.

Takacs, L. (1995). On the Local Time of the Brownian Motion. *Annals of Applied Probability,* **5**, 741–756.

Taqqu, M. S. (1975). Weak convergence to fractional Brownian motion and to the Rosenblatt process. *Z. Wahrsch. Verw. Geb.* **31**, 287–302.

Tong, H. (1990). Nonlinear Time Series: A Dynamical System Approach. *Oxford University Press.*

Wang, Q. (2014). Martingale limit theorem revisited and nonlinear cointegrating regression, *Econometric Theory,* **30**, 509–535.

Wang, Q. (2015). Convergence to local time: new results with weakly dependent random variables, manuscript.

Wang, Q. and Chan, N. (2014). Uniform convergence rates for a class of martingales with application in nonlinear co-integrating regression. *Bernoulli,* **1**, 207–230.

Wang, Q., Lin, X., Gulati, C. (2001). Asymptotics for moving average processes with dependent innovations. Statistics and Probability Letters, 54(4), 347–356.

Wang, Q., Lin, Y. X. and Gulati, C. M. (2003). Asymptotics for general fractionally integrated processes with applications to unit root tests. *Econometric Theory,* **19**, 143–164.

Wang, Q. and Phillips, P. C. B. (2009a). Asymptotic Theory for Local Time Density Estimation and Nonparametric Cointegrating Regression. *Econometric Theory* **25**, 710–738.

Wang, Q. and Phillips, P. C. B. (2009b). Structural Nonparametric Cointegrating Regression. *Econometrica* **77**, 1901–1948.

Wang, Q. and Phillips, P. C. B. (2011). Asymptotic Theory for zero energy Functionals with Nonparametric Regression Applications. *Econometric Theory,* **27**, 235–259.

Wang, Q. and Phillips, P. C. B. (2012). A Specification Test for Nonlinear Nonstationary Models. *The Annals of Statistics,* **40**, 727–758.

Wang, Q. and Phillips, P. C. B. (2015). Nonparametric cointegrating regression with endogeneity and long memory. *Econometric Theory*, In Press.

Wang, Q. and Wang, R. (2013). Nonparametric cointegrating regression with NNH errors. *Econometric Theory*, **29**, 1–27.

Wu, C. F. (1981). Asymptotic theory of nonlinear least square estimations. *Annals of Statistics*, **9**, 501–513.

Wu, W. B., Huang, Y. and Huang, Y. (2010). Kernel estimation for time series: An asymptotic theory. *Stochastic Processes and their Applications* **120**, 2412–2431.

Wu, W. B., Shao, X. (2004). Limit theorems for iterated random functions. *Journal of Applied Probability* **41**, 425–436.

Xiao, Y. (2006). Properties of local-nondeterminism of Gaussian and stable random fields and their applications *Annales de la faculté des sciences de Toulouse Sér.* 6, 15 no. 1 (2006), 157–193.

Xiao, Z. (2009). Functional-coefficient cointegration models. *Journal of Econometrics*, **152**, 81–92.

Index

Printed in the United States
By Bookmasters